Singular and Degenerate Cauchy Problems

ACADEMIC PRESS RAPID MANUSCRIPT REPRODUCTION

This is Volume 127 in
MATHEMATICS IN SCIENCE AND ENGINEERING
A Series of Monographs and Textbooks
Edited by RICHARD BELLMAN, *University of Southern California*

The complete listing of books in this series is available from the Publisher upon request.

Singular and Degenerate Cauchy Problems

R. W. Carroll

Department of Mathematics
University of Illinois at Urbana-Champaign

R. E. Showalter

Department of Mathematics
University of Texas at Austin

ACADEMIC PRESS New York San Francisco London 1976

A Subsidiary of Harcourt Brace Jovanovich, Publishers

ACADEMIC PRESS, INC.
111 Fifth Avenue, New York, New York 10003

United Kingdom Edition published by
ACADEMIC PRESS, INC. (LONDON) LTD.
24/28 Oval Road, London NW1

Library of Congress Cataloging in Publication Data

Carroll, Robert Wayne, Date
Singular and degenerate Cauchy problems.
(Mathematics in science and engineering ;)
Bibliography: p.
Includes index.
1. Cauchy problem. I. Showalter, Ralph E., joint author. II. Title. III. Series.
QA374.C38 515'.353 76-46563
ISBN 0-12-161450-6

PRINTED IN THE UNITED STATES OF AMERICA

Contents

3
Degenerate Equations with Operator Coefficients

4
Selected Topics

Preface

In this monograph we shall deal primarily with the Cauchy problem for singular or degenerate equations of the form ($u' = u_t = \partial u/\partial t$)

$$A(t)u_{tt} + B(t)u_t + C(t)u = g \tag{1.1}$$

where $u(\cdot)$ is a function of t, taking values in a separated locally convex space E, while $A(t)$, $B(t)$, and $C(t)$ are families of linear or nonlinear differential type operators acting in E, some of which become zero or infinite at $t = 0$. Appropriate initial data $u(0)$ and $u_t(0)$ will be prescribed at $t = 0$, and g is a suitable E-valued function. Similarly some equations of the form

$$(A(\cdot)u)_{tt} + (B(\cdot)u)_t + C(\cdot)u = g \tag{1.2}$$

will be considered. Problems of the type (1.1) for example will be called singular if at least one of the operator coefficients tends to infinity in some sense as $t \to 0$. Such problems will be called degenerate if some operator coefficient tends to zero as $t \to 0$ in such a way as to change the type of the problem. Similar considerations apply to (1.2). We shall treat only a subclass of the singular and degenerate problems indicated and will concentrate on well-posed problems. If (1.1) or (1.2) can be exhibited in a form where the highest order derivative appears with coefficient one, then the problems treated will be of parabolic or hyperbolic type; if the highest order derivative cannot be so isolated, the equation will be said to be of Sobolev type. We have included a discussion of not necessarily singular or degenerate Sobolev equations for completeness, since it is not available elsewhere. The distinction between singular and degenerate can occasionally be somewhat artificial when some of the operators (or factors thereof) are invertible or when a suitable change of variable can be introduced (see, e.g., Example 1.3); however, we shall not usually resort to such artifices. In practice E frequently will be a space of functions or distributions in a region $\Omega \subset \mathbb{R}^n$ and $0 \leqslant t \leqslant b < \infty$.

The study of singular and degenerate Cauchy problems is partially motivated by problems in physics, geometry, applied mathematics, etc., many examples of which are given in the text, and there is an extensive literature. We have tried to give a fairly complete bibliography and apologize for any omissions. The material is aimed basically at analysts, and standard theorems from functional analysis will be assumed (cf. Bourbaki [1; 2], Carroll [14], Horváth [1], Köthe [1], Schaeffer [1], Treves [1]); in particular Schwartz distributions will be used with the standard notations (cf. Schwartz [1]). There is also some material involving the abstract theory of Lie groups which is assumed but clearly indicated, with references, in Chapter 2; and there are enough concrete examples worked out in the text to illustrate the matter completely for the reader unfamiliar with Lie theory. The chapters are essentially independent; Chapters 1 and 3 begin with introductory material outlining content, motivation, and objectives, whereas Chapters 2 and 4 are organized somewhat differently.

Some remarks on notation should perhaps be made here. If, in a given chapter, there appears a reference to formula $(x.y)$ it means $(x.y)$ of that chapter. If a reference $(x.y.z)$ appears, it means formula $(y.z)$ of Chapter x. Theorems, Lemmas, Examples, Remarks, etc. will be labeled consecutively; thus one might have in order: Lemma $x.y$, Theorem $x.y+1$, Remark $x.y+2$, Lemma $x.y+3$, etc. in a given section.

The first author (R.W.C.) would like to acknowledge a professional debt to A. Weinstein, who initiated systematic work on Euler-Poisson-Darboux (EPD) equations (cf. Weinstein [1; 2]) and whose further contributions to this theory have been essential to its development; his encouragement motivated this author's early work in the area. He was also instrumental in the undertaking of the present monograph. Thus this book is dedicated to Alexander Weinstein.

The authors would like to acknowledge the aid of the National Science Foundation from time to time in supporting some of the research presented here.

Chapter 1

Singular Partial Differential Equations of EPD Type

1.1 Examples. Let us first give some typical examples of singular and related degenerate problems; their solutions, in various spaces via diverse techniques, will appear in the course of the book, usually as special cases of more general results. Further examples of degenerate problems will appear in Chapter 3.

Example 1.1 The singular Cauchy problem for EPD equations. Let $t \geq 0$ and $x \in \mathbb{R}^n$; we consider the problem

$$(1.1) \qquad u^m_{tt} + \frac{2m+1}{t} u^m_t = \Delta_x u^m$$

$$(1.2) \qquad u^m(x,o) = f(x); \qquad u^m_t(x,o) = 0$$

where the parameter $m \in \mathbb{C}$ is arbitrary for the moment, Δ_x denotes the Laplace operator in $\mathbb{R}^n$, and classically f and u^m were numerical functions (note that the left hand side of (1.1) denotes the radial form of a Laplace operator in $\mathbb{R}^n$ if $2m + 1 = n - 1$). These equations, for integral m, provide a model for some very interesting scales of canonical singular Cauchy problems with their natural origins in Lie theory (see Chapter 2). They also arise in physics as is indicated in Example 1.3 and of course for $m = -1/2$ we have the wave equation; their connection with mean values is indicated in Section 2 and in Chapter 2. In various forms one can trace their origins back to Euler [1], Poisson [1], and Darboux [1]. In this chapter we will present five general techniques for the solution of these problems and of similar abstract singular problems; more general singularities in the u_t term and some nonlinear terms will be treated. The first is based on a Fourier method in distribution spaces developed by one of the authors (cf. Carroll [2; 3; 4; 5; 6; 8; 10; 11; 14]). The second involves a spectral technique in a Hilbert space related to the Fourier method (cf. Carroll, loc. cit.) while the third is based on transmutation methods in various spaces (cf. Carroll [24], Carroll-Donaldson [20], Delsarte [1], Delsarte-Lions [2; 3] Donaldson [1; 3; 5], Hersh [2], Lions [1; 2; 3; 4; 5], Thyssen [1; 2]). The fourth involves the idea of related differential equations (cf. Bragg [2; 3; 4; 5; 9], Bragg-Dettman [1; 6; 7; 8], Carroll [24], Carroll-Donaldson [20], Dettman [1], Donaldson [1; 3; 5], Donaldson-Hersh [6], Hersh [1; 2; 3]). There are certain relations between the method of transmutation and that of related equations and this is brought out in Carroll [24], Carroll-Donaldson [20], and Hersh [2] (cf. also Carroll [18; 19; 25], Donaldson [4; 7], and Hersh [1; 2]). The fifth technique involves "energy" methods following Lions [5] and yields some results of Lions indicated in Carroll [8].

Remark 1.2 There are many additional papers devoted to problems of the form (1.1) - (1.2) and we will list some of them here; further references can be found in the bibliographies to these articles. Some of this material will appear in this or other chapters. In giving such references we have tried to avoid repetition of the citations above and there will be considerable relevant material and further references in the work on the Cauchy problem for abstract Tricomi type problems to be treated later in Chapter 3 (cf. Example 1.3), in Remark 2.4 on mean values, and in Chapter 4. Thus, let us cite here Agmon [2], Babenko [1], Baouendi-Goulaouic [1], Baranovskij [1; 2; 3; 4; 5; 6; 7], Barantsev [1], Berezanskij [1], Berezin [1], Bers [1; 2], Bitsadze [1; 2], Blumkina [1], Bresters [1], Bureau [1; 2], Carroll [7; 9; 13; 18; 19; 21; 22], Carroll-Silver [15; 16; 17], Carroll-Wang [12], Chi [1; 2], Cibrario [1; 2; 3], Cinquini-Cibrario [1;2], Conti [1; 2; 3], Copson [1], Copson-Erdelyi [2], Courant-Hilbert [1], Delache-Leray [1], Diaz [5; 6], Diaz-Kiwan [7], Diaz-Ludford [3; 4], Diaz-Martin [9], Diaz-Weinberger [2], Diaz-Young [1; 8], Ferrari-Tricomi [1], Filipov [1], Fox [1], Frank'l [1], Friedlander-Heins [1], B. Friedman [1], Fusaro [2; 3; 4; 5; 6], Garabedian [1], Germain [1], Germain-Bader [2; 3], Gilbert [1], Gordeev [1], Günther [2; 3], Haack-Hellwig [1], Hariullina [1; 2; 3], Hairullina-Nikolenko [1], Hellwig [1; 2; 3], Kapilevič [1; 2; 3; 4; 5; 6], Karapetyan [1], Karimov-Baikuziev [1; 2], Karmanov [1], Karol [1; 2], Kononenko [1], Krasnov [1; 2; 3], Krivenko [1; 2], Ladyženskaya [1], Lavrentiev-Bitsadze [1], Lieberstein [1; 2], Makarov [1; 2], Makusina [1], Martin [1], Mikhlin [1], Miles-Williams [1], Miles-Young [2; 3], Morawetz [1; 2], Nersesyan [1], Nevostruev [1], Nosov [1], Olevskij [2], Ossicini [1; 2], Ovsyanikov [1], Payne [1], Payne-Sather [2], Protter [2; 3; 4; 5;

6], Protter-Weinberger [1], Rosenbloom [1; 2; 3], Sather-Sather [4], Silver [1], Smirnov [1; 2; 3; 4], Solomon [1; 2], Sun [1], Suschowk [1], Tersenov [1; 2; 3; 4], Tong [1], Travis [1], Tricomi [1], Volkodarov [1], Walker [1; 2; 3], Walter [1], Wang [1; 2], Weinstein [1; 2; 3; 4; 5; 6; 7; 8; 9; 10; 11; 13; 14; 15; 16; 17; 18; 19; 20], Williams [1], Young [1; 2; 3; 4; 5; 6; 7; 8; 9; 10; 11], Žitomirskij [1].

Example 1.3 The degenerate Cauchy problem for the Tricomi equation. The original Tricomi equation is of the form

$$(1.3) \qquad u_{tt} - tu_{xx} = 0$$

and we will be primarily concerned with the Cauchy problem for abstract versions of the more general equation

$$(1.4) \qquad u_{tt} + a(x,t)u_t + b(x,t)u_x + c(x,t)u - K(x,t)u_{xx} = 0$$

in the hyperbolic region $t > 0$ where it is assumed that $tK(x,t) > 0$ for $t \neq 0$ while $K(x,0) = 0$; suitable initial data $u(x,0)$ and $u_t(x,0)$ will be prescribed on the parabolic line $t = 0$. This example will be treated as a degenerate problem in Chapter 3 but is displayed here because of the connection of (1.3) to a (singular) EPD equation of index $2m + 1 = 1/3$ under the change of variable $\tau = 2/3\ t^{3/2}$ for $t > 0$ which yields

$$(1.5) \qquad u_{\tau\tau} + \frac{1}{3\tau} u_\tau = u_{xx}$$

Similarly in the region $t < 0$ the change of variable $\tau = 2/3(-t)^{3/2}$

transforms (1.3) into $u_{\tau\tau} + \frac{1}{3\tau} u_\tau + u_{xx} = 0$ which is a singular elliptic equation of the type studied by Weinstein and others in the context of generalized axially symmetric potential theory (GASPT); cf. Remark 1.2 for some references to this. The references above in Remark 1.2 also deal with mixed problems where suitable data is prescribed on various types of curves in the elliptic ($t < 0$) and hyperbolic ($t > 0$) regions; such problems occur for example in transonic gas dynamics. We will not deal with the elliptic region in general but some other types of singular hyperbolic problems for EPD equations will be discussed briefly in Chapter 4.

1.2 Mean values in $\mathbb{R}^n$. We refer now to the EPD problem (1.1) - (1.2) for $x \in \mathbb{R}^n$ and $2m + 1 = n - 1$ which is a pivotal case in the theory. Let $\langle T,\phi\rangle$ denote the action in $\mathbb{R}^n$ of a distribution $T \in D' \equiv D'_x$ on a "test function" $\phi \in D \equiv D_x$; the same bracket notation will also be used for pairings $E' - E$, $S' - S$, etc. One defines a surface (or spherical) mean value operator $\mu_x(t)$ (t fixed for the moment) by the rule

$$(2.1)\qquad \langle\mu_x(t),\phi\rangle = \frac{1}{t^{n-1}\omega_n}\int_{|x|=t} \phi(x)\,d\sigma_n$$

where $\phi \in E$, $|x|^2 = \sum_1^n x_i^2$, $\omega_n = 2\pi^{n/2}/\Gamma(n/2)$ is the surface area of the n-dimensional unit ball, and $d\sigma_n = t^{n-1}d\Omega_n$ represents the surface area measure on the ball of radius $|x| = t$. The surface mean value measure $\mu_x(t)$ averages ϕ over a sphere of radius t centered at the origin. Similarly one defines a solid mean value operator $A_x(t) \in E'$ (t still fixed) by the rule

$$(2.2) \qquad \langle A_x(t), \phi\rangle = \frac{n}{t^n \omega_n} \int_{|x|\le t} \phi(x)dx$$

which averages $\phi \in E$ over the entire ball $|x| \le t$. Now we observe that $\langle \mu_x(t), \phi(x)\rangle = \langle \mu_y(1), \phi(ty)\rangle$ and, allowing t to vary, the following calculations are easily checked (ν denotes the exterior normal)

$$(2.3) \qquad \frac{d}{dt} \langle \mu_x(t), \phi\rangle = \frac{1}{\omega_n} \int_{|y|=1} (\sum y_i \frac{\partial \phi}{\partial x_i})\, d\Omega_n$$

$$= \frac{1}{t^{n-1}\omega_n} \int_{|x|=t} (\frac{\partial \phi}{\partial \nu}) d\sigma_n = \frac{1}{t^{n-1}\omega_n} \int_{|x|\le t} \Delta\phi dx$$

$$= \frac{t}{n} \langle A_x(t), \Delta\phi\rangle = \frac{t}{n} \langle \Delta A_x(t), \phi\rangle$$

$$(2.4) \qquad \frac{d}{dt} \langle A_x(t), \phi\rangle = \frac{-n^2}{t^{n+1}\omega_n} \int_{|x|\le t} \phi(x)dx$$

$$+ \frac{n}{t^n \omega_n} \int_{|x|=t} \phi(x) d\sigma_n = \frac{n}{t} \langle \mu_x(t) - A_x(t), \phi\rangle$$

Note in (2.3) that if $\psi \in D$ with $\psi = 1$ in a neighborhood of $|x| \le t$ then the E-E$'$ pairing $\langle A_x(t), \Delta\phi\rangle_E = \langle A_x(t), \Delta(\phi\psi)\rangle_D$ in a D-D$'$ pairing so that $\langle A_x(t), \Delta(\phi\psi)\rangle_D = \langle \Delta A_x(t), \phi\psi\rangle_D = \langle \Delta A_x(t), \phi\rangle_E$ (i.e., $\Delta A_x(t) \in E'$ is computed in D').

By Lemma 2.3 below, (2.3) and (2.4) imply that the functions $t \to \mu_x(t)$ and $t \to A_x(t)$ belong to $C^\infty(E_x')$ for $t > 0$ with

$$(2.5) \qquad \frac{d}{dt} \mu_x(t) = \frac{t}{n} \Delta A_x(t)$$

(2.6) $\frac{d}{dt} A_x(t) = \frac{n}{t}(\mu_x(t) - A_x(t))$

<u>Theorem 2.1</u> Let $T \varepsilon D'_x$ be arbitrary. Then $u^m_x(t) = \mu_x(t) * T$ satisfies the EPD equation (1.1) with $2m + 1 = n - 1$ (i.e., $m = \frac{n}{2} - 1$) and the initial conditions $u^m_x(o) = T$ with $\frac{d}{dt} u^m_x(o) = 0$.

Proof: We recall that $(S, T) \to S * T$: $E' \times D' \to D'$ is separately continuous so equation (1.1) results immediately by differentiating (2.5) and using (2.5) - (2.6). Evidently $\mu_x(t) \to \delta$ (Dirac measure at $x = 0$) when $t \to 0$ and we note that $\frac{d}{dt}(\mu_x(t) * T) = \frac{t}{n} \Delta A_x(t) * T = (\frac{t}{n})[\Delta\delta * A_x(t) * T] = (\frac{t}{n})(A_x(t) * \Delta T) \to 0$ as $t \to 0$. QED

For information about distributions and the differentiation of vector valued functions let us refer to Carroll [14], A. Friedman [2], Garsoux [1], Gelfand-Šilov [5; 6; 7], Horváth [1], Nachbin [1], Schwartz [1; 2], Treves [1].

<u>Definition 2.2</u> A function $f: \mathbb{R} \to E$, E a general locally convex space, has a scalar derivative $f'_s(t) \varepsilon (E')^*$ if $\frac{d}{dt}\langle f(\cdot), e' \rangle = \langle f'_s(\cdot), e' \rangle$ for all $e' \varepsilon E'$ (F^* denotes the algebraic dual of F). On the other hand if $\frac{df}{dt} = f'(t) \varepsilon E$ in the topology of E then f' is called the strong derivative of f.

The following lemma is a special case of results in Carroll [2; 5; 14] and Schwartz [2].

Lemma 2.3 If $f:\mathbb{R} \to E'$ has a scalar derivative f_s' then f is strongly differentiable with $f' = f_s'$.

Proof: Fix t_0 and set $g(t) = \frac{f(t)-f(t_0)}{t - t_0}$. In discussing limits as $t \to t_0$ it suffices to consider sequences $t_n \to t_0$ since any t_0 has a countable fundamental system of neighborhoods. By hypothesis $g(t_n) \to f_s'(t_0)$ weakly. Using the fact that E is a Montel space and a version of the Banach-Steinhaus theorem (cf. Carroll [14]) it follows that $g(t_n) \to f_s'(t_0)$ strongly in E' so that $f_s'(t_0) = f'(t_0)$. QED

Remark 2.4 A great deal of information is available relating "spherical" mean values and differential equations, both in $\mathbb{R}^n$ and on certain Riemannian manifolds. In addition to references included in the Introduction, Example 1.1, Remark 1.2, and work on GASPT, relative to such mean values, let us mention Asgeirsson [1], Carroll [1], Fusaro [1], Ghermanesco [1; 2], Günther [1], Helgason [1; 2; 3; 4; 5; 6; 7; 8; 9; 10; 11], John [1], Olevskij [1], Poritsky [1], Walter [2], Weinstein [12], Willmore [1] and their bibliographies. This will be discussed further in a Lie group context in Chapter 2. For some references to complex function theory and various mean values and measures related to differential equations see Zalcman [1; 2] and the bibliographies there.

1.3 The Fourier method. We will work with a more general equation than (1.1) which includes (1.1) as a special case. Thus let $D_k = \frac{1}{i}\frac{\partial}{\partial x_k}$ $(1 \leq k \leq n)$, $D_\alpha = (D_1)^{\alpha_1} \ldots (D_n)^{\alpha_n}$, and

$$(3.1) \qquad A_x = \sum a_\alpha D_\alpha$$

where the a_α are constant (and real) and the sum in (3.1) is over a finite set of multi-indices $\alpha = (\alpha_1, \ldots, \alpha_n)$. The Fourier transform $F{:}T \to \hat{T} = FT{:}S' \to S'$ is a topological isomorphism (cf. Schwartz [1]) and for $\phi \in S$ for example we use the form $F\phi = \hat{\phi} = \int_{-\infty}^{\infty} \phi(x) \exp - i\langle x,y\rangle dx$ where $\langle x,y\rangle = \sum_1^n x_i y_i$. Hence in particular $F(D_k\phi) = y_k F\phi = y_k\hat{\phi}$. We recall also that $A_x T = A_x\delta * T$ with $F\delta = 1$ and will write $A(y) = F(A_x\delta) = \sum a_\alpha y^\alpha = \sum a_\alpha y_1^{\alpha_1} \ldots y_n^{\alpha_n}$. It will be assumed that $A(y)$ satisfies

<u>Condition 3.1</u> $A(y) = \sum a_\alpha y^\alpha$ is real with $A(y) \geq a > 0$ for $|y| > R_0$.

In particular Condition 3.1 holds when $A_x = -\Delta_x = -\sum i^2 D_k^2 = \sum D_k^2$ so that $F(-\Delta\delta) = \sum y_k^2 = |y|^2$. It can also hold for certain hypoelliptic operators A since then $|A(y)| \to \infty$ as $|y| \to \infty$ (cf. Hörmander [1]) or even nonhypoelliptic operators (e.g., $A(y) = a + y_2^2 + \ldots + y_n^2$). One might also envision a parallel theory when $\text{Re}\, A(y) \geq a > 0$ for $|y| > R_0$ but we will not investigate this here. Let us first consider the problem for $w^m(\cdot) \in C^2(S_x')$ on $0 \leq t \leq b < \infty$ (b arbitrary)

$$(3.2) \qquad w_{tt}^m + \frac{2m+1}{t} w_t^m + A_x w^m = 0$$

$$(3.3) \qquad w^m(o) = T \in S' \; ; \qquad w_t^m(o) = 0$$

where $m \geq -1/2$ is real (complex m with Re $m > -1/2$ can also be treated by the methods of this section). Taking Fourier transforms in x one obtains

$$(3.4) \qquad \hat{w}^m_{tt} + \frac{2m+1}{t}\hat{w}^m_t + A(y)\,\hat{w}^m = 0$$

$$(3.5) \qquad \hat{w}^m(o) = \hat{T}; \quad \hat{w}^m_t(o) = 0$$

and when $T = \delta$ we denote the solution of (3.2) - (3.3) by $R^m(t)$ so that $\hat{R}^m(t)$ satisfies (3.4) - (3.5) with $\hat{T} = 1$. The expressions $R^m(t)$ and $\hat{R}^m(t)$ will be called resolvants and one notes that if $R^m(t) \in O'_C$ (i.e., $\hat{R}^m(t) \in O_M$), with suitable differentiability in O'_C, then $R^m(t) * T = w^m(t)$ satisfies (3.2) - (3.3) (see Theorem 3.4). We recall here that O_M is the space of C^∞ functions f on $\mathbb{R}^n$ such that $D_\alpha f$ is bounded by a polynomial (depending perhaps on α) while $f_\beta \to 0$ in O_M whenever $gD_\alpha b_\beta \to 0$ uniformly in $\mathbb{R}^n$ for any α and any $g \in S$. $O'_C = F^{-1}O_M \subset S'$ and is given the topology transported by the Fourier transform (i.e., $T_\beta \to 0$ in $O'_C \leftrightarrow FT_\beta \to 0$ in O_M). It is known that O'_C is the space of distributions mapping $S' \to S'$ under convolution and the map $(S,T) \to S * T$: $O'_C \times S' \to S'$ is hypocontinuous.

Now the solution of (3.4) - (3.5) when $\hat{T} = 1$ is given by $(\hat{R}^m(t) = \hat{R}^m(\cdot,t))$

$$(3.6) \qquad \hat{R}^m(y,t) = 2^m\Gamma(m+1)z^{-m}J_m(z)$$

$$= \Gamma(m+1)\sum_{k=0}^{\infty}\frac{(-1)^k}{k!}\left(\frac{z}{2}\right)^{2k}\frac{1}{\Gamma(k+m+1)}$$

where $z = t\sqrt{A(y)}$ (cf. B. Friedman [1] and Rosenbloom [1]). One observes that $\hat{R}^m(y,t)$ is in fact a function of z^2 so that it is not necessary to define the square root $\sqrt{A(y)}$; evidently $\hat{R}^m(\cdot,t)$ is an entire function of the y_k for $y_k \in \mathbb{C}$ (and hence of $y \in \mathbb{C}^n$ by Hartog's theorem - cf. Hörmander [2]).

Lemma 3.2 The function $t \to \hat{R}^m(\cdot,t):[o,b] \to O_M$ is continuous ($m \geq -1/2$).

Proof: Clearly $t \to \hat{R}^m(\cdot,t):[o,b] \to E_y$ is continuous and one knows that on bounded sets in O_M the topology of O_M coincides with that induced by E (see Schwartz [2] and recall that a set $B \subset O_M$ is bounded if $D_\alpha f$ is bounded by the same polynomial, depending perhaps on α, for all $f \in B$). Consequently one need only show that each $D_\alpha \hat{R}^m(y,t)$ is bounded by a polynomial in y, independent of t for $t \in [o,b]$. We choose here a straightforward method of doing this and refer to subsequent material for another, much more general, technique (cf. Carroll [5; 8]). Thus, one knows that for $m > -1/2$ (cf. Watson [1])

$$(3.7) \qquad z^{-m}J_m(z) = \frac{2^{1-m}\pi^{-\frac{1}{2}}}{\Gamma(m+\frac{1}{2})} \int_0^{\pi/2} \cos(z\cos\theta)\sin^{2m}\theta\, d\theta$$

(the case $m = -1/2$ is trivial since $z^{1/2}J_{-1/2}(z) = (2/\pi)^{1/2}\cos z$). Writing $|\alpha| = \alpha_1 + \ldots + \alpha_n$ and $c_m = 2\Gamma(m+1)/\sqrt{\pi}\,\Gamma(m+\frac{1}{2})$ we have from (3.7)

$$(3.8)\qquad \left|\frac{\partial^{|\alpha|}\hat{R}^m(y,t)}{\partial z^{|\alpha|}}\right| = c_m\left|\int_0^{\pi/2}(-1)^{|\alpha|}\right.$$

$$\left.\cdot\left\{\begin{matrix}\cos\,(z\,\cos\theta)\\ \sin\,(z\,\cos\theta)\end{matrix}\right\}\cos^{|\alpha|}\theta\,\sin^{2m}\theta d\theta\right| \leq \frac{\pi c_m}{2}$$

Now for $|y| > R_0$ we write

$$(3.9)\qquad \frac{\partial}{\partial y_k}\hat{R}^m(y,t) = \left(t\frac{\partial A(y)}{\partial y_k}/2\sqrt{A(y)}\right)\frac{\partial}{\partial z}\hat{R}^m(y,t)$$

and since $A(y) \geq a$ with $\partial A(y)/\partial y_k = B_k(y)$ a polynomial we have from (3.8) - (3.9)

$$(3.10)\qquad \left|\frac{\partial\hat{R}^m(y,t)}{\partial y_k}\right| \leq \frac{\pi bc_m}{4\sqrt{a}}|B_k(y)| = c|B_k(y)| \leq 1 + c^2B_k^2(y)$$

A similar argument obviously applies to majorize any $D_\alpha\hat{R}^m(y,t)$ for $|y| > R_0$ by a polynomial in y not depending on t. For $|y| \leq R_0$ the power series in (3.6) may be used, since $(y,t) \to D_\alpha\hat{R}^m(y,t)$ is continuous on the compact set $\{|y| \leq R_0\} \times [0,b]$, to obtain a bound $|D_\alpha\hat{R}^m(y,t,)| \leq M_\alpha$. Consequently $|D_\alpha\hat{R}^m(y,t)| \leq M_\alpha$ + polynomial = polynomial in y, for all y, independent of $t \in [0,b]$.

QED

<u>Lemma 3.3</u> The function $t \to \hat{R}^m(\cdot,t)$ belongs to $C^\infty(O_M)$ for $t \in [0,b]$ $(m \geq -1/2)$ and

$$(3.11)\qquad \frac{\partial}{\partial t}\hat{R}^m(y,t) = \frac{-tA(y)}{2(m+1)}\hat{R}^{m+1}(y,t)$$

$$(3.12)\qquad \frac{\partial}{\partial t}\hat{R}^m(y,t) = \frac{2m}{t}[\hat{R}^{m-1}(y,t) - \hat{R}^m(y,t)]$$

Proof: The formulas (3.11) - (3.12) are easily seen to hold in E_y by using well-known recursion formulas for Bessel functions (cf. Watson [1], Carroll [5]). Further we know by Schwartz [3; 4] that if $t \to \hat{R}^m(\cdot,t) \in C^0(O_M) \cap C^1(E)$ with $t \to \hat{R}^m_t(\cdot,t) \in C^0(O_M)$ then $t \to \hat{R}^m(\cdot,t) \in C^1(O_M)$. Using (3.11) and Lemma 3.2 however we see that $t \to \hat{R}^m_t(\cdot,t) \in C^0(O_M)$ and in particular (3.11) - (3.12) hold in O_M. By iteration of this argument it is seen that $t \to \hat{R}^m(\cdot,t)$ belongs to $C^\infty(O_M)$. QED

<u>Theorem 3.4</u> Let $T \in S'$ and $R^m(t) = F^{-1}\hat{R}^m(\cdot,t)$ with $\hat{R}^m(y,t)$ given by (3.6). Then $R^m(t) \in O'_C$ and $w^m(t) = R^m(t) * T$ satisfies (3.2) - (3.3).

Proof: We know $t \to \hat{R}^m(\cdot,t)\hat{T}:[o,b] \to S'$ is continuous since the map $(\hat{S},\hat{T}) \to \hat{S}\hat{T}:O_M \times S' \to S'$ is hypocontinuous (cf. Schwartz [1]). Using (3.11), $t \to \frac{2m+1}{t}\hat{R}^m_t(\cdot,t)\hat{T}:[o,b] \to S'$ is also continuous. From (3.4) in E we see (using (3.11) again) that $t \to \hat{R}^m_{tt}(\cdot,t) = A(\cdot)[\frac{2m+1}{2(m+1)}\hat{R}^{m+1}(\cdot,t) - \hat{R}^m(\cdot,t)]$ belongs to $C^0(O_M)$ which shows explicitly that $t \to \hat{R}^m(\cdot,t) \in C^2(O_M)$ on $[o,b]$ (cf. Lemma 3.3) and that $t \to \hat{R}^m(\cdot,t)\hat{T}$ satisfies (3.4) in S'. Consequently (3.2) holds in S' and $w^m_t(o) = 0$ since by (3.11) $\hat{R}^m_t(y,o) = 0$. QED

To deal with uniqueness we make first a few formal calculations following Carroll [8; 11]; this technique is somewhat different than that used in Carroll [5] but by contrast it can be generalized. Thus for $o < \tau \leq t \leq b < \infty$ let $\hat{R}^m(y,t,\tau)$ and

$\hat{S}^m(y,t,\tau)$ be two linearly independent solutions of (3.4) with $\hat{R}^m(y,\tau,\tau) = 1$, $\hat{R}^m_t(y,\tau,\tau) = 0$, $\hat{S}^m(y,\tau,\tau) = 0$, and $\hat{S}^m_t(y,\tau,\tau) = 1$. Such (unique) numerical solutions exist since the problem is non-singular on $[\tau,b]$ for $t > 0$. Let

$$(3.13)\qquad \hat{G}^m(y,t,\tau) = \begin{pmatrix} \hat{R}^m(y,t,\tau) & \hat{S}^m(y,t,\tau) \\ \hat{R}^m_t(y,t,\tau) & \hat{S}^m_t(y,t,\tau) \end{pmatrix}$$

so that $\hat{G}^m$ is a fundamental matrix for the first order system corresponding to (3.4) written in the form

$$(3.14)\qquad \vec{V}^m_t + \hat{M}\vec{V}^m = 0;\ \hat{M}(y,t) = \begin{pmatrix} 0 & -1 \\ A(y) & \frac{2m+1}{t} \end{pmatrix};\ \vec{V}^m = \begin{pmatrix} \hat{V}^m \\ \hat{V}^m_t \end{pmatrix}$$

As in Schwartz [3] it is easily shown that

$$(3.15)\qquad \frac{\partial}{\partial\tau}\hat{G}^m(y,t,\tau) = \hat{G}^m(y,t,\tau)\,\hat{M}(y,\tau)$$

from which follows in particular

$$(3.16)\qquad \hat{R}^m_\tau(y,t,\tau) = A(y)\,\hat{S}^m(y,t,\tau)$$

$$(3.17)\qquad \hat{S}^m_\tau(y,t,\tau) = -\hat{R}^m(y,t,\tau) + \frac{2m+1}{\tau}\hat{S}^m(y,t,\tau)$$

Given that $\hat{R}^m$ and $\hat{S}^m$ have suitable properties (stated and verified below) it follows that if $\hat{w}^m$ is a solution of (3.4) in S' with $\hat{w}^m(\tau) = \hat{T}$ and $\hat{w}^m_t(\tau) = 0$ then using (3.16) - (3.17) the formulas (3.18) - (3.20) below make sense.

$$(3.18)\qquad \int_\tau^t \hat{S}^m(y,t,\xi)A(y)\hat{w}^m(\xi)d\xi = \int_\tau^t \hat{R}^m_\xi(y,t,\xi)\hat{w}^m(\xi)d\xi$$

$$= \hat{w}^m(t) - \hat{R}^m(y,t,\tau)\hat{T} - \int_\tau^t \hat{R}^m(y,t,\xi)\hat{w}^m_\xi(\xi)\,d\xi$$

$$(3.19)\qquad \int_\tau^t \hat{S}^m(y,t,\xi)\hat{w}^m_{\xi\xi}(\xi)d\xi = -\int_\tau^t \hat{S}^m_\xi(y,t,\xi)\hat{w}^m_\xi(\xi)d\xi$$

$$(3.20)\qquad 0 = \int_\tau^t \hat{S}^m(y,t,\xi)[\hat{w}^m_{\xi\xi}(\xi) + \frac{2m+1}{\xi}\hat{w}^m_\xi(\xi) + A(y)\hat{w}^m(\xi)]d\xi$$

$$= -\int_\tau^t [\hat{S}^m_\xi(y,t,\xi) - \frac{2m+1}{\xi}\hat{S}^m(y,t,\xi) + \hat{R}^m(y,t,\xi)]w^m_\xi(\xi)d\xi$$

$$+ \hat{w}^m(t) - \hat{R}^m(y,t,\tau)\hat{T} = \hat{w}^m(t) - \hat{R}^m(y,t,\tau)\hat{T}$$

From (3.20) it then follows formally that any solution $\hat{w}^m$ of (3.4) in S' with $\hat{w}^m(\tau) = \hat{T}$ and $\hat{w}^m_t(\tau) = 0$ $(\tau > 0)$ must be of the form $\hat{w}^m(t) = \hat{R}^m(y,t,\tau)\hat{T}$. We will now proceed to justify this and to attend to the case $\tau = 0$ as well. We remark also that if one considers

$$(3.21)\qquad \hat{w}^m_{tt} + \frac{2m+1}{t}\hat{w}^m_t + A(y)\hat{w}^m = \hat{f}$$

with initial conditions given at $t = \tau$ as above $(\hat{f}(\cdot)\varepsilon C^0(S'))$ then the procedure of (3.20) formally yields the unique solution as

$$(3.22)\qquad \hat{w}^m(t) = \hat{R}^m(y,t,\tau)\hat{T} + \int_\tau^t \hat{S}^m(y,t,\xi)\hat{f}(\xi)d\xi$$

We will develop a procedure now which can also be used in other situations (see Section 1.5). It is helpful to have an explicit example of the technique however and hence we introduce it here; in particular this will make the considerable details more easily visible and shorten the development later. Thus for $m \geq -1/2$ let $P(t) = \exp\left(-\int_t^b \frac{(2m+1)}{\xi}d\xi\right) = (t/b)^{2m+1}$ so that $P(\tau)/P(t) = (\tau/t)^{2m+1} \leq 1$ for $0 \leq \tau \leq t \leq b$ and $P(t) \to 0$ as $t \to 0$. Let us consider equation (3.4) for $\hat{R}^m(y,t,\tau)$ with initial data at

$t = \tau > 0$ as in (3.13) and then, multiplying by $P(t)$, we obtain

$$\frac{d}{dt}(P\hat{R}^m_t) + PA\hat{R}^m = 0 \tag{3.23}$$

This leads to the integral equation ($\tau \le \xi \le \sigma \le t$), equivalent to (3.4), with initial conditions at $t = \tau$ as indicated,

$$\begin{aligned}\hat{R}^m(y,t,\tau) &= 1 - \int_\tau^t \frac{1}{P(\sigma)} \int_\tau^\sigma P(\xi)A(y)\hat{R}^m(y,\xi,\tau)\,d\xi d\sigma \\ &= 1 - A(y)\int_\tau^t\int_\tau^\sigma \left(\frac{\xi}{\sigma}\right)^{2m+1}\hat{R}^m(y,\xi,\tau)\,d\xi d\sigma = 1 - J \cdot \hat{R}^m\end{aligned} \tag{3.24}$$

and evidently $\tau = 0$ is permitted in (3.24). A solution to (3.24) is given formally by the series

$$\hat{R}^m(y,t,\tau) = \sum_{k=0}^{\infty} (-1)^k J^k \cdot 1 \tag{3.25}$$

(cf. Hille [1]) where J^k denotes the kth iterate of the operation J and J^0 is the identity. Writing $\hat{R}^{m,0} = 1$ and

$$\hat{R}^{m,p} = 1 - J\hat{R}^{m,p-1} = \sum_{k=0}^{p} (-1)^k J^k \cdot 1 \tag{3.26}$$

we have $\hat{R}^{m,p} - \hat{R}^{m,p-1} = (-1)^p J^p \cdot 1$ and if $y \in K \subset \mathbb{C}^n$, with K compact, then $|A(y)|_C \le c_K$ and in particular

$$\begin{aligned}|\hat{R}^{m,1} - \hat{R}^{m,0}|_C &= |J \cdot 1|_C \le |A(y)|_C \int_\tau^t\int_\tau^\sigma \left(\frac{\xi}{\sigma}\right)^{2m+1} d\xi d\sigma \\ &\le c_K F(t,\tau) \le c_K F(t,0) = c_K F(t)\end{aligned} \tag{3.27}$$

where $F(t,\tau)$ is the double integral in (3.27) and $F(t_1) \le F(t) \le F(b)$ for $0 \le t_1 \le t \le b$ (note that $F(t) = \frac{t^2}{4(m+1)}$). Continuing we have, using (3.27),

$$(3.28)\qquad |\hat{R}^{m,2} - \hat{R}^{m,1}|_C = |J(\hat{R}^{m,1} - \hat{R}^{m,0})|_C$$
$$\leq c_K^2 \int_\tau^t \int_\tau^\sigma (\tfrac{\xi}{\sigma})^{2m+1} F(\xi)\, d\xi\, d\sigma$$
$$\leq c_K^2 \int_\tau^t F(\sigma) \int_\tau^\sigma (\tfrac{\xi}{\sigma})^{2m+1} d\xi\, d\sigma$$

so that $|\hat{R}^{m,2} - \hat{R}^{m,1}|_C \leq c_K^2 F^2(t)/2!$ since if $\tilde{F}(t) = \int_0^t F(\sigma) \int_0^\sigma (\frac{\xi}{\sigma})^{2m+1} d\xi d\sigma$ then $\tilde{F}' = F\,F'$ with $\tilde{F}(o) = F(o) = 0$. By iteration one obtains

$$(3.29)\qquad |\hat{R}^{m,p} - \hat{R}^{m,p-1}|_C = |J^p \cdot 1|_C \leq \frac{c_K^p F^p(b)}{p!}$$

Hence the series (3.25) converges absolutely and uniformly on $[0 \leq \tau \leq t \leq b] \times K$ and since the terms $J^p \cdot 1$ are continuous in (y,t,τ) and analytic in y the same is true of $\hat{R}^m(y,t,\tau)$. It is clear that $\hat{R}^m$ given by (3.25) satisfies (3.24) and we can state

Lemma 3.5 The series (3.25) represents a solution of (3.24) with $\hat{R}^m(y,\tau,\tau) = 1$ and $\hat{R}_t^m(y,\tau,\tau) = 0$ ($0 \leq \tau \leq t \leq b$ and $m > -1/2$). The maps $(y,t,\tau) \to \hat{R}^m(y,t,\tau)$ and $(y,t,\tau) \to \hat{R}_t^m(y,t,\tau)$ are continuous numerical functions while $\hat{R}^m(\cdot,t,\tau)$ and $\hat{R}_t^m(\cdot,t,\tau)$ are analytic.

Proof: Everything has been proved for $\hat{R}^m$ and the statements for $\hat{R}_t^m$ follow immediately upon differentiating (3.24) in t. QED

We examine now (cf. (3.24))

$$(3.30)\qquad \frac{2m+1}{t}\hat{R}^m_t(y,t,\tau) = -\frac{(2m+1)}{t}A(y)\int_\tau^t (\frac{\xi}{t})^{2m+1}\hat{R}^m(y,\xi,\tau)\,d\xi$$

For $\tau > 0$ there is obviously no problem in proving continuity and analyticity as in Lemma 3.5 so we consider $\tau \to 0$. Let

$$(3.31)\qquad H(t,\tau) = \frac{2m+1}{t}\int_\tau^t (\frac{\xi}{t})^{2m+1}d\xi = \frac{2m+1}{2(m+1)}[1 - (\frac{\tau}{t})^{2(m+1)}]$$

Clearly $(t,\tau) \to H(t,\tau)$ is not continuous as $(t,\tau) \to (0,0)$ so there is no hope that $(y,t,\tau) \to \frac{2m+1}{t}\hat{R}^m_t(y,t,\tau)$ will be continuous on $[0 \le \tau \le t \le b] \times K$ (cf. Lemma 3.5). We write $H(t,o) = H(t) = \frac{2m+1}{2(m+1)}$ and then clearly, from (3.30) - (3.31), $(y,t) \to \frac{2m+1}{t}\hat{R}^m_t(y,t,o)$ is continuous on $[o,b] \times K$ with limit as $(y,t) \to (y_0,o)$ equal to $-A(y_0)\frac{2m+1}{2(m+1)}$. Further from (3.30) $y \to \frac{2m+1}{t}\hat{R}^m_t(y,t,\tau)$ is analytic for $0 \le \tau \le t \le b$. These properties are transported to $\hat{R}^m_{tt}(y,t,\tau)$ after differentiating (3.24) twice in t and equation (3.4) is satisfied for $0 \le \tau \le t \le b$. Summarizing we have

<u>Lemma 3.6</u> For $[0 \le \tau \le t \le b]$ the maps $(y,t) \to \hat{R}^m_{tt}(y,t,\tau)$ and $(y,t) \to \frac{2m+1}{t}\hat{R}^m_t(y,t,\tau)$ are continuous while $y \to \hat{R}^m_{tt}(y,t,\tau)$ and $y \to \frac{2m+1}{t}\hat{R}^m_t(y,t,\tau)$ are analytic. The numerical function $\hat{R}^m(y,t,\tau)$ satisfies (3.4) with $\hat{R}^m(y,\tau,\tau) = 1$ and $\hat{R}^m_t(y,\tau,\tau) = 0$.

Next we will derive some bounds for $\hat{R}^m$, assuming that y is real. If we multiply (3.23) (with t replaced by ξ) by $P^{-1}(\xi)\hat{R}^m_\xi(y,\xi,\tau)$ and integrate by parts it follows that

$$(3.32)\qquad |\hat{R}^m_t(y,t,\tau)|^2 + 2(2m+1)\int_\tau^t \frac{1}{\xi}|\hat{R}^m(y,\xi,\tau)|^2 d\xi + A(y)|\hat{R}^m(y,t,\tau)|^2 = A(y)$$

since $\frac{d}{dt}|\hat{R}^m|^2 = 2\hat{R}^m\hat{R}^m_t$ and $\frac{d}{dt}|\hat{R}^m_t|^2 = 2\hat{R}^m_t\hat{R}^m_{tt}$ (cf. also Section 1.4). In particular if $|y| > R_0$ we can say that $|\hat{R}^m_t(y,t,\tau)|^2 \leq A(y)$ and $|\hat{R}^m(y,t,\tau)|^2 \leq 1$. Since $|\hat{R}^m(y,t,\tau)|$ and $|\hat{R}^m_t(y,t,\tau)|$ can be bounded on $[0 \leq \tau \leq t \leq b] \times \{0 \leq |y| \leq R_0\}$ by continuity (cf. Lemma 3.5) it follows that $|\hat{R}^m(y,t,\tau)|^2 \leq c_1$ and $|\hat{R}^m_t(y,t,\tau)|^2 \leq c_2 + A(y)$ for $0 \leq \tau \leq t \leq b$ and all y. To bound the term $\frac{2m+1}{t}\hat{R}^m_t(y,t,\tau)$ we refer to (3.30) - (3.31) to obtain

$$\text{(3.33)} \qquad \frac{2m+1}{t}|\hat{R}^m_t(y,t,\tau)| \leq |A(y)|c_1^{1/2}H(t,\tau) \leq c_3|A(y)|$$

For $|y| > R_0$, $|A(y)| = A(y)$ and for $|y| \leq R_0$, $|A(y)|$ is bounded; this leads to

<u>Lemma 3.7</u> The function $\hat{R}^m$ satisfies $|\hat{R}^m(y,t,\tau)|^2 \leq c_1$, $|\hat{R}^m_t(y,t,\tau)|^2 \leq c_2 + A(y)$, and $\frac{2m+1}{t}|\hat{R}^m_t(y,t,\tau)| \leq c_4 + c_3 A(y)$ for $0 \leq \tau \leq t \leq b$ and all $y \in \mathbb{R}^n$.

Let us examine now some similar properties of $\hat{S}^m$. If we differentiate (3.24) in τ there results (since $\hat{R}^m(y,\tau,\tau) = 1$)

$$\text{(3.34)} \qquad \hat{R}^m_\tau = -J \cdot \hat{R}^m_\tau + \int_\tau^t \frac{P(\tau)}{P(\sigma)} A(y)\, d\sigma$$

$$= -J \cdot \hat{R}^m_\tau + A(y)U(t,\tau)$$

where for $m \geq -1/2$, $U(t,\tau) = \frac{\tau}{2m}[1 - (\frac{\tau}{t})^{2m}]$ is continuous in (t,τ) for $0 \leq \tau \leq t \leq b$ (note that $U(t,\tau) \to -\tau \log(\tau/t)$ as $m \to 0$). From (3.16) we have then

$$\text{(3.35)} \qquad \hat{S}^m = U(t,\tau) - J \cdot \hat{S}^m$$

(except perhaps where $A(y) = 0$). A solution of (3.35) is given by

$$(3.36) \qquad \hat{S}^m = \sum_{k=0}^{\infty} (-1)^k J^k \cdot U$$

and the terms $J^k \cdot U$ in (3.36) are continuous in (y,t,τ) and analytic in y with $|U(t,\tau)| \leq c$. It follows by our previous analysis (cf. (3.26) - (3.29)) that the series in (3.36) converges absolutely and uniformly on $[0 \leq \tau \leq t \leq b] \times K$ ($K \subset \mathbb{C}^n$ being compact). Defining now $\hat{S}^m$ by this series (whether or not $A(y) = 0$) there results

<u>Lemma 3.8</u> For $m \geq -1/2$ the expression $\hat{S}^m(y,t,\tau)$ defined by (3.36) satisfies (3.35) and 3.4); it is continuous in (y,t,τ) and analytic in y for $[0 \leq \tau \leq t \leq b]$ along with $\hat{S}^m_t(y,t,\tau)$ for $0 < \tau \leq t \leq b$. One has $\hat{R}^m_\tau = A(y)\hat{S}^m$ (i.e., (3.16)) and, for $\tau > 0$, (3.17) holds while $\hat{S}^m(y,\tau,\tau) = 0$ with $\hat{S}^m_t(y,\tau,\tau) = 1$; finally $\hat{S}^m(y,t,o) = 0$ for $m > -1/2$.

Proof: One notes first that the series obtained by termwise differentiation of (3.25) in τ is the same as $A(y)\hat{S}^m(y,t,\tau)$ where $\hat{S}^m$ is given by (3.36). Since $J^k \cdot U = 0$ at $t = \tau$ for $k \geq 1$ with $U(\tau,\tau) = 0$ it follows that $\hat{S}^m(y,\tau,\tau) = 0$. On the other hand $\hat{S}^m(y,t,o) = 0$ for $m > -1/2$ since $U(t,o) = 0$ while (3.17) holds for $\tau > 0$ since $\dot{\hat{S}}^m$ clearly satisfies (3.35) and hence (since $\hat{S}^m(y,\tau,\tau) = 0$) $\hat{S}^m_\tau = U_\tau - \frac{d}{d\tau} J \cdot \hat{S}^m = U_\tau - J \cdot \hat{S}^m_\tau$. But $U_\tau(t,\tau) = \frac{2m+1}{\tau} U(t,\tau) - 1$ and hence for $\tau > 0$

$$(3.37)\qquad \hat{S}^m_\tau = \sum_{k=0}^{\infty} (-1)^k J^k \cdot U_\tau = \frac{2m+1}{\tau}\hat{S}^m - \hat{R}^m$$

which is (3.17). Further, since $U_t = (\frac{\tau}{t})^{2m+1}$,

$$(3.38)\qquad \hat{S}^m_t(y,t,\tau) = (\frac{\tau}{t})^{2m+1} - \int_\tau^t (\frac{\xi}{t})^{2m+1} A(y)\hat{S}^m(y,\xi,\tau)d\xi$$

(from (3.35) and the definition of J), from which follows $\hat{S}^m_t(y,\tau,\tau) = 1$ for $\tau > 0$ and the continuity and analyticity properties of $\hat{S}^m_t$. To check (3.4) one can use (3.17) to obtain for $\tau > 0$

$$(3.39)\qquad \hat{S}^m(y,t,\tau) = \int_\tau^t (\frac{\tau}{\eta})^{2m+1}\hat{R}^m(y,t,\eta)d\eta$$

which leads to (3.4) upon differentiation in t. QED

Now $\hat{S}^m_\tau$ is determined by (3.17) and as $\tau \to 0$ a priori $\lim \hat{S}^m_\tau$ may not exist. However let us look first at

$$(3.40)\qquad \frac{1}{\tau}\hat{S}^m(y,t,\tau) = \sum_{k=0}^{\infty}(-1)^k J^k \cdot \frac{1}{\tau}U(\cdot,\tau)$$

where $\frac{1}{\tau}U(\xi,\tau) = \frac{1}{2m}[1 - (\frac{\tau}{\xi})^{2m}] \le \frac{1}{2m}$ for $m > 0$ and $0 \le \tau \le \xi \le t \le b$. Hence, as in (3.26) - (3.29), the series in (3.40) converges uniformly and absolutely for $0 \le \tau \le t \le b$ and $y \in K \subset \mathbb{C}^n$. There is little hope that $(y,t,\tau) \to \frac{1}{\tau}\hat{S}^m(y,t,\tau)$ will be continuous on $[0 \le \tau \le t \le b]$ since e.g., $\frac{1}{\tau}U(t,\tau)$ is not so continuous (nor is $J \cdot \frac{1}{\tau}U(\cdot,\tau)$, etc.). We can however let $\tau \to 0$ in (3.40) for $m > 0$ to obtain for $t > 0$

$$(3.41)\qquad \lim_{\tau\to 0}\frac{1}{\tau}\hat{S}^m(y,t,\tau) = \frac{1}{2m}\sum_{k=0}^{\infty}(-1)^k J^k \cdot 1 = \frac{1}{2m}\hat{R}^m(y,t,o)$$

Consequently from (3.17) for $t > 0$ and $m > 0$

$$(3.42)\qquad \hat{S}^m_\tau(y,t,o) = \lim_{\tau\to 0} S^m_\tau(y,t,\tau) = \left(\frac{2m+1}{2m} - 1\right)\hat{R}^m(y,t,o)$$

$$= \frac{1}{2m}R^m(y,t,o)$$

Lemma 3.9 For $m > 0$ and $m = -1/2$ the function $(y,t) \to \hat{S}^m_\tau(y,t,o) = \frac{1}{2m}\hat{R}^m(y,t,o)$ is continuous for $0 \leq t \leq b$ and $y \in K$ while $y \to \hat{S}^m_\tau(y,t,o)$ is analytic. Further, $\hat{S}^m(y,t,\cdot)$ is absolutely continuous for $m \geq -1/2$ with $\hat{S}^m(y,t,\beta) - \hat{S}^m(y,t,\alpha) = \int_\alpha^\beta \hat{S}^m_\tau(y,t,\tau)d\tau$ and for $0 \leq \tau \leq t \leq b$ and $m \geq -1/2$ the expression $\tau\hat{S}^m_\tau(y,t,\tau)$ is continuous in (y,t,τ) and analytic in y.

Proof: The continuity and analyticity indicated for $\hat{S}^m_\tau(y,t,o)$ and $\tau\hat{S}^m_\tau(y,t,\tau)$ follow immediately from (3.42) and (3.17). For $0 < \alpha \leq \beta \leq t \leq b$ the formula $\hat{S}^m(y,t,\beta) - \hat{S}^m(y,t,\alpha) = \int_\alpha^\beta \hat{S}^m_\tau(y,t,\tau)d\tau$ is obvious and by continuity one can take limits as $\alpha \to 0$. In the cases $-1/2 < m \leq 0$ we note that $\hat{S}^m_\tau$ will have an integrable singularity at $\tau = 0$ since $\frac{1}{\tau}U(t,\tau) \sim \tau^{2m}$ for $-1/2 < m < 0$ and $\sim \log\tau$ for $m = 0$. QED

In order to obtain some bounds for $\hat{S}^m$ when $y \in \mathbb{R}^n$ we multiply (3.17) (with τ replaced by ξ) by $\hat{S}^m(y,t,\xi)$ and since $\frac{d}{d\xi}|\hat{S}^m|^2 = 2\hat{S}^m\hat{S}^m_\xi$ one obtains upon integration

$$(3.43) \qquad \frac{1}{2}|\hat{S}^m(y,t,\tau)|^2 + (2m+1)\int_\tau^t \frac{1}{\xi}|\hat{S}^m(y,t,\xi)|^2\, d\xi = \int_\tau^t \hat{R}^m(y,t,\xi)\,\hat{S}^m(y,t,\xi)\, d\xi$$

Since $|\hat{R}^m\hat{S}^m| \le \frac{1}{2}(|\hat{R}^m|^2 + |\hat{S}^m|^2)$ this leads to (cf. Lemma 3.7)

$$(3.44) \qquad |\hat{S}^m|^2 \le c_1 b + \int_\tau^t |\hat{S}^m|^2\, d\xi$$

By Gronwall's lemma (cf. Section 5 and Carroll [14]) there results $|\hat{S}^m(y,t,\tau)|^2 \le c_5$ for $0 \le \tau \le t \le b$ and $y \in \mathbb{R}^n$. Consequently by (3.17) and Lemma 3.7 $|\tau\hat{S}^m_\tau(y,t,\tau)| \le c_6$ on the same domain. Finally from (3.38) we have $|\hat{S}^m_t| \le 1 + bc_5^{1/2}|A(y)| \le c_7 + c_8A(y)$.

<u>Lemma 3.10</u> For $y \in \mathbb{R}^n$ and $0 \le \tau \le t \le b$ one has $|\hat{S}^m(y,t,\tau)|^2 \le c_5$ and $|\tau\hat{S}^m_\tau(y,t,\tau)| \le c_6$ while $|\hat{S}^m_t(y,t,\tau)| \le c_7 + c_8A(y)$.

<u>Lemma 3.11</u> The function $t \to \hat{R}^m(\cdot,t,\tau) \in C^2(E_y)$ for $0 \le \tau \le t \le b$ while $(t,\tau) \to \hat{R}^m(\cdot,t,\tau)$ and $(t,\tau) \to \hat{R}^m_t(\cdot,t,\tau)$: $[0 \le \tau \le t \le b] \to E_y$ are continuous. Similarly $\xi \to \hat{S}^m(\cdot,t,\xi) \in C^0(E_y)$, $\xi \to \xi\hat{S}^m_\xi(\cdot,t,\xi) \in C^0(E_y)$, $t \to \hat{S}^m(\cdot,t,\xi) \in C^2(E_y)$, $(t,\xi) \to \hat{S}^m(\cdot,t,\xi) \in C^0(E_y)$, and $\xi \to \hat{R}^m(\cdot,t,\xi) \in C^1(E_y)$ for $0 \le \xi \le t \le b$ while $\xi \to \hat{S}^m(\cdot,t,\xi) \in C^1(E_y)$ and $(t,\xi) \to \hat{S}^m_t(\cdot,t,\xi) \in C^0(E_y)$ for $0 < \xi_0 \le \xi \le t \le b$.

Proof: We will indicate the proof for some sample cases and the rest follows in a similar manner. To show $(t,\tau) \to$

$\hat{R}^m(\cdot,t,\tau)$ and $(t,\tau) \to \hat{R}^m_t(\cdot,t,\tau)$ continuous with values in E_y while $t \to \hat{R}^m(\cdot,t,\tau) \in C^2(E_y)$ for instance let $y \in K \subset \mathbb{R}^n$ and enclose K in the interior of a compact polydisc $\tilde{K} \subset \mathbb{C}^n$. If $\alpha = (\alpha_1, \ldots, \alpha_n)$ write $D^\alpha_y = (\partial/\partial y_1)^{\alpha_1} \ldots (\partial/\partial y_n)^{\alpha_n}$ and let $\Gamma = \partial\tilde{K}$ be the distinguished boundary of $\tilde{K}$. We have a Cauchy integral formula (cf. Gunning-Rossi [1]) which leads to

$$(3.45) \qquad D^\alpha_y\hat{R}^m(y,t,\tau) - D^\alpha_y\hat{R}^m(y,t_0,\tau_0)$$

$$= \frac{\Pi\alpha_k!}{(2\pi i)^n} \int_\Gamma \frac{[\hat{R}^m(\zeta,t,\tau) - \hat{R}^m(\zeta,t_0,\tau_0)]}{\Pi(\zeta_k - y_k)^{\alpha_k+1}} \Pi d\zeta_k$$

By uniform continuity in (ζ,t,τ) on $\tilde{K} \times [0 \le \tau \le t \le b]$ if $(t,\tau) \to (t_0,\tau_0)$ then $\hat{R}^m(\zeta,t,\tau) \to \hat{R}^m(\zeta,t_0,\tau_0)$ uniformly on Γ and consequently $D^\alpha_y\hat{R}^m(y,t,\tau) \to D^\alpha_y\hat{R}^m(y,t_0,\tau_0)$ uniformly for $y \in K$. It follows that $(t,\tau) \to \hat{R}^m(\cdot,t,\tau) \in C^0(E_y)$. A similar argument can obviously be applied to $\hat{R}^m_t$ etc. For differentiability in t of $\hat{R}^m(\cdot,t,\tau) \in E_y$ for example one wants to show that $\Delta\hat{R}^m/\Delta t = [\hat{R}^m(\cdot,t,\tau) - \hat{R}^m(\cdot,t_0,\tau)]/\Delta t \to \hat{R}^m_t(\cdot,t_o,\tau)$ in E_y as $\Delta t = t - t_0 \to 0$. This means that we must prove that $D^\alpha_y(\Delta\hat{R}^m/\Delta t) \to D^\alpha_y\hat{R}^m_t(y,t_0,\tau)$ uniformly for $y \in K$ $\mathbb{R}^n$. This may be written (assuming $t > t_0$ for convenience)

$$(3.46) \qquad D^\alpha_y \frac{\Delta\hat{R}^m}{\Delta t} - D^\alpha_y\hat{R}^m_t = \frac{1}{\Delta t}\int_{t_0}^t D^\alpha_y[\hat{R}^m_t(y,\xi,\tau) - \hat{R}^m_t(y,t_0,\tau)]d\xi$$

Byt by the continuity of $\xi \to \hat{R}^m_t(\cdot,\xi,\tau)$ we can make $|D^\alpha_y\hat{R}^m_t(y,\xi,\tau) - D^\alpha_y\hat{R}^m(y,t_0,\tau)| \le \varepsilon$ for $|\xi - t_0| \le \delta(\varepsilon)$, uniformly for $y \in K$, and hence $t \to \hat{R}^m(\cdot,t,\tau) \in C^1(E_y)$. A similar argument may be applied to $\Delta\hat{R}^m_t/\Delta t - \hat{R}^m_{tt}$, etc. QED

<u>Lemma 3.12</u> For $0 \leq \tau \leq t \leq b$ the formula (3.22) holds pointwise or in E_y if $\hat{w}^m$ is replaced by any numerical function $\hat{\phi}$ of argument (y,t,τ) satisfying (3.21) with $\hat{f} \in C^0(E_y)$ where $t \to \hat{\phi}(\cdot,t,\tau) \in C^2(E_y)$ while initial conditions $\hat{\phi}(\cdot,\tau,\tau) = \hat{T}(\cdot) \in E_y$ and $\hat{\phi}_t(\cdot,\tau,\tau) = 0$ are stipulated. Such a function $\hat{\phi}$ is necessarily unique by (3.20).

Proof: Pointwise for (y,t,τ) fixed everything is trivial since our lemmas concerning properties of $\hat{R}^m$ and $\hat{S}^m$ make all calculations in (3.18) - (3.20) legitimate for $\tau > 0$ and one can let $\tau \to 0$ in the resulting formulas (cf. Lemmas 3.5, 3.6, 3.8, 3.9 and note that terms $\frac{1}{\xi}\hat{S}^m(y,t,\xi)\hat{\phi}(y,\xi,\tau)$ are continuous since $\hat{\phi}$ satisfies (3.21) with $\frac{1}{\xi}\hat{\phi}(y,\xi,\tau)$ continuous). To work in E_y we recall that $(S,T) \to ST : E_y \times E_y \to E_y$ is separately continuous with E_y a Frechet space and hence this map is continuous (see Bourbaki [2] and for vector valued integration see Bourbaki [3; 4], Carroll [14]). Hence all of the operations in (3.18) - (3.20) and (3.22) make sense in E_y. QED

<u>Lemma 3.13</u> For $0 \leq \tau \leq t \leq b$ we have $(t,\tau) \to \hat{R}^m(\cdot,t,\tau) \in C^0(O_M)$, $(t,\tau) \to \hat{R}^m_t(\cdot,t,\tau) \in C^0(O_M)$, $t \to \hat{R}^m(\cdot,t,\tau) \in C^2(O_M)$, $\tau \to \hat{S}^m(\cdot,t,\tau) \in C^0(O_M)$, $\tau \to \tau\hat{S}^m_\tau(\cdot,t,\tau) \in C^0(O_M)$, and $\tau \to \hat{R}^m(\cdot,t,\tau) \in C^1(O_M)$ while, for $0 < \tau \leq t \leq b$, $\tau \to \hat{S}^m(\cdot,t,\tau) \in C^1(O_M)$.

Proof: Again we will prove this for some sample cases to

indicate a procedure which can then be applied in general. We recall that on bounded sets in O_M the topology of O_M coincides with that induced by E and that if $t \to \psi(\cdot,t) \in C^0(O_M) \cap C^1(E)$ with $t \to \psi_t(\cdot,t) \in C^0(O_M)$ then $t \to \psi(\cdot,t) \in C^1(O_M)$ (cf. Schwartz [2; 3; 4] and Lemma 3.3). In order to study $D_y^\alpha \hat{R}^m$ for example we replace $\hat{w}^m$ by $\hat{R}^m$ in (3.4) and differentiate in y to obtain

$$(3.47) \qquad (D_y^\alpha \hat{R}^m)_{tt} + \frac{2m+1}{t}(D_y^\alpha \hat{R}^m)_t + A(y) D_y^\alpha \hat{R}^m$$

$$= \sum_{|\gamma| \leq |\alpha|-1} P_\gamma(y) D_y^\gamma \hat{R}^m$$

where the $P_\gamma(y)$ are polynomials and $|\alpha| = \sum \alpha_k$. The initial conditions are $D_y^\alpha \hat{R}^m(y,\tau,\tau) = D_y^\alpha \hat{R}_t^m(y,\tau,\tau) = 0$ and by Lemma 3.12 the unique solution of (3.47) is given by (3.22) in the form

$$(3.48) \qquad D_y^\alpha \hat{R}^m(y,t,\tau) = \int_\tau^t \hat{S}^m(y,t,\xi) \sum P_\gamma(y) D_y^\gamma \hat{R}^m(y,\xi,\tau)\, d\xi$$

where the $D_y^\gamma \hat{R}^m$ may be regarded as known functions (by an induction procedure). Now, referring to Lemmas 3.7 and 3.10, we know $|\hat{S}^m(y,t,\xi)|^2 \leq c_5$ and by induction the $D_y^\gamma \hat{R}^m(y,\xi,\tau)$ for $|\gamma| \leq |\alpha| - 1$ will be bounded by polynomials in y independent of (ξ,τ). Consequently all $D_y^\alpha \hat{R}^m(y,t,\tau)$ will be bounded by polynomials in y depending only on α so that $\hat{R}^m(\cdot,t,\tau)$ is bounded in O_M for $0 \leq \tau \leq t \leq b$ and hence from Lemma 3.11 we can say that $(t,\tau) \to \hat{R}^m(\cdot,t,\tau) \in C^0(O_M)$. Further if one differentiates (3.48) in t there results

$$(3.49) \qquad D_y^\alpha \hat{R}_t^m(y,t,\tau) = \int_\tau^t \hat{S}_t^m(y,t,\xi) \sum P_\gamma(y) D_y^\gamma \hat{R}^m(y,\xi,\tau)\, d\xi$$

for $\tau > 0$, with $|\hat{S}^m_t(y,t,\xi)| \le c_7 + c_8 A(y)$ for $0 \le \tau \le t \le b$, from which follows the desired polynomial bound for $D^\alpha_y \hat{R}^m_t$ since one can let $\tau \to 0$ in the resulting inequality for $|D^\alpha_y \hat{R}^m_t|$. Finally we have from (3.49) for $\tau > 0$

$$(3.50) \qquad D^\alpha_y \hat{R}^m_{tt}(y,t,\tau) = \sum P_\gamma(y) D^\gamma_y \hat{R}^m(y,t,\tau) + \int_\tau^t \hat{S}^m_{tt}(y,t,\xi) \sum P_\gamma(y) D^\gamma_y \hat{R}^m(y,\xi,\tau)\, d\xi$$

since $\hat{S}^m_t(y,t,t) = 1$ while $\hat{S}^m_{tt}$ is continuous in (y,t,τ) for $\tau > 0$. But $\hat{S}^m_{tt} = -A(y)\hat{S}^m - \frac{2m+1}{t}\hat{S}^m_t$ and we need only check

$$(3.51) \qquad \frac{1}{t}\int_\tau^t |\hat{S}^m_t(y,t,\xi)|\, d\xi \le (c_7 + c_8 A(y))\,(1 - \frac{\tau}{t}) \le c_7 + c_8 A(y)$$

Again the resulting estimate on $|D^\alpha_y \hat{R}^m_{tt}|$ will hold as $\tau \to 0$. Hence $D^\alpha_y \hat{R}^m_{tt}$ can be bounded by polynomials in y independent of (t,τ) for $0 \le \tau \le t \le b$ and consequently $t \to \hat{R}^m(\cdot,t,\tau) \in C^2(O_M)$. The rest follows in a similar manner. QED

We can now prove uniqueness (and well posedness) for the solution $w^m(t) = R^m(t) * T$ of (3.2) - (3.3) given by Theorem 3.4. As will be seen later there are quicker ways to prove uniqueness when one is working in a Hilbert or Banach space but in order to study the growth of solutions for example it is absolutely necessary to work in "big" spaces such as S' or D' and for this a more elaborate uniqueness technique is required, such as we have developed. It is sufficient now to deal with the solution

$\hat{w}^m(t,\tau) = \hat{R}^m(y,t,\tau)\hat{T}$ of (3.4) with initial conditions of the form (3.5) given at $t = \tau$. We recall that the map $(S,T) \to ST$: $0_M \times S' \to S'$ is hypocontinuous and then, given any solution $\hat{w}^m$ of (3.4) (with t replaced by ξ), multiply by $\hat{S}^m(y,t,\xi)$ as in (3.20) to conclude that $\hat{w}^m$ is necessarily of the form $\hat{R}^m(y,t,\tau)\hat{T}$. The same uniqueness argument works for any solution $\hat{w}^m$ of (3.21) so that (3.22) holds in S'. Hence we have proved

<u>Theorem 3.14</u> Assume $m \geq -1/2$ and that $A(y) = F(A_x\delta)$ satisfies Condition 3.1. Then the unique solution of the equation

$$(3.52) \qquad w^m_{tt} + \frac{2m+1}{t} w^m_t + A_x w^m = f$$

with $f(\cdot) \in C^0(S')$ and initial data $w^m(\tau) = T \in S'$ and $w^m_t(\tau) = 0$ $(0 \leq \tau \leq t \leq b)$ is given by

$$(3.53) \qquad w^m(t,\tau) = R^m(\cdot,t,\tau) * T + \int_\tau^t S^m(\cdot,t,\xi) * f(\xi)\, d\xi$$

<u>Definition 3.15</u> Following Schwartz [3] we will say that (3.53) with $f(\cdot) \in C^0(S')$, $w^m(\tau) = T \in S'$, and $w^m_t(\tau) = 0$ is uniformly well posed if the solution (3.53) depends continuously in S' on (t,τ,T,f).

<u>Theorem 3.16</u> The problem (3.52) with $f(\cdot) \in C^0(S')$, $w^m(\tau) = T \in S'$, and $w^m_t(\tau) = 0$ is uniformly well posed in S' with solution (3.53).

Proof: This follows by hypocontinuity. QED

Remark 3.17 The formulas of Lemma 3.3 can be composed with $\hat{T} \in S'$ and after an inverse Fourier transformation we have

$$(3.54) \qquad w_t^m(t) = -\frac{t}{2(m+1)} A_x w^{m+1}(t)$$

$$(3.55) \qquad w_t^m(t) = \frac{2m}{t}[w^{m-1}(t) - w^m(t)]$$

These recursion formulas are generalizations of formulas first developed by Weinstein [1; 3; 4; 6; 8; 9; 10; 11; 16; 18] when $A_x = -\Delta_x$. They were systematically exploited by Carroll [2; 5] in a general existence-uniqueness theory and it turns out that they have a group theoretic significance (see Chapter 2).

1.4 Connection formulas and properties of solutions for EPD equations in D' and S'. Let us return now to the case when $A_x = -\Delta_x$ with $A(y) = |y|^2 = \sum y_k^2$ and look at the resolvant $\hat{R}^m(y,t)$ given by (3.6). We denote by supp S the support of a distribution S.

Lemma 4.1 The resolvant $\hat{R}^m(y,t) = 2^m\Gamma(m+1)z^{-m}J_m(z)$, $z = t|y|$, of (3.6) is an entire function of $y \in \mathbb{C}^n$ of exponential type so that $R^m(\cdot,t) \in E'_x$ and supp $R^m(\cdot,t)$ is contained in a fixed compact set $|y_k| \leq B$ for $0 \leq t \leq b$.

Proof: The order ρ of an entire function $g(y) = \sum a_\alpha y^\alpha$, $y \in \mathbb{C}^n$, is given by (cf. Fuks [1])

$$(4.1) \qquad \rho = \rho(g) = \limsup_{|\alpha| \to \infty} \frac{|\alpha| \log |\alpha|}{-\log|a_\alpha|_C}$$

where $|\alpha| = \sum \alpha_k$ and if $\rho(g) = 1$ with

$$\text{(4.2)} \qquad \limsup_{\|y\|_C \to \infty} \frac{\log |g(y)|_C}{\|y\|_C} < \gamma$$

where $\|y\|_C = \sum |y_k|_C$ then g is said to be of exponential type $\leq \gamma$. Now looking at $\hat{R}^m(y,t) = \sum a_{2k} z^{2k}$ as a function of one variable $z = t|y|$ we have by Stirling's formula (cf. Titchmarsh [1]) that $\rho = 1$. This means that $|\hat{R}^m(y,t)| \leq \delta \exp(\beta+\varepsilon)|z|_C$ for any $\varepsilon > 0$ where β is the type of $\hat{R}^m$ as a function of z. Since $|y|_C = |(\sum y_k^2)^{1/2}|_C \leq (\sum |y_k|_C^2)^{1/2} \leq \sum |y_k|_C = \|y\|_C$ one has then

$$\text{(4.3)} \qquad \limsup_{\|y\|_C \to \infty} \frac{\log |\hat{R}^m(y,t)|_C}{\|y\|_C} \leq \limsup_{|z|_C \to \infty} \frac{t|\hat{R}^m(y,t)|_C}{|z|_C}$$

$$\leq \beta t \leq \beta b = B$$

The conclusion of the Lemma now follows from the Paley-Wiener-Schwartz theorem (see Schwartz [1]). QED

<u>Theorem 4.2</u> If $A_x = -\Delta_x$, $m \geq -1/2$, $0 \leq t \leq b$, and $T \in D'$ then $w^m(t) = R^m(\cdot,t) * T$ is a solution of (3.2) - (3.3) in D' depending continuously in D' on (t,T).

Proof: By Lemma 4.1 supp $R^m(\cdot,t)$ is contained in a fixed compact set for $0 \leq t \leq b$ so that $R^m(\cdot,t) * T$ makes sense for $T \in D'$. Furthermore the set $\{R^m(\cdot,t);\ 0 \leq t \leq b\}$ is bounded in E' (cf. Ehrenpreis [3], Schwartz [1]) and one knows by Schwartz [1] that the map $(S,T) \to S * T : E' \times D' \to D'$ is hypocontinuous. To check the differentiability of $t \to R^m(\cdot,t)$ in E' we recall that on bounded sets in E' the topology coincides with that induced by O'_C (cf. Schwartz [1], p. 274, where the Fourier transform is shown to establish a topological isomorphism between E' and Exp O_M and note that on sets $B \subset$ Exp O_M of bounded exponential type the topology is that induced by O_M). The result follows immediately. QED

For uniqueness in Theorem 4.2 we can go to a direct expression for $\hat{R}^m(y,t,\tau)$ derived in Carroll [2; 5] and then follow the procedure involving (3.20). Thus one can write for $m \geq -1/2$ not an integer

$$\hat{R}^m(y,t,\tau) = \gamma_m\Big[\frac{J_m(z)}{z^m}\,\frac{J_{-m-1}(z_1)}{z_1^{-m-1}} + z_1^2\Big(\frac{\tau}{t}\Big)^{2m}\frac{J_{-m}(z)}{z^{-m}}\,\frac{J_{m+1}(z_1)}{z_1^{m+1}}\Big] \tag{4.4}$$

where $z = t\sqrt{A(y)}$, $z_1 = \tau\sqrt{A(y)}$, and $\gamma_m = \Gamma(m+1)\Gamma(-m)/2$, while for $m \geq -1/2$ an integer we have

$$R^m(y,t,\tau) = \frac{\pi}{2}\Big[\Big(\frac{\tau}{t}\Big)^{2m} z_1^2 z^m N_m(z)\frac{J_{m+1}(z_1)}{z_1^{m+1}} - z_1^{m+1}N_{m+1}(z_1)\frac{J_m(z)}{z^m}\Big] \tag{4.5}$$

where N_m denotes the Neumann function (cf. Courant-Hilbert [2], Watson [1]). Some analysis of the Bessel and Neumann functions, which we omit here (see Carroll [2; 5] for details), leads to

<u>Lemma 4.3</u> Wnen $A(y) = |y|^2$ the resolvants $\hat{R}^m(y,t,\tau)$ are

entire functions of bounded exponential type for $0 \le \tau \le t \le b$ with $\{R^m(\cdot,t,\tau)\}$ a bounded set in E'. Similarly $\{R^m_\xi(\cdot,t,\xi)\} \subset E'$ is bounded for $0 \le \xi \le t \le b$.

Using for example (3.39) we can say that given $A(y) = |y|^2$, $\{S^m(\cdot,t,\xi)\} \subset E'$ is bounded for $0 \le \xi \le t \le b$ and certainly Lemmas 3.5, 3.6, 3.7, 3.8, 3.9, 3.10, 3.11, and 3.12 remain valid. Lemma 3.13 holds with 0_M replaced by $\text{Exp}\ 0_M = FE' \subset 0_M$ and if we take inverse Fourier transforms in (3.18) - (3.20) the resulting convolution formulas will be valid for $T \in D'$. Hence we have

<u>Theorem 4.4</u> The solution $w^m(t) = R^m(\cdot,t) * T \in D'$ of (3.2)-(3.3) when $A_x = -\Delta_x$ and $T \in D'$, given by Theorem 4.2, is unique. Similarly the problem (3.52) with $f(\cdot) \in C^0(D')$, $w^m(\tau) = T \in D'$, and $w^m_t(\tau) = 0$ is uniformly well posed in D' with solution given by (3.53).

Going back now to the resolvant $\hat{R}^m(y,t)$ in 0_M or $\text{Exp}\ 0_M$ we exploit the Sonine integral formula (cf. Rosenbloom [1], Watson [1]) to obtain for $m > p \gtrsim -1/2$

$$(4.6) \qquad \hat{R}^m(y,t) = \frac{2\Gamma(m+1}{\Gamma(p+1)\Gamma(m-p)} \int_0^1 \hat{R}^p(y,\xi t)\xi^{2p+1}(1-\xi^2)^{m-p-1}d\xi$$

$$= \frac{2\Gamma(m+1)t^{-2m}}{\Gamma(p+1)\Gamma(m-p)} \int_0^t \hat{R}^p(y,\eta)\eta^{2p+1}(t^2-\eta^2)^{m-p-1}d\eta$$

We note in passing that for m-p integral (4.6) is also a direct consequence of the recursion relation (3.12) (cf. Carroll-Silver [15; 16; 17], Silver [1]) and this will be utilized in Chapter 2 where an explicit derivation is given; analogous formulas in other group situations will also be derived. We recall next the formula (cf. Watson [1])

$$(4.7)\qquad \left(\frac{1}{z}\frac{d}{dz}\right)^p (z^k J_k(z)) = z^{k-p} J_{k-p}(z)$$

for $p \geq 0$ an integer. This leads to the formula

$$(4.8)\qquad \hat{R}^m(y,t) = \frac{\Gamma(m+1)t^{-2m}}{2^p \Gamma(m+p+1)} \left(\frac{1}{t}\frac{\partial}{\partial t}\right)^p [t^{2(m+p)} \hat{R}^{m+p}(y,t)]$$

provided $m \neq -1, -2, \ldots, -p$. We observe that (4.8) gives resolvants for values of $m < -1/2$ and it should be noted here that $j_m(z) = z^m J_{-m}(z)$ satisfies $j_m'' + (\frac{-2m+1}{z}) j_m' + j_m = 0$ so that (cf. (3.6)) $\hat{R}^{-m}(y,t) = 2^{-m}\Gamma(1-m) z^m J_{-m}(z)$, for $z = t\sqrt{A(y)}$ and any nonintegral $m \geq 0$, satisfies the equation $\hat{R}^{-m}_{tt} + (\frac{-2m+1}{t})\hat{R}^{-m}_t + A(y)\hat{R}^{-m} = 0$ with initial data $\hat{R}^{-m}(y,o) = 1$ and $\hat{R}^{-m}_t(y,o) = 0$. Now from (4.8) with nonintegral $m < -1/2$ we can choose an integer p so that $m + p \geq -1/2$ and express $\hat{R}^m$ in terms of derivatives of $\hat{R}^{m+p}$, whose properties are known from Section 1.3. In particular (4.8) holds in E_y and also in O_M or Exp O_M(cf. Lemma 3.13 and the remarks before Theorem 4.4) so we can state, referring to (3.11),

<u>Theorem 4.5</u> For nonintegral $m < -1/2$ resolvants $\hat{R}^m(y,t)$ of the form (3.6) obey (4.8) and thus inherit the properties of the $\hat{R}^m$ for $m \geq -1/2$. Also (4.6) holds in O_M or Exp O_M.

Taking inverse Fourier transforms in (4.6) and (4.8) and composing the result with $T \in S'$ or $T \in D'$ (when $A_x = -\Delta_x$) we obtain

Theorem 4.6 For $m > p \geq -1/2$ the (unique) solution of (3.2)-(3.3) in S' (or D') for index m is given by

$$(4.9) \qquad w^m(t) = \frac{2\Gamma(m+1)t^{-2m}}{\Gamma(p+1)\Gamma(m-p)} \int_0^t \eta^{2p+1}(t^2-\eta^2)^{m-p-1} w^p(\eta)\, d\eta$$

where w^p satisfies (3.2) - (3.3) in S' (or D') for index p. Similarly if w^{m+p} is the (unique) solution of (3.2) - (3.3) in S' (or D') for index $m + p \geq -1/2$ (p an integer) then

$$(4.10) \qquad w^m(t) = \frac{\Gamma(m+1)t^{-2m}}{2^p\Gamma(m+p+1)} \left(\frac{1}{t}\frac{\partial}{\partial t}\right)^p [t^{2(m+p)} w^{m+p}(t)]$$

is a solution of (3.2) - (3.3) in S' (or D') for index m provided $m \neq -1, \ldots, -p$.

Remark 4.7 Formulas of the form (4.9) - (4.10) were first discovered by Weinstein (loc. cit.), in a classical setting, using different methods. In particular Weinstein observed that if w^{-m} satisfies (3.2) with index -m then $t \to w^m(t) = t^{-2m} w^{-m}(t)$ satisfies (3.2) with index m. This fact, plus a version of (3.54), leads to (4.10) for example. Weinstein's version of (4.9) involves $p = n - 1$ and is obtained by a generalized method of descent (see e.g., Weinstein [3]). We note also here that there is no uniqueness theorem for solutions of (3.2) - (3.3) when $m < -1/2$ since if $m = -s$ ($s > 1/2$) then $t \to u^m(t) = t^{2s} w^s(t)$ satisfies (3.2) for index m with $u^m(o) = u^m_t(o) = 0$; here w^s is the solution of (3.2) - (3.3) for index s.

Remark 4.8 The exceptional cases m = -1, -2, . . . have

been treated by Blum [1; 2], Diaz-Weinberger [2], Diaz-Martin [10], B. Friedman [1], Martin [1], and Weinstein (loc. cit.) and we refer to these papers for details. Here for $m = -p$ with $p \geq 1$ an integer we may take

$$\hat{R}^{-p}(y,t) = -\frac{\pi}{(p-1)!}\left(\frac{z}{2}\right)^p N_p(z) \tag{4.11}$$

(cf. B. Friedman [1]). This resolvant $\hat{R}^{-p}$ satisfies (3.4) - (3.5) for index $m = -p$ with $\hat{T} = 1$ and one may write

$$z^p N_p(z) = \sum_{k=0}^{\infty} c_k z^{2k} + z^{2p} \log z \sum_{k=0}^{\infty} \tilde{c}_k z^{2k} \tag{4.12}$$

Consequently the $2p^{\text{th}}$ derivative of $\hat{R}^{-p}$ in t will have a logarithmic singularity at $t = 0$. However, writing formally

$$\begin{aligned} w^{-p}(t) = R^{-p}(\cdot,t) * T &= \sum_{k=0}^{\infty} c_k t^{2k}(A_x\delta * T) \\ &+ t^{2p}A_x^p\delta * T * F^{-1}(\log z \sum_{k=0}^{\infty} \tilde{c}_k z^{2k}) \end{aligned} \tag{4.13}$$

we see that the logarithmic term varnishes if $A_x^p T = 0$, in which event w^{-p} will depend smoothly on t.

<u>Remark 4.9</u> We note that if one takes $p = 1$ in (4.8), with m replaced by $m - 1$, the result is exactly (3.12).

We will now discuss some growth and convexity theorems for solutions of (3.2) - (3.3) when $A_x = -\Delta_x$. Such results were first discovered by Weinstein [9; 10; 15] in a classical setting and subsequently generalized and improved in certain directions by Carroll [1; 2; 5] in a distribution framework; we will follow the latter approach. (Other kinds of growth and convexity theorems for singular Cauchy problems were developed by Carroll [18; 19; 25] and will be treated later.) Referring to (4.6), the basic fact one requires for these theorems is that $R^p(\cdot,\eta) \geq 0$ for some $p \geq -1/2$ (i.e., $R^p(\cdot,\eta)$ should be a positive measure), in which case all $R^m(\cdot,t) \geq 0$ for $m \geq p$. For general A_x not too much is known in this direction but when $A_x = -\Delta_x$ it follows from Section 2 that $R^{n/2-1}(\cdot,\eta) = \mu_x(\eta) \geq 0$. Another case of interest involves the metaharmonic operator $A_x = -\Delta_x - \alpha^2$ where $R^m(\cdot,\eta) \geq 0$ for $m > \frac{n}{2} - 1$. Indeed for $m > \frac{n}{2} - 1$ and $r = (\sum x_i^2)^{1/2} \leq t$ one has (see Carroll [5] and cf. Ossicini [1])

$$(4.14) \qquad R^m(\cdot,t) = \frac{\Gamma(m+1)2^{m-\frac{n}{2}}}{\pi^{n/2}\, t^{2m}} \, \frac{(i\alpha\sqrt{t^2-r^2})^{m-n/2}}{(i\alpha)^{2m-n}} J_{m-n/2}(i\alpha\sqrt{t^2-r^2})$$

$$= \frac{\Gamma(m+1)(t^2-r^2)^{m-n/2}}{\pi^{n/2}\, t^{2m}} \sum_{k=0}^{\infty} \frac{[\alpha^2(t^2-r^2)]^k}{2^{2k} k!\; \Gamma(m-\frac{n}{2}+k+1)} \geq 0$$

while $R^m(\cdot,t) = 0$ for $|r| \geq t$; in particular then $R^m(\cdot,t) \in E'$.

Now, general theorems of growth and convexity involving assumptions of the form $(-A_x)^k T \geq 0$ were developed in Carroll [2; 5] using the notions of value and section of a distribution introduced by Lojasiewicz [1]. This can be simplified somewhat in dealing with $A_x = -\Delta_x$ since $\Delta T \geq 0$ implies that T is an almost subharmonic function (see Schwartz [1]) and hence T automatically has a "value" almost everywhere (a.e.). We recall briefly some notations and results of Lojasiewicz [1].

Definition 4.10 A distribution T is said to have the value c at x_0 if $\lim T_{x_0+\lambda x} = c$ as $\lambda \to 0^+$ (i.e., $\langle T_{x_o+\lambda x}, \phi\rangle = \langle T_\xi, \frac{1}{\lambda^n}\phi\left(\frac{\xi - x_o}{\lambda}\right)\rangle \to c$ for $\phi \in D$ with $\int \phi(x)dx = 1$); when T is a positive measure the notion of value coincides with that of density. One may fix x_0 in a distribution $T_{x,t}$ if $\lim T_{x_0+\lambda x,t} = S_t \in D'$ as $\lambda \to 0^+$ and S_t is said to be the section $T_{x_0,t}$ of T. Thus one must have $\langle T_{\xi,t}, \frac{1}{\lambda^n}\phi\left(\frac{\xi - x_o}{\lambda}\right)\psi(t)\rangle \to \langle S,\psi\rangle$ when $\phi \in D_x$, $\psi \in D_t$, and $\int\phi(x)dx = 1$.

Remark 4.11 It is proved in Lojasiewicz [1] that if $\sum_{x,t}$ is a measure then a.e. in x we may fix $x = x_0$ and the section $\sum_{x_0,t}$ is a measure in D_t'. Further if x_0 may be fixed for $T_{x,t}$ then x_0 may be fixed for $(\partial/\partial t)T_{x,t}$ and $(\partial/\partial t)T_{x,t}\,|_{x_0,t} = \partial/\partial t T_{x_0,t}$.

For the following theorems see Carroll [5].

Theorem 4.12 Let $w_x^m(t) = R^m(\cdot,t) * T \in D_x'$ be the solution of (3.2) - (3.3) for $A_x = -\Delta_x$ with $m \geq \frac{n}{2} - 1$ and $T \in D'$ satisfying $\Delta T \geq 0$. Then a.e. in x the section $t \to w_{x_0}^m(t)$ is a nondecreasing function of t with $w_{x_0}^m(t) \geq T_{x_0}$.

Proof: Let $m > \frac{n}{2} - 1$ and set $p = \frac{n}{2} - 1$ in (4.6) to obtain $(R^{\frac{n}{2}-1}(\cdot,t) = \mu_x(t))$

$$(4.15) \qquad R^m(\cdot,t) = \frac{2\Gamma(m+1)}{\Gamma(\frac{n}{2})\Gamma(m - \frac{n}{2} + 1)}\int_0^1 \xi^{n-1}(1-\xi^2)^{m-\frac{n}{2}}\mu_x(\xi t)d\xi$$

Using (3.11) (transformed) and composing with T we have (cf. (2.5) where $R^{n/2}(\cdot,t) = A_x(t)$)

$$(4.16)\qquad w_t^m(t) = \frac{\partial}{\partial t}(R^m(\cdot,t) * T)$$
$$= \frac{\Gamma(m+1)t}{\Gamma(\frac{n}{2}+1)\Gamma(m-\frac{n}{2}+1)}\int_0^1 \xi^{n+1}(1-\xi^2)^{m-n/2}(A_x(\xi t)*\Delta T)d\xi$$

Now we may consider $w^m(t)$ as a distribution $w^m_{x,t}$ in two variables (since $D'_t(D'_x) = D'_{x,t}$ algebraically and topologically with $E_t(D'_x) \subset D'_t(D'_x)$ - cf. Schwartz [5]) and here one may extend $\mu_x(t)$ and $A_x(t)$ as even functions of t for $t < 0$. In D'_x one has $w^m(t) - T = \int_0^t w^m_\tau(\tau)d\tau$ which, by (4.16) with $\Delta T \geq 0$, is a positive measure for $t \geq 0$. Writing $\tilde{T} = T \times 1_t$ in $D'_{x,t}$ it follows that $w^m_{x,t} - \tilde{T}$ will be a positive measure in $D'_{x,t}$ for $t \geq 0$ and hence we can fix $x = x_0$ a.e. in x for $w^m_{x,t}$ and $\tilde{T}$ together (cf. Remark 4.11 and recall that T is an almost subharmonic function). For such x_0, $(\partial/\partial t)w^m_{x_o,t} = (w^m_t)_{x_o,t}$ by Remark 4.11, and, by (4.6), $(w^m_t)_{x_0,t} \geq 0$ for $t \geq 0$. Hence $t \to w^m_{x_0,t} = w^m_{x_o}(t)$ will be a nondecreasing function for $t \geq 0$ of bounded variation (cf. Schwartz [1]). We conclude that $w^m_{x_0,t} \geq T_{x_0}$ for $t \geq 0$ and to spell this out let $\phi \in D_x$, $\phi \geq 0$, $\int\phi(x)dx = 1$ and set $\psi_\lambda(\xi) = \lambda^{-n}\phi\left(\frac{\xi - x_0}{\lambda}\right)$ (cf. Definition 4.10). Then from (4.16) the continuous function $t \to \langle w^m(t),\psi_\lambda\rangle = \langle w^m_{\xi,t},\psi_\lambda(\xi)\rangle$ is nondecreasing for $t \geq 0$. Consequently for $t > 0$ fixed $\langle w^m_{\xi,t},\psi_\lambda(\xi) \geq \langle T_\xi,\psi_\lambda(\xi)\rangle$ and letting $\lambda \to 0^+$ we obtain $w^m_{x_0,t} \geq T_{x_0}$. The result for $m = \frac{n}{2} - 1$ is now immediate from (2.5) and the remarks above. QED

Let us now combine (4.8) with Lemma 3.3 when $A_x = -\Delta_x$ in order to obtain a formula for $R^m(\cdot,t)$ $(m \geq -1/2)$ in terms of

$R^{m+p}(\cdot,t)$ and $R^{m+p+1}(\cdot,t)$, where the integer p is chosen so that $m + p \geq \frac{n}{2} - 1$. Some simple but tedious calculations yield (see Carroll [2; 5] for details)

Lemma 4.13 For $-1/2 \leq m < n/2 - 1$ the resolvant $R^m(\cdot,t)$ for (3.2) - (3.3) (i.e., $T = \delta$ in (3.3)) given by F^{-1} applied to (4.8) can be written

$$(4.17) \qquad R^m(\cdot,t) = \Big(\sum_{k=0}^{q-1} \alpha_k(m)t^{2k}(\Delta\delta)^k\Big) * R^{m+p}(\cdot,t)$$

$$+ \Big(\sum_{k=0}^{q-1} \beta_k(m)t^{2k}(\Delta\delta)^k\Big) * R^{m+p+1}(\cdot,t)$$

where p is an integer chosen so that $m + p \geq \frac{n}{2} - 1$, $q = 1 + [\frac{p+1}{2}]$ ($[u]$ denotes the largest integer in u), $\alpha_k(m) \geq 0$, $\beta_k(m) \geq 0$, and $\alpha_o(m) = 1$.

Theorem 4.14 For $-1/2 \leq m < \frac{n}{2} - 1$ let the integer p be chosen so that $m + p \geq \frac{n}{2} - 1$. Let $q = 1 + [\frac{p+1}{2}]$ and assume $T \in D'$ satisfies $(\Delta)^k T \geq 0$ for $1 \leq k \leq q$. If $w^m(t)$ denotes the solution of (3.2) - (3.3) for $A_x = -\Delta (t \geq 0)$ then a.e. in x the section $t \to w^m_{x_o}(t)$ is a nondecreasing function of t with $w^m_{x_o}(t) \geq T_{x_o}$ for $t \geq 0$.

Proof: We convolute (4.17) with T to get $w^m(t)$ and differentiate, using (3.11). Since $R^{m+p}(\cdot,t)$, $R^{m+p+1}(\cdot,t)$, and $R^{m+p+2}(\cdot,t)$ are positive measures we have again that $w^m_t(t) \geq 0$ in D'_x for $t \geq 0$. Hence, as in Theorem 4.12, we may fix $x = x_o$ a.e. for $w^m_{x,t} \in D'_{x,t}$ and $t \to w^m_{x_o,t} = w^m_{x_o}(t)$ will be a nondecreasing function of bounded variation ($t \geq 0$). Further $w^m(t) = R^{m+p}(\cdot,t) * T + \sum(\cdot,t)$ where $\sum(\cdot,t) \geq 0$ in D'_x for $t \geq 0$ (we recall here that $\alpha_0(m) = 1$). From Theorem 4.12 we know that $R^{m+p}(\cdot,t) * T = w^{m+p}(t)$ satisfies $w^{m+p}_{x_0}(t) \geq T_{x_0}$ a.e. in x when $\Delta T \geq 0$ (which holds since $q \geq 2$). Hence, a.e. in x, $w^m_{x_0}(t) \geq w^{m+p}_{x_0}(t) \geq T_{x_0}$. QED

Corollary 4.15 For $-1/2 \leq m < \frac{n}{2} - 1$ choose p as in Theorem 4.14 and assume $(\Delta)^k T \geq 0$ for $1 \leq k \leq q - 1$. Given $w^m(t)$ as in Theorem 4.14 then a.e. in x the section $t \to w^m_{x_0}(t)$ satisfies $w^m_{x_0}(t) \geq T_{x_0}$ for $t \geq 0$.

Proof: We note first that $p \geq 1$ so that $q - 1 = [\frac{p+1}{2}] \geq 1$. Then from (4.17) one writes again $w^m(t) = R^{m+p}(\cdot,t) * T + \sum(\cdot,t)$ where $\sum(\cdot,t) \geq 0$ in D'_x for $t \geq 0$ under our hypotheses. As in the proof of Theorem 4.14 it follows that $w^m_{x_0}(t) \geq T_{x_0}$ a.e. in x.

QED

Let us write now $L^t_m = \partial^2/\partial t^2 + \frac{2m+1}{t}\, \partial/\partial t$; then as noted by Weinstein, for $m \neq 0$, $L^t_m = (2m)^2 t^{-2(2m+1)}\, \partial^2/\partial^2(t^{-2m})$, while for $m = 0$, $L^t_0 = t^{-2}\, \partial^2/\partial^2(\log t)$. Referring now to (3.2) with $A_x = -\Delta$ we have $L^t_m w^m = \Delta w^m$ and using (4.15), composed with $T \in D'$, for $m > n/2 - 1$ it follows that

$$(4.18) \qquad L^t_{n/2-1} w^m(t) = \frac{2\Gamma(m+1)}{\Gamma(\frac{n}{2})\Gamma(m - \frac{n}{2}+1)} \int_0^1 \xi^{n+1}(1-\xi^2)^{m-\frac{n}{2}} \cdot (\mu_x(\xi t) * \Delta T)\, d\xi$$

since $L^t = \xi^2 L^{\xi t}$. Thus if $\Delta T \geq 0$ then $L^t_{\frac{n}{2}-1} w^m(t)$ is a positive measure for $t \geq 0$. Similarly from (4.17) with $-1/2 \leq m < \frac{n}{2} - 1$ we convolute with $T \in D'_x$ to obtain

$$(4.19) \qquad L^t_m w^m(t) = \sum_{k=0}^{q-1} \alpha_k(m) t^{2k} (R^{m+p}(\cdot,t) * \Delta^{k+1} T)$$

$$+ \sum_{k=1}^{q-1} \beta_k(m) t^{2k} (R^{m+p+1}(\cdot,t) * \Delta^{k+1} T)$$

Hence if $\Delta^k T \geq 0$ for $1 \leq k \leq q$ then $L^t_m w^m(t)$ is a positive measure in D'_x for $t \geq 0$.

<u>Theorem 4.16</u> Let $w^m(t)$ denote the solution of (3.2) - (3.3) for $A_x = -\Delta$ and let $T \in D'$ satisfy $\Delta T \geq 0$ for $m \geq \frac{n}{2} - 1$ while, for $-1/2 \leq m < \frac{n}{2} - 1$, $\Delta^k T \geq 0$ for $1 \leq k \leq q = 1 + [\frac{p+1}{2}]$ where the integer p is chosen so that $m + p \geq \frac{n}{2} - 1$. Then for $t > 0$, a.e. in x, the section $t \to w^m_{x_0}(t)$ is a convex function of t^{2-n} for $m \geq \frac{n}{2} - 1$, of t^{-2m} for $-1/2 \leq m < \frac{n}{2} - 1$ $(m \neq 0)$, and of $\log t$ for $m = 0$.

Proof: By (4.18) - (4.19) we know that for $t > 0$, $(\partial^2/\partial\tau^2) w^m(t) \geq 0$ in D'_x under the hypotheses given, where $\tau = t^{2-n}$ for $m \geq \frac{n}{2} - 1$, $\tau = t^{-2m}$ for $-1/2 \leq m < \frac{n}{2} - 1$ $(m \neq 0)$, and $\tau = \log t$ for $m = 0$. Now by Theorems 4.12 and 4.14 our hypotheses insure that $x = x_0$ may be fixed a.e. in x for $w^m_{x,t}$ $(t \geq 0)$ and by Definition 4.10 with Remark 4.11 it follows that $\partial^2 w^m_{x_0,t}/\partial\tau^2 \geq 0$ in D'_x for $t > 0$. Hence by Schwartz [1] $t \to w^m_{x_0}(t)$ is a convex function of t. QED

Remark 4.17 Of particular interest in the EPD frameword of course is the wave equation when $m = -1/2$ and we see that in the hypotheses of Corollary 4.15 $-1/2 + p \geq \frac{n}{2} - 1$ means $p \geq \frac{n-1}{2}$. Note here that in dimension $n = 1$, $-1/2 \leq m < -1/2$ is impossible so that we are in the situation of Theorem 4.12. Now $q - 1 = [\frac{p+1}{2}]$ so that, given n even, $p = n/2$, while, for n odd $(n \geq 3)$, $p = \frac{n-1}{2}$; thus for n even $q - 1 = [\frac{n+2}{4}]$ while for n odd $q - 1 = [\frac{n+1}{4}]$. These results on the number of positive Laplacians of T sufficient for a minimum (or maximum) principle improve Weinstein's estimates and are comparable with results of D. Sather [1; 2; 3]. Sather in [3] for example sets $N = \frac{n-2}{2}$ for n even and $N = \frac{n-3}{2}$ for n odd with $a = [\frac{N+2}{2}] = [\frac{n+2}{4}]$ for n even while $a = [\frac{n+1}{4}]$ for n odd; similarly he sets $b = [\frac{N+1}{2}] = [\frac{n}{4}]$ for n even while $b = [\frac{n-1}{4}]$ for n odd. Then he requires that $\Delta^k T \geq 0$ for $1 \leq k \leq a$ and $\Delta^k w_t^{-1/2}(o) \geq 0$ for $0 \leq k \leq b$ in order that $w^{-1/2}(x,t) \geq T(x)$ for $t \geq 0$. Thus our results seem at least equivalent and certainly more concise than those of Sather. For further information on maximum and minimum principles for Cauchy type problems we refer to Protter-Weinberger [1] and the bibliography there; cf. also Agmon-Nirenberg-Protter [1], Bers [1], Protter [7], Weinberger [1]. For growth and convexity properties of solutions of parabolic equations with subharmonic initial data see Pucci-Weinstein [1].

Remark 4.18 The existence of positive elementary solutions (or resolvants) for second order hyperbolic equations is in general somewhat nontrivial and we mention in this regard Duff [1; 2], in addition to Carroll [2], where the problem is discussed.

1.5 Spectral techniques and energy methods in Hilbert space. First we will present some typical theorems of Lions for EPD type equations, using "energy" methods, which were discussed in Carroll [8]. Lions proved these results in 1958 in lectures at the University of Maryland (unpublished). Thereafter follow some theorems in Hilbert space based on a spectral technique developed by Carroll [4; 6; 8; 10; 11] (suggested by Lions as a variation on Carroll's Fourier technique). The two approaches (energy and spectral) are not directly connected and lead to weak and strong solutions respectively.

Thus, regarding energy methods (cf. Carroll [14] and Lions [5] for a more complete exposition), let $(u,v) \to a(t,u,v)$: $V \times V \to \mathbb{C}$ be a family of continuous sesquilinear forms, $V \subset H$, where V and H are Hilbert spaces with V dense in H having a finer topology. One can define a linear operator $A(t)$ such that $a(t,u,v) = (A(t)u,v)_H$ for $u \in D(A(t)) \subset V$ and $v \in V$ (i.e., $u \in D(A(t))$ if $v \to a(t,u,v) : V \to \mathbb{C}$ is continuous in the topology of H). Assume $t \to a(t,u,v) \in C^1[o,b]$ with $a(t,u,v) = \overline{a(t,v,u)}$ and consider

<u>Problem 5.1</u> Find $u(\cdot) \in C^2(H)$ on $[o,b]$, $u(t) \in D(A(t))$, $t \to A(t)u(t) \in C^0(H)$ on $[o,b]$, and

(5.1) $$u_{tt} + kq(t)u_t + A(t)u = f; \quad u(o) = u_t(o) = 0$$

where $k \in ¢$, $q(\cdot) \in C^0(o,b]$ is real valued with $q(t) \to \infty$ as $t \to 0$, and $f \in C^0(H)$ on $[o,b]$. Note here that if $u(o) = T \in D(A(t))$ for $t \in [o,b]$ then $v(t) = u(t) - T$ satisfies (5.1) with f replaced by $f(\cdot) - A(\cdot)T$ (assuming the latter is continuous on $[o,b]$ with values in H - which holds if $A(t) = A$).

This can be transformed into a weak problem (which is more tractable) as follows ($\frac{d}{dt} \equiv {}'$).

Problem 5.2 Find $u \in L^2(V)$ on $[o,b]$, $u(o) = 0$, $u'\sqrt{q} \in L^2(H)$ on $[o,b]$, such that given $f/\sqrt{q} \in L^2(H)$

(5.2) $$\int_0^b \{a(t,u,h) + kq(u',h)_H - (u',h')_H - (f,h)_H\}dt = 0$$

for all $h \in L^2(V)$ with $h\sqrt{q} \in L^2(H)$, $h'/\sqrt{q} \in L^2(H)$, and $h(b) = 0$.

Theorem 5.3 Let $t \to a(t,u,v) \in C^1[o,b]$, $a(t,u,v) = \overline{a(t,v,u)}$, $q \in C^0(o,b]$, $q > 0$, $q(t) \to \infty$ as $t \to 0$, Re $k > 0$, and $a(t,u,u) \geq \alpha\|u\|_V^2$ for $\alpha > 0$. Then there exists a solution of Problem 5.2.

Proof: Set $h = e^{-\gamma t}\phi'$ for γ real (to be determined) and

(5.3) $$E_\gamma(u,\phi) = \int_0^b \{a(t,u,e^{-\gamma t}\phi') + kq(t)(u',e^{-\gamma t}\phi')_H - (u',(e^{-\gamma t}\phi')')_H\}dt$$

Let F be the space of functions $u \in L^2(V)$, $u'\sqrt{q} \in L^2(H)$, $u(o) = 0$, while G is the space of $\phi \in L^2(V)$, $\phi' \in L^2(V)$, $\phi'\sqrt{q} \in L^2(H)$, $\phi''/\sqrt{q} \in L^2(H)$, $\phi(o) = 0$, and $\phi'(b) = 0$. Clearly

$G \subset F$ and we put on G the induced topology of F defined by $\int_0^b (\|u\|_V^2 + q\|u'\|_H^2)dt = \|u\|_F^2$ so that G is a prehilbert space. Then if $u \in F$ satisfies $E_\gamma(u,\phi) = \int_0^b (f,e^{-\gamma t}\phi')_H dt$ for all $\phi \in G$ it follows that u is a solution of Problem 5.2. Now $E_\gamma(u,\phi)$ is a sesquilinear form on $F \times G$ with $u \to E_\gamma(u,\phi) : F \to \mathbb{C}$ continuous for $\phi \in G$ fixed. Further, $a(t,\phi,e^{-\gamma t}\phi)' = a_t(t,\phi,e^{-\gamma t}\phi) + 2a(t,\phi,e^{-\gamma t}\phi') - \gamma a(t,\phi,e^{-\gamma t}\phi)$ so that,

$$(5.4) \qquad 2\text{Re}\int_0^b a(t,\phi,e^{-\gamma t}\phi')dt = a(b,\phi(b),\phi(b))e^{-\gamma b} - \int_0^b a_t(t,\phi,\phi)e^{-\gamma t}dt + \gamma\int_0^b a(t,\phi,\phi)e^{-\gamma t}dt \geq \int_0^b (\gamma\alpha - c)e^{-\gamma t}\|\phi\|_V^2 dt$$

where $|a_t(t,\phi,\phi)| \leq c\|\phi\|_V^2$. Since $2\text{Re}\,(\phi',\phi'')_H = \frac{d}{dt}\|\phi'\|_H^2$ we have in addition

$$(5.5) \qquad -2\text{Re}\int_0^b (\phi',(e^{-\gamma t}\phi')')_H dt = \int_0^b [-e^{-\gamma t}\frac{d}{dt}\|\phi\|_H^2 + 2\gamma e^{-\gamma t}\|\phi'\|_H^2]dt = \|\phi'(0)\|_H^2 + \gamma\int_0^b e^{-\gamma t}\|\phi'\|_H^2\, dt \geq 0$$

Hence, for $\gamma > c/\alpha$, $|E_\gamma(\phi,\phi)| \geq \beta\|\phi\|_G^2$ (note the second term in $2\text{Re}\, E_\gamma(\phi,\phi)$) and by the Lions projection theorem (cf. Carroll [14], Lions [5], Mazumdar [1; 2; 3; 4]) there exists $u \in F$ satisfying $E_\gamma(u,\phi) = \int_0^b (f,e^{-\gamma t}\phi')_H dt$, since $\phi \to \int_0^b (f,e^{-\gamma t}\phi')_H dt$ is a continuous semilinear form on G. QED

<u>Remark 5.4</u> For an interesting generalization of the Lions

projection theorem and applications see Mazumdar [1; 2; 3; 4]. We note that in Problem 5.2 q could become infinite at other points on [o,b] provided the "weight" functions h are suitably chosen; if $q > 0$ but $q \not\to \infty$ the problem can be solved directly as in Lions [5].

Theorem 5.5 Let $t \to a(t,u,v) \in C^1[o,b]$, $a(t,u,v) = \overline{a(t,v,u)}$, $q \in C^1(o,b]$, $q > 0$, $q(t) \to \infty$ as $t \to 0$, $\mathrm{Re}\, k > 0$, and $a(t,u,u) \geq \alpha \|u\|_V^2$ for $\alpha > 0$. Then a solution of Problem 5.2 is unique.

Proof: Let h be given by $h(t) = 0$ for $t \geq s$ with $h' - kqh = u$ on $[o,s)$; then some calculation shows that h is admissable in Problem 5.2 and, as $\varepsilon \to 0$, $\lim a(\varepsilon,h(\varepsilon),h(\varepsilon)) = \theta^2 \geq 0$. Taking real parts in (5.2), with $f = 0$, one obtains with this h,

$$(5.6) \qquad \theta^2 + \| u(s) \|_H^2 + 2\mathrm{Re}\, k \int_0^s q\, a(t,h,h)dt + \int_0^s a_t(t,h,h)dt = 0$$

where $a(t,h,h)' = 2\mathrm{Re}\, a(t,h,h') + a_t(t,h,h)$. Consequently

$$(5.7) \qquad \int_0^s (2\alpha q\ \mathrm{Re}\, k - c)\|h\|_V^2 dt \leq 0$$

and since $q(t) \to \infty$ as $t \to 0$ one may choose s_0 small enough so that $2\alpha q\ \mathrm{Re}\, k - c > 0$ in $[o,s_0]$ which will imply that $h = 0$ and $u = 0$ in $[o,s_0]$. In this step only $q \in C^0(o,s_0]$ is needed; to extend the result $u = 0$ to $[s_0,b]$ one can resort to standard methods (cf. Lions [5]) where $q \in C^1[s_0,b]$ is used. QED

Remark 5.6 There are also versions of Theorems 5.3 and 5.5 due to Lions when $a(t,v,v) + \lambda\|v\|_H^2 \geq \alpha\|v\|_V^2$ (cf. Carroll [8]) but we will omit the details here. We note that no restrictions have been placed on the growth of $q(t)$ as $t \to 0$; however the solution given by Theorem 5.3 is a weak solution and not necessarily a solution of (5.1).

We consider now the operator

$$(5.8) \qquad L = \frac{\partial^2}{\partial t^2} + (\alpha(t) + \beta(t))\frac{\partial}{\partial t} + \gamma(t)$$

for $0 \leq \tau \leq t \leq b < \infty$, where $\alpha \in C^0(o,b]$ with $\int_\tau^b \operatorname{Re} \alpha(\xi)\, d\xi \to \infty$ as $\tau \to 0$, $\beta \in C^0[o,b]$, and $\gamma \in C^0[o,b]$. Let Λ be a self adjoint (densely defined) operator in a separable Hilbert space H with $(\Lambda u,u)_H + \delta\|u\|_H^2 \geq 0$ and consider for $Q \in C^0[o,b]$

Problem 5.7 Find $w \in C^2(H)$ on $[\tau,b]$ $(0 \leq \tau \leq t \leq b < \infty)$, $w(t) \in D(\Lambda)$, and $t \to \Lambda w(t) \in C^0(H)$ on $[\tau,b]$, such that

$$(5.9) \qquad (L + Q(t)\Lambda)w = 0; \quad w(\tau) = T \in H; \quad w_t(\tau) = 0$$

It is to be noted that one may always assume $\Lambda \geq 0$ since a change of variables $\tilde{\Lambda} = \Lambda + \delta$ and $\tilde{\gamma} = \gamma - \delta Q$ can be made. Now from the von Neumann spectral decomposition theorem (cf. Dixmier [1], Carroll [14]) it follows that there is a measure ν, a ν-measurable family of Hilbert sapces $\lambda \to H(\lambda)$ (cf. Chapter 3), and an isometric isomorphism $\theta : H \to \mathcal{H} = \int^{\oplus} \mathcal{H}(\lambda)d\nu(\lambda)$, such that Λ is transformed into the diagonalizable operator of multiplication by λ. Here $\theta T \in \mathcal{H}$ means $\int_0^\infty \|\theta T\|^2_{H(\lambda)} d\nu < \infty$ ($\|\theta T\|_{H(\lambda)} = \|\theta T(\lambda)\|_{H(\lambda)}$) and $T \in D(\Lambda)$ if and only if in addition $\int_0^\infty \lambda^2 \|\theta T\|^2_{H(\lambda)} d\nu < \infty$ where θT is the image of $T \in H$ in $\mathcal{H}$; we write here θT for the family $(\theta T)(\lambda)$ determined up to a set of ν-measure zero. For various other applications to differential equations see e.g., Berezanskij [2], Brauer [1], Gårding [1], Lions [6], Maurin [1; 2; 3]. Now by use of the map θ (5.9) is transformed into the equivalent problem

$$(5.10) \qquad (L + \lambda Q)\theta w = 0; \quad \theta w(\tau) = \theta T \in \mathcal{H}; \quad \theta w_t(\tau) = 0$$

with $t \to \theta w(t) \in C^2(\mathcal{H})$ and $t \to \lambda\theta w(t) \in C^0(\mathcal{H})$ on $[o,b]$ (note that $\theta\Lambda w = \theta\Lambda\theta^{-1}\theta w = \lambda\theta w$). This essentially reduces Problem 5.7 to a numerical problem, as in the case of the Fourier transform, and we will construct suitable "resolvants" $Z(\lambda,t,\tau)$ and $Y(\lambda,t,\tau)$ analogous to the $\hat{R}^m(y,t,\tau)$ and $\hat{S}^m(y,t,\tau)$ respectively of Section 3 so that for example

$$(5.11) \qquad (L + \lambda Q)Z = 0; \quad Z(\lambda,\tau,\tau) = 1; \quad Z_t(\lambda,\tau,\tau) = 0$$

in which event $\theta w = Z(\lambda,t,\tau)\theta T$ will be a solution of (5.10). Many of the techniques of Section 3 will be employed and this will enable us to shorten somewhat the exposition here.

Let us assume $\mathrm{Re}\,\alpha \geq 0$ for $0 < t \leq s$ and consider $P(t) = \exp\left(-\int_t^b (\alpha(\xi) + \beta(\xi))d\xi\right)$ (cf. (3.23)). Then $P(t) \to 0$ as $t \to 0$ and

$$(5.12)\qquad |P(\xi)/P(\sigma)| = \exp\left(-\int_\xi^\sigma \mathrm{Re}[\alpha(\eta) + \beta(\eta)]d\eta\right) \le M$$

for $0 \le \xi \le \sigma \le b$. (In the terminology of Feller [2] $t = 0$ is an entrance boundary.) As in (3.23) - (3.24) we obtain from (5.11) $(PZ_t)_t + P(\gamma+\lambda Q)Z = 0$ and

$$(5.13)\qquad Z(\lambda,t,\tau) = 1 - \int_\tau^t \frac{1}{P(\sigma)} \int_\tau^\sigma P(\xi)[\gamma(\xi)+\lambda Q(\xi)]Z(\lambda,\xi,\tau)d\xi d\sigma$$

which we write again in the form $Z = 1 - J \cdot Z$. Setting $F(t,\tau) = \int_\tau^t \int_\tau^\sigma |P(\xi)/P(\sigma)| d\xi d\sigma$ (cf. (3.27)) we have $F(t,\tau) < \infty$ with $F(t,\tau) \le F(t,o) = F(t) \le F(\hat{t}) \le F(b)$ for $t \le \hat{t} \le b$. The solution of (5.13) is given again formally by $Z(\lambda,t,\tau) = \sum_{k=0}^{\infty} (-1)^k J^k \cdot 1$ (cf. (3.25)) and one may proceed as in Section 1.3 to prove

<u>Lemma 5.8</u> Given $\mathrm{Re}\,\alpha \ge 0$ on $(o,s]$ the series $\sum_{k=0}^{\infty} (-1)^k J^k \cdot 1$ converges absolutely and uniformly on $\{0 \le \tau \le t \le b\} \times \Gamma$, $\Gamma \subset \mathbb{C}$ compact ($\lambda \in \Gamma$), and represents a continuous function $Z(\lambda,t,\tau)$ of (λ,t,τ), analytic in λ, which satisfies (5.13). Similarly $Z_t(\lambda,t,\tau)$ is continuous in (λ,t,τ) and analytic in λ.

Proof: The statements for $Z_t(\lambda,t,\tau)$ follow upon differentiating (5.13) in t (cf. Lemma 3.5). We note also that if $\tilde{F}(t) = \int_0^t F(\sigma) \int_0^\sigma \left|\frac{P(\xi)}{P(\sigma)}\right| d\xi d\sigma$ then $\tilde{F}' = FF'$ with $\tilde{F} = F^2/2$ as in Section 3. QED

Now, as in (3.30), we must examine the behavior of $\alpha(t)Z_t(\lambda,t,o)$ as $t \to 0$. Thus (cf. (3.31)) we set

$\alpha(t)\int_\tau^t (P(\xi)/P(t))d\xi = H(t,\tau)$ with $H(t) = H(t,o)$. Let us assume now that $\alpha \in C^1(o,b]$ with $\phi = \alpha'/\alpha^2 \in C^o[o,s]$, $|\alpha| \leq k\mathrm{Re}\,\alpha$ on $[o,s]$, and $|\alpha(t)/\alpha(\xi)| \leq N$ for $0 \leq \xi \leq t \leq s$. If $\lim H(t) = H(o)$ exists as $t \to 0$ then clearly (cf. (5.13) upon differentiation) $\lim \alpha(t)\, Z_t(\lambda,t,o) = -H(o)[\gamma(o) + \lambda Q(o)]$. Now consider $\alpha(t)\int_\tau^t \exp\left(-\int_\xi^t \alpha(\eta)\,d\eta\right) d\xi = \hat{H}(t,\tau)$ with $\hat{H}(t) = \hat{H}(t,o)$. Since $\beta \in C^o[o,b]$ it is easily seen that $|\hat{H}(t) - H(t)|$ can be made arbitrarily small for t sufficiently small. Indeed if $B(t,\xi) = \exp\left(-\int_\xi^t \beta(\eta)\,d\eta\right)$ then, by continuity, given ε there exists $\delta(\varepsilon)$ such that $|B(t,\xi) - 1| \leq \varepsilon$ for $t \leq \delta(\varepsilon)$. Hence for $t \leq \min(s,\delta(\varepsilon))$ one has

$$(5.14)\qquad |\hat{H}(t)-H(t)| \leq |\alpha(t)| \int_0^t |1 - B(t,\xi)|\, e^{-\int_\xi^t \mathrm{Re}\alpha(\eta)d\eta}\, d\xi$$

$$\leq \varepsilon \int_0^t \left|\frac{\alpha(t)}{\alpha(\xi)}\right| |\alpha(\xi)| e^{-\int_\xi^t \mathrm{Re}\alpha(\eta)d\eta}\, d\xi$$

$$\leq \varepsilon k N \int_0^t \mathrm{Re}\alpha(\xi) e^{-\int_\xi^t \mathrm{Re}\alpha(\eta)d\eta}\, d\xi = \varepsilon k N$$

To find now $\hat{H}(o) = \lim \hat{H}(t)$ as $t \to 0$ we integrate the definition of $\hat{H}(t,\tau)$ by parts to obtain

$$(5.15)\qquad \hat{H}(t,\tau) = 1 - \frac{\alpha(t)}{\alpha(\tau)} \exp\left(-\int_\tau^t \alpha(\eta)d\eta\right)$$

$$+ \int_\tau^t \frac{\alpha(t)}{\alpha(\xi)} \alpha(\xi)\phi(\xi) \exp\left(-\int_\xi^t \alpha(\eta)d\eta\right)d\xi$$

Letting $\tau \to 0$ we have

$$(5.16)\qquad \hat{H}(t) = 1 + \int_0^t f(t,\xi)\phi(\xi)d\xi$$

where $f(t,\xi) = \alpha(t) \exp(-\int_{\xi}^{t} \alpha(\eta)d\eta)$. But $\hat{H}(t) = \int_0^t f(t,\xi)d\xi$ so (5.16) can be written as $\hat{h}(t) \equiv 1 = \int_0^t f(t,\xi)(1-\phi(\xi))d\xi$. Evidently $\hat{h}(\cdot)$ is continuous as $t \to 0$ and choosing t small enough so that $|\phi(\xi)-\phi(o)| \leq \varepsilon$ for $0 \leq \xi \leq t$ one has (cf. (5.14))

$$(5.17) \qquad |\hat{h}(t) - (1-\phi(o)) \int_0^t f(t,\xi)d\xi|$$

$$\leq \int_0^t |f(t,\xi)||\phi(o)-\phi(\xi)|d\xi \leq \varepsilon Nk$$

Hence $(1-\phi(o))\hat{H}(t) \to 1$ as $t \to 0$ and if $\phi(o) \neq 1$ this means $\lim \hat{H}(t) = \lim H(t) = H(o) = 1/(1-\phi(o))$ as $t \to 0$ from which follows that $t \to \alpha(t)Z_t(\lambda,t,o)$ is continuous in t as $t \to 0$ with $\lim \alpha(t)Z_t(\lambda,t,o) = \frac{\gamma(o) + \lambda Q(o)}{\phi(o) - 1}$. It may then also be shown easily that $\alpha(t)Z_t(\lambda,t,\tau)$ is continuous in (λ,t) for $\tau \geq 0$ fixed and analytic in λ for (t,τ) fixed $(o \leq \tau \leq t \leq b)$; similar properties are then transported to $Z_{tt}(\lambda,t,\tau)$ by (5.11). Consequently

<u>Lemma 5.9</u> Let $\alpha \in C^1(o,b]$ with $\phi = \alpha'/\alpha^2 \in C^0[o,s]$, $|\alpha| \leq k\text{Re}\alpha$ on $[o,s]$, and $|\alpha(t)/\alpha(\xi)| \leq N$ for $0 \leq \xi \leq t \leq s$. Then $Z(\lambda,\cdot,\tau) \in C^2[o,b]$ and satisfies (5.11) for $0 \leq \xi \leq t \leq b$ while $\alpha(t)Z_t(\lambda,t,\tau)$ and $Z_{tt}(\lambda,t,\tau)$ are continuous in (λ,t) and analytic in λ.

Proof: It remains to prove that $\phi(o) \neq 1$ and we suppose the contrary. Then $\text{Re}\,\phi \to 1$ and $\text{Im}\,\phi \to 0$ and we can make $|\text{Re}\phi(t) - 1| \leq \varepsilon$ for $t \leq \delta(\varepsilon)$. Now $\frac{1}{\alpha(t)} - \frac{1}{\alpha(\tau)} =$

$-\int_\tau^t \phi(\xi)d\xi$ $(\tau < t)$ with $\lim \frac{1}{\alpha(\tau)} = 0$ as $\tau \to 0$ since $\text{Re}\alpha(\tau) \to \infty$. Hence $1/\alpha(t) = -\int_0^t \phi(\xi)d\xi$ and for $t \le \min(s,\delta(\varepsilon))$ and $\varepsilon < 1$ we can write

$$(5.18) \qquad \text{Re}\,\frac{1}{\alpha(t)} = -t + \int_0^t (1-\text{Re}\phi(\xi))d\xi \le -(1-\varepsilon)t < 0$$

which violates the fact that $\text{Re}\alpha(t) > 0$ for t small. QED

We can now derive some bounds for the resolvant Z as in Section 3 (cf. (3.32) - (3.33) and Lemma 3.7). Here one multiplies $(PZ_\xi)_\xi + P(\gamma+\lambda Q)Z = 0$ by $P^{-1}(\xi)\overline{Z}_\xi(\lambda,\xi,\tau)$ and integrates from τ to t to obtain

$$(5.19) \qquad |Z_t(\lambda,t,\tau)|^2 - \int_\tau^t Z_\xi \overline{Z}_{\xi\xi} d\xi + \int_\tau^t (\alpha+\beta)|Z_\xi|^2 d\xi + \int_\tau^t (\gamma+\lambda Q) Z\overline{Z}_\xi d\xi = 0$$

Assume now that $Q \in C^1[o,b]$ with Q real and $0 < q \le Q(t)$. Adding (5.19) to its complex conjugate and noting that $2\text{Re}(Z_\xi \overline{Z}_{\xi\xi}) = \frac{d}{d\xi}|Z_\xi|^2$ with $2\text{Re}\, Z\overline{Z}_\xi = \frac{d}{d\xi}|Z|^2$ we have

$$(5.20) \qquad |Z_t(\lambda,t,\tau)|^2 + 2\int_\tau^t \text{Re}(\alpha+\beta)|Z_\xi|^2 d\xi + 2\int_\tau^t \text{Re}(\gamma Z\overline{Z}_\xi)d\xi + \lambda Q(t)|Z(\lambda,t,\tau|^2 = \lambda Q(\tau) + \lambda\int_\tau^t Q'|Z|^2 d\xi$$

It is convenient now to write $\alpha + \beta = \alpha_1 + \alpha_2$ where $\alpha_1 = \alpha$ on $(o,\tilde{s}]$, $\tilde{s} < s$, $\alpha_1 = 0$ on $[s,b]$, and $\alpha_1 = \alpha[(s-t)/(s-\tilde{s})]$ on $[\tilde{s},s]$. Then $\alpha_1 \in C^0(o,b]$ with $|\alpha_1| \le k_1\text{Re}\alpha_1$ on $(o,s]$ and $|\alpha_1(t)/\alpha_1(\xi)| \le$

N_1 for $0 \leq \xi \leq t \leq b$. Then from (5.20) we obtain

$$(5.21) \qquad |Z_t|^2 + \lambda Q(t)|Z|^2 \leq 2\int_\tau^t |\mathrm{Re}\alpha_2||Z_\xi|^2 d\xi + \lambda Q(\tau) + \lambda\int_\tau^t |Q'||Z|^2 d\xi + 2\int_\tau^t |\mathrm{Re}(\gamma Z\bar{Z}_\xi)|d\xi$$

We note that $|\mathrm{Re}(\gamma Z\bar{Z}_\xi)| \leq |\gamma Z\bar{Z}_\xi| \leq \frac{1}{2}|\gamma|(|Z|^2 + |Z_\xi|^2)$ and set $c_1 = 2 \sup |\mathrm{Re}\alpha_2|$, $c_2 = \sup |\gamma|$, and $c_3 = \sup |Q'|$ (on $[o,b]$). It may be assumed that $\lambda \geq 1/q$ since this can always be assured by adding if necessary a constant multiple of Q to γ (e.g., set $\lambda' = \lambda + 1/q$ and $\gamma' = \gamma - (\frac{1}{q})Q$ so that $\gamma' + \lambda' Q = \gamma + \lambda Q$ with $\lambda' \geq 1/q$). Then from (5.21) we have

$$(5.22) \qquad |Z_t|^2 + \lambda Q(t)|Z|^2 \leq \lambda Q(\tau) + (\lambda c_3 + c_2)\int_\tau^t |Z|^2 d\xi + (c_2+c_1)\int_\tau^t |Z_\xi|^2 d\xi \leq \lambda Q(\tau) + (\lambda c_3 + c_2)\int_\tau^t |Z|^2 d\xi + c_4\int_\tau^t \{|Z_\xi|^2 + \lambda Q(\xi)|Z|^2\}d\xi$$

where $c_4 = c_1 + c_2$. Now let us state a version of Gronwall's lemma which will be useful (see Carroll [14] or Sansone-Conti [1] for proof).

<u>Lemma 5.10</u> Let $u \in C^0$, $\ell \in L^1$ with $\ell \geq 0$, and ϕ absolutely continuous. Assume $u(t) \leq \phi(t) + \int_{t_0}^t \ell(\eta)u(\eta)d\eta$. Then

$$(5.23) \qquad u(t) \leq \int_{t_0}^t \phi'(\xi) \exp\left(\int_\xi^t \ell(\theta)d\theta\right) d\xi + \phi(t_0) \exp\left(\int_{t_0}^t \ell(\theta)d\theta\right)$$

Applying Lemma 5.10 to (5.22) with $\ell = c_4$, $t_o = \tau$, $u(t) = |Z_t|^2 + \lambda Q(t)|Z|^2$, and $\phi(t) = \lambda Q(\tau) + (\lambda c_3+c_2)\int_\tau^t |Z|^2 d\xi$ it follows that

$$(5.24) \qquad |Z_t|^2 + \lambda Q(t)|Z|^2 \leq \lambda Q(\tau) \exp c_4(t-\tau)$$

$$+ \int_\tau^t (\lambda c_3+c_2)|Z|^2 \exp c_4(t-\xi)d\xi$$

$$\leq c_5\lambda Q(\tau) + c_5(\lambda c_3+c_2)\int_\tau^t |Z|^2 \, d\xi$$

where $c_5 = \exp c_4 b$. In particular

$$(5.25) \qquad |Z|^2 \leq \frac{c_5 Q(\tau)}{q} + \frac{c_5}{q}(c_3+\frac{c_2}{\lambda}) \int_\tau^t |Z|^2 d\xi$$

Since $\lambda \geq 1/q$ we have $\frac{1}{\lambda q} \leq 1$ and using Lemma 5.10 again there results from (5.25)

$$(5.26) \qquad |Z|^2 \leq \frac{c_5 Q(\tau)}{q} \exp c_6 b = c_7$$

where $c_6 = \frac{c_5}{q}(c_3+c_2)$. Putting this in (5.24) we obtain

$$(5.27) \qquad |Z_t|^2 \leq c_8\lambda + c_9$$

<u>Lemma 5.11</u> Assume $Q \in C^1[o,b]$ with Q real and $0 < q \leq Q(t)$ while $|\alpha| \leq k\mathrm{Re}\alpha$ on $[o,s]$, $\phi = \alpha'/\alpha^2 \in C^o[o,s]$, and $|\alpha(t)/\alpha(\xi)| \leq N$ on $[o,s]$. Then Z given by Lemma 5.8 satisfies (5.26) - (5.27) for $0 \leq \tau \leq t \leq b$ while $|\alpha(t)Z_t| \leq c_{10}\lambda + c_{11}$.

Proof: There remains only the last statement. Let $c_{12} = \sup |Q|$ on $[o,b]$, $c_{13} = \sup |\alpha_2|$ on $[o,b]$ (recall that $\alpha + \beta =$

$\alpha_1 + \alpha_2)$, and $c_{14} = \sup \exp(-\int_\tau^t \mathrm{Re}\, \alpha_2(\eta)d\eta)$ for $0 \leq \xi \leq t \leq b$. Then differentiate (5.13) in t and for $t \leq s$ we have (cf. (5.12))

$$(5.28) \qquad |(\alpha+\beta)Z_t| \leq (c_2+\lambda c_{12})\{c_{13}M \int_\tau^t |Z|d\xi$$

$$+ c_{14}N_1k_1 \int_\tau^t \mathrm{Re}\alpha_1(\xi) \exp(-\int_\xi^t \mathrm{Re}\alpha_1(\eta)d\eta)|Z|d\xi\}$$

$$\leq c_{15}\lambda + c_{16}$$

where (5.26) has been used. QED

<u>Remark 5.12</u> Under certain mild additional hypotheses the assumption $\phi \in C^0[o,s]$ is sufficient to insure that $|\alpha(t)/\alpha(\tau)| \leq N$ for $0 \leq \tau \leq t \leq s$ (cf. Lemmas 5.9 and 5.11). In particular this holds if $\phi(o) \neq 0$ or if $\phi/|\phi| \in C^0[o,\tilde{s}]$ for some $\tilde{s} \leq s$ (see Carroll [8] for details).

We construct an "associate" resolvant $Y(\lambda,t,\tau)$, corresponding to $\hat{S}^m(y,t,\tau)$ in Section 3, by the same technique to obtain (cf. (3.16) - (3.17))

$$(5.29) \qquad Z_\tau(\lambda,t,\tau) = [\gamma(\tau)+\lambda Q(\tau)]\, Y(\lambda,t,\tau)$$

$$(5.30) \qquad Y_t(\lambda,t,\tau) = -Z(\lambda,t,\tau) + [\alpha(\tau)+\beta(\tau)]\, Y(\lambda,t,\tau)$$

Then given $\mathrm{Re}\alpha \geq 0$ on $(o,s]$, (5.12) holds and one can differentiate (5.13) in τ to obtain $Y(\lambda,t,\tau) = U(t,\tau) - (J \cdot Y)(\lambda,t,\tau)$ where U is given as in (3.34) by $U(t,\tau) = \int_\tau^t (P(\tau)/P(\sigma))d\sigma =$

$\int_\tau^t I(\sigma,\tau)d\sigma$ and J is defined by (5.13). The formal solution is again $Y = \sum_{k=0}^{\infty} (-1)^k J^k \cdot U$ (cf. (3.36)) and to show that $U(t,\tau)$ is continuous in (t,τ) we need only consider the critical point $\tau = 0$. Thus defining evidently $U(t,o) = 0$ for $t \geq 0$ (note that our assumptions preclude a case where $\alpha(t) = 0$ as when $m = -1/2$ in Section 3) one has $|I(\sigma,\tau)| \leq M$ for $0 \leq \tau \leq \sigma \leq b$ (cf. (5.12)) and we write $U(t,\tau) = \int_0^t \tilde{I}(\sigma,\tau)d\sigma$ where $\tilde{I}(\sigma,\tau) = 0$ for $\sigma < \tau$. Now let $(t,\tau) \to (t_0,o)$ for $t_0 \geq 0$; then $\tilde{I}(\sigma,\tau) \to 0$ for any $\sigma \neq 0$ as $\tau \to 0$ with $|\tilde{I}(\sigma,\tau)| \leq M$ and we may invoke the Lebesque bounded convergence theorem (cf. Bourbaki [3]) to conclude that $U(t,\tau) \to 0$. Hence the terms $J^k \cdot U$ will be continuous in (λ,t,τ) and analytic in λ with $|U(t,\tau)| \leq Mb$. By previous analysis (cf. Lemmas 3.8 and 5.8) we have

<u>Lemma 5.13</u> Given $\text{Re}\alpha \geq 0$ on $(o,s]$ the series $Y = \sum_{k=0}^{\infty} (-1)^k J^k \cdot U$ converges absolutely and uniformly on $\{0 \leq \tau \leq t \leq b\} \times \Gamma$, $\Gamma \subset \mathbb{C}$ compact $(\lambda \in \Gamma)$, and represents a continuous function $Y(\lambda,t,\tau)$ of (λ,t,τ), analytic in λ, which satisfies $Y = U - J \cdot Y$ (J given as in (5.13)). One has (5.29) and for $\tau > 0$ (5.30) holds. Moreover $Y(\lambda,\tau,\tau) = 0$, $Y(\lambda,t,o) = 0$, and, for $\tau > 0$, $Y_t(\lambda,\tau,\tau) = 1$.

Proof: This goes as in the proof of Lemma 3.8 (note here that $U_t = P(\tau)/P(t)$, $U_\tau = -1 + (\alpha+\beta)U$, and from (5.30), cf. (3.39), $Y(\lambda,t,\tau) = \int_\tau^t Z(\lambda,t,\eta)(P(\tau)/P(\eta))d\eta$). QED

Now Y_τ is defined by (5.30) and does not necessarily tend to infinity as $\tau \to 0$ (cf. (3.42)). We need only examine $\alpha(\tau)Y(\lambda,t,\tau)$, or $\alpha_1(\tau)Y(\lambda,t,\tau)$, as $\tau \to 0$. As in (3.40) $\alpha_1(\tau)Y(\lambda,t,\tau) = \sum_{k=0}^{\infty} (-1)^k J^k \cdot \alpha_1(\tau)U(\cdot,\tau)$ so we look first at $\alpha_1(\tau)U(t,\tau)$. For $\tau > 0$ one has

$$\alpha_1(\tau)U(t,\tau) = \alpha_1(\tau) \int_\tau^t B_2(\sigma,\tau) \exp\left(-\int_\tau^\sigma \alpha_1(\eta)d\eta\right) d\sigma \tag{5.31}$$

$$= 1 - B_2'(t,\tau)\frac{\alpha_1(\tau)}{\alpha_1(t)} \exp\left(-\int_\tau^t \alpha_1(\eta)d\eta\right)$$

$$+ \alpha_1(\tau) \int_\tau^t \left[\frac{B_2'(\sigma,\tau)}{\alpha_1(\tau)} - B_2(\sigma,\tau)\phi_1(\sigma)\right]e^{(-\int_\tau^\sigma \alpha_1(\eta)d\eta)} d\sigma$$

where $B_2'(\sigma,\tau) = \dfrac{\partial B_2(\sigma,\tau)}{\partial\sigma}$ and $\phi_1 = \alpha_1'/\alpha_1^2 = \phi$ for sufficiently small arguments (e.g., $t \le \tilde{s}$); here $B_2(\sigma,\tau) = \exp\left(-\int_\tau^\sigma \alpha_2(\eta)d\eta\right)$ is C^∞ in (σ,τ) with $|B_2(\sigma,\tau)| \le c_{14}$ (cf. Lemma 5.11). We examine first the term $\Lambda(t,\tau) = \dfrac{\alpha_1(\tau)}{\alpha_1(t)} \exp\left(-\int_\tau^t \alpha_1(\eta)d\eta\right)$ as $\tau \to 0$ and one can compare here with Section 3 where $\alpha(t) = \dfrac{2m+1}{t}$ and $\Lambda(t,\tau) = (\tau/t)^{2m} \to 0$ as $\tau \to$ for $m > 0$ ($B_2(t,\tau) \equiv 1$ in Section 3). Thus assume $\Lambda(t,\tau) \to 0$ as $\tau \to 0$ in which event, from (5.31) there results (since $B_2'/B_2 = -\alpha_2(\sigma)$)

$$\alpha_1(\tau) \int_\tau^t B_2(\sigma,\tau)\left[1 + \frac{\alpha_2(\sigma)}{\alpha_1(\sigma)} + \phi(\sigma)\right]e^{(-\int_\tau^t \alpha_1(\eta)d\eta)} d\sigma \to 1 \tag{5.32}$$

as $\tau \to 0$, independently of t. But $\alpha_2(\sigma)/\alpha_1(\sigma) \to 0$ as $\sigma \to 0$ and taking t small enough so that $1 + \dfrac{\alpha_2(\sigma)}{\alpha_1(\sigma)} + \phi(\sigma) \doteq 1 + \phi(o)$ (and $B_2(\sigma,\tau) \doteq 1$) we have

$$\lim_{\tau\to 0} \alpha_1(\tau)U(t,\tau) = \frac{1}{1 + \phi(o)} \tag{5.33}$$

provided $\phi(o) \neq -1$. This checks with Section 1.3 (for $m > 0$) where $1/(1+\phi(o)) = \frac{2m+1}{2m}$ or $\frac{2m+1}{\tau} U(t,\tau) \to \frac{2m+1}{2m}$. Thus when $\Lambda(t,\tau) \to 0$ as $\tau \to 0$ with $\phi(o) \neq -1$ we have by (5.30) as $\tau \to 0$ $Y_\tau(\lambda,t,o) = -(\phi(o)/(1+\phi(o))Z(\lambda,t,o)$ (cf. (3.42) and the definition of Z in Lemma 5.8).

In general we cannot expect of course that $\Lambda(t,\tau) \to 0$ as $\tau \to 0$ or that $\phi(o) \neq -1$ (cf. Section 1.3). However, as with $\hat{S}^m(y,t,\tau)$, we can show that $Y_\tau(\lambda,t,\tau) \in L^1$ in τ (with $Y(\lambda,t,\cdot)$ absolutely continuous) and we follow again Carroll [8]. Thus consider $\alpha_1(\tau)Y(\lambda,t,\tau) = \sum_{k=0}^{\infty} (-1)^k \alpha_1(\tau) J^k \cdot U(\cdot,\tau)$ and define for some upper limit $\tilde{t} \leq \tilde{s}$

$$K_n = \int_0^{\tilde{t}} \alpha_1(\tau)\, J^n \cdot U(\cdot,\tau) d\tau \tag{5.34}$$

where $J^n \cdot U(\cdot,\tau) = (J^n \cdot U)(\tilde{t},\tau)$ (note here that $(J^n \cdot U)(t,\tau) = 0$ by definition if $\tau > t$). We set again $|B_2(\sigma,\tau)| \leq c_{14}$ and $|\alpha_1(\tau)| \leq k_1 \text{Re}\alpha_1(\tau)$ so that

$$\begin{aligned} |K_0| &\leq c_{14}k_1 \int_0^{\tilde{t}} \text{Re}\alpha_1(\tau) \int_\tau^{\tilde{t}} e^{-\int_\tau^\sigma \text{Re}\alpha_1(\eta)d\eta} d\sigma d\tau \\ &\leq c_{14}k_1 \int_0^{\tilde{t}} \int_0^{\sigma} \text{Re}\alpha_1(\tau) e^{-\int_\tau^\sigma \text{Re}\alpha_1(\eta)d\eta} d\tau d\sigma \leq c_{14}k_1 b \end{aligned} \tag{5.35}$$

(for Fubini-Tonelli theorems see McShane [1]). Similarly, setting $|\lambda|_C \leq R$ and $(c_2 + Rc_{12}) = c$, we have (cf. (5.13) for J)

(5.36) $$|K_1| \le \int_0^{\tilde{t}} |\alpha_1(\tau)||(J \cdot U)(\tilde{t},\tau)|d\tau$$

$$\le c_{14}k_1c \int_0^{\tilde{t}} \text{Re}\alpha_1(\tau) \int_\tau^{\tilde{t}} \int_\tau^{\sigma} |\frac{P(\xi)}{P(\sigma)}|$$

$$\cdot \int_\tau^{\xi} e^{-\int_\tau^{\eta}\text{Re}\alpha_1(s)ds} d\eta d\xi d\sigma a\tau \le$$

$$c_{14}k_1c \int_0^{\tilde{t}} \int_0^{\sigma} |\frac{P(\xi)}{P(\sigma)}| \int_0^{\xi} \int_0^{\eta} \text{Re}\alpha_1(\tau)e^{-\int_\tau^{\eta}\text{Re}\alpha_1(s)ds} d\tau d\eta d\xi d\sigma$$

$$\le c_{14}k_1cbF(\tilde{t}) \le c_{14}k_1cbF(b)$$

Upon iteration we obtain (cf. Section 3)

(5.37) $$|K_n| \le c_{14}k_1b\,\frac{c^nF^n(b)}{n!}$$

and thus $\sum_{n=0}^{\infty}(-1)^nK^n = \sum_{n=0}^{\infty}(-1)^n \int_0^{\tilde{t}} \alpha_1(\tau)(J^n \cdot U)(\tilde{t},\tau)d\tau$ converges uniformly and absolutely. An elementary argument (cf. Carroll [8]) then shows that $\alpha_1(\cdot)Y(\lambda,t,\cdot) \in L^1$ with

(5.38) $$\int_0^{\tilde{t}} \alpha_1(\tau)Y(\lambda,\tilde{t},\tau)d\tau = \sum_{n=0}^{\infty}(-1)^nK_n$$

<u>Lemma 5.14</u> If $|\alpha_1| \le k_1\text{Re}\alpha_1$ on $[o,\tilde{s}]$ then $\tau \to Y_\tau(\lambda,t,\tau) = -Z(\lambda,t,\tau) + (\alpha+\beta)(\tau)Y(\lambda,t,\tau) \in L^1$ for $0 \le \tau \le t \le b$ and $Y(\lambda,t,\cdot)$ is absolutely continuous with $Y(\lambda,t,\tau) = \int_0^{\tau} Y_\xi(\lambda,t,\xi)d\xi$. If $\phi = \alpha'/\alpha^2 \in C^0[o,s]$ and $\phi(o) \ne -1$ with $\Lambda(t,\tau) = (\alpha_1(\tau)/\alpha_1(t))\exp(-\int_\tau^t \alpha_1(\xi)d\xi) \to 0$ as $\tau \to 0$ then $Y_\tau(\lambda,t,\cdot)$ is continuous with $Y_\tau(\lambda,t,o) = -(\phi(o)/(1+\phi(o)))Z(\lambda,t,o)$.

Proof: The absolute continuity follows as in Lemma 3.9 and we recall that $Y(\lambda,t,o) = 0$ by Lemma 5.13. QED

Bounds for Y can be obtained as for Z. We simply multiply (5.30), with τ replaced by ξ, by $\overline{Y}(\lambda,t,\xi)$, add this to its complex conjugate, and integrate in ξ from τ to t, using the relation $\frac{d}{d\xi}|Y|^2 = 2\mathrm{Re}(Y_\xi\overline{Y})$, to obtain (recall that $\alpha + \beta = \alpha_1 + \alpha_2$)

$$(5.39)\quad \frac{1}{2}|Y(\lambda,t,\tau)|^2 + \int_\tau^t \mathrm{Re}\alpha_1(\xi)|Y|^2(\lambda,t,\xi)d\xi$$

$$\leq -\int_\tau^t \mathrm{Re}\alpha_2(\xi)|Y|^2(\lambda,t,\xi)d\xi$$

$$+\frac{1}{2}\int_\tau^t (|Z|^2 + |Y|^2)(\lambda,t,\xi)d\xi$$

Recalling that $c_1 = 2\sup|\mathrm{Re}\alpha_2|$ we have, assuming Lemma 5.11,

$$(5.40)\quad |Y|^2 \leq (c_1+1)\int_\tau^t |Y|^2 d\xi + \int_\tau^t |Z|^2 d\xi$$

$$\leq c_7 b + (c_1+1)\int_\tau^t |Y|^2(\lambda,t,\xi)d\xi$$

Lemma 5.15 Under the hypotheses of Lemma 5.11 we have $|Y|^2(\lambda,t,\tau) \leq \tilde{c}$ for $0 \leq \tau \leq t \leq b$ and $\lambda > 0$.

Proof: One applies Gronwall's lemma (Lemma 5.10) to (5.40). QED

Now let us go to the solution of the Cauchy problem (5.9) - (5.10), recalling that λ can always be made larger than zero.

Let $L_s(E,F)$ be the space of continuous linear maps $E \to F$ with the topology of simple convergence, i.e., $A_\nu \to 0$ in $L_s(E,F)$ if $A_\nu e \to 0$ in F for each $e \in E$. We write $L_s(E)$ for $L_s(E,E)$ and recall that $\theta H = \mathcal{H} = \int^\oplus \mathcal{H}(\lambda) d\nu(\lambda)$. Now $Z(\cdot,t,\tau)$ and $Y(\cdot,t,\tau)$ belong to $L_s(\theta D(\Lambda),\mathcal{H})$, $L_s(\mathcal{H})$, and, $L_s(\theta D(\Lambda))$ since they are bounded by Lemmas 5.11 and 5.15 (on $\theta D(\Lambda)$ we put the norm $\|\theta T\|^2 = \int_0^\infty \|\theta T\|^2_{\mathcal{H}(\lambda)} d\nu + \int_0^\infty \lambda^2 \|\theta T\|^2_{\mathcal{H}(\lambda)} d\nu$, where $\|\theta T\|_{\mathcal{H}(\lambda)}$ denotes $\|\theta T(\lambda)\|_{\mathcal{H}(\lambda)}$). If now $\theta T \in \theta D(\Lambda)$ consider for example $\Delta(\theta w)(\lambda) = [Z(\lambda,t,\tau) - Z(\lambda,t_o,\tau)](\theta T)(\lambda) = (\Delta Z)(\theta T)(\lambda)$. To show $t \to Z(\cdot,t,\tau)\theta T : [o,b] \to \theta D(\Lambda)$ is "strongly" continuous, and hence that $t \to Z(\cdot,t,\tau) : [o,b] \to L_s(\theta D(\Lambda))$ is "simply" continuous, one must show that as $t \to t_o$ $\|\Delta(\theta w)\|_{\mathcal{H}} \to 0$ and $\|\lambda\Delta(\theta w)\|_{\mathcal{H}} \to 0$. But, for example, $\|\Delta(\theta w)\|_{\mathcal{H}} \to 0$ means $\|\Delta(\theta w)(\lambda)\|^2_{\mathcal{H}(\lambda)} \to 0$ in $L^1(\nu)$ and since $\Delta(\theta w)(\lambda) \to 0$ ν almost everywhere (cf. Lemma 5.8) with $\|\Delta(\theta w)(\lambda)\|^2_{\mathcal{H}(\lambda)} \leq 2c_7 \|\theta T\|^2_{\mathcal{H}(\lambda)} \in L^1(\nu)$ (cf. (5.26) in Lemma 5.11) one can apply the Lebesgue dominated convergence theorem (cf. Bourbaki [3]) to conclude that $\|\Delta(\theta w)\|_{\mathcal{H}} \to 0$ as $t \to t_o$ (note that $\|\theta T\|^2_{\mathcal{H}(\lambda)}$ is finite ν almost everywhere). A similar argument applies to $\|\lambda\Delta(\theta w)\|_{\mathcal{H}}$ since $\lambda^2 \|\theta T\|^2_{\mathcal{H}(\lambda)} \in L^1(\nu)$; thus one has shown that $t \to Z(\cdot,t,\tau) \in C^o(L_s(\theta D(\Lambda)))$ on $[o,b]$. Analogous reasoning, using (5.29) with Lemmas 5.8, 5.11, 5.13, and 5.15, leads to (cf. also Remark 5.22)

<u>Lemma 5.16</u> Under the hypotheses of Lemma 5.11 we have, for $0 \leq \tau \leq t \leq b$, $t \to Z(\cdot,t,\tau) \in C^o(L_s(\theta D(\Lambda))$ or $C^o(L_s(\mathcal{H}))$ (and

$C^0(L_s(\theta D(\Lambda),H))$ trivially); $\tau \to Y(\cdot,t,\tau)$ and $\tau \to Z(\cdot,t,\tau) \in C^0(L_s(H))$ or $C^0(L_s(\theta D(\Lambda)))$; while $t \to Z_t(\cdot,t,\tau)$, $t \to \alpha(t)Z_t(\cdot,t,\tau)$, and $\tau \to Z_\tau(\cdot,t,\tau) \in C^0(L_s(\theta D(\Lambda),H))$.

We wish to show now that in fact $\partial/\partial t(\theta w) = \partial/\partial t(Z\theta T) = Z_t\theta T$ in H. For fixed (λ,τ) one has evidently $\Delta Z/\Delta t - Z_t(\lambda,t_0\tau) = (\int_{t_o}^{t} [Z_\eta(\lambda,\eta,\tau) - Z_\eta(\lambda,t_0,\tau)]d\eta)/\Delta t$. Hence in $H(\lambda)$ we have, ν almost everywhere, in an obvious notation, $\tilde{\Delta}_\lambda = (\Delta(\theta w)/\Delta t) - Z_t(\lambda,t_0,\tau)\theta T(\lambda) = [(1/\Delta t)\int_{t_0}^{t} \Delta Z_\eta d\eta]\theta T(\lambda)$ and consequently

$$(5.41) \qquad \|\tilde{\Delta}_\lambda\|^2 \le [\frac{1}{\Delta t}\int_{t_0}^{t} |\Delta Z_\eta|d\eta]^2\|\theta T(\lambda)\|^2_{H(\lambda)}$$

Now $\|\theta T\|_{H(\lambda)} = \|\theta T(\lambda)\|_{H(\lambda)}$ is finite ν almost everywhere and for (λ,τ) fixed we can make (by Lemma 5.8) $|\Delta Z_\eta| \le \varepsilon$ for $|\eta - t_0| \le \delta(\varepsilon)$. Hence, for $|t-t_0| \le \delta(\varepsilon)$, $\|\tilde{\Delta}_\lambda\|^2 \le \varepsilon^2\|\theta T(\lambda)\|^2_{H(\lambda)}$ so that $\|\tilde{\Delta}_\lambda\| \to 0$ ν almost everywhere as $t \to t_0$. Also by (5.27) we know that $\|\tilde{\Delta}_\lambda\|^2 \le [(2/\Delta t)(\int_{t_0}^{t} (c_8\lambda + c_9)^{1/2}d\eta]^2\|\theta T\|^2_{H(\lambda)} \le (c_{17}\lambda^2 + c_{18})\|\theta T(\lambda)\|^2_{H(\lambda)}$. Therefore $\|\tilde{\Delta}_\lambda\|^2 \to 0$ in $L^1(\nu)$ and $\partial/\partial t(\theta w) = Z_t\theta T$ in H. Similarly one checks $\partial^2/\partial t^2(\theta w)$, using estimates for $\alpha(t)Z_t(\lambda,t,\tau)$ and $\lambda Z(\lambda,t,\tau)$ indicated above (cf. also Lemma 5.9), to establish

<u>Theorem 5.17</u> Under the hypotheses of Lemma 5.11, $t \to Z(\cdot,t,\tau) \in C^2(L_s(\theta D(\Lambda)),H)$ and there exists a solution of the Cauchy problem 5.9 (Problem 5.7) given by $w(t) = \theta^{-1}[Z(\cdot,t,\tau)\theta T(\cdot)]$ for $T \in D(\Lambda)$ $(0 \le \tau \le t \le b)$.

One notes also that by (5.30), and Lemma 5.16 with references thereto, $\tau \to (1/\alpha(\tau))Y_\tau(\cdot,t,\tau) \in C^0(L_s(H))$ or $C^0(L_s(\theta D(\Lambda))$ for τ small with $\tau \to Y(\cdot,t,\tau) \in C^1(L_s(H))$ or $C^1(L_s(\theta D(\Lambda))$ for $\tau > 0$. Now we note that if F is barreled then any separately continuous bilinear map $E \times F \to G$ is hypocontinuous relative to bounded sets in E (see Bourbaki [2]). But a Hilbert space is barreled so the map $u : (A,h) \to Ah : L_s(F,G) \times F \to G$ is continuous on bounded sets $B \subset L_s(F,G)$. Since $0 \le \tau \le \xi \le t \le b$ is compact and the continuous image of a compact set is bounded we may state for example that $\xi \to Y(\cdot,t,\xi)\theta w(\xi)$, $\xi \to Y(\cdot,t,\xi)\theta w_\xi(\xi)$, $\xi \to (\alpha(\xi) + \beta(\xi))Y(\cdot,t,\xi)\theta w_\xi(\xi)$, $\xi \to Y_\xi(\cdot,t,\xi)\theta w_\xi(\xi)$, and $\xi \to Y(\cdot,t,\xi)\theta w_{\xi\xi}(\xi)$ are continuous (note that $Y_\xi \theta w_\xi = (1/\alpha(\xi))Y_\xi \; \alpha \; \theta w_\xi$ and $\theta w(\xi) = \theta w(\xi,\tau)$ is given by Theorem 5.17). The following formulas are then easily verified using Lemma 5.14 (cf. (3.18) - (3.20) and see Carroll [8] for details).

$$(5.42) \qquad \int_\tau^t \{Y\theta w_{\xi\xi} + Y_\xi \theta w_\xi\} d\xi = 0$$

$$(5.43) \qquad \theta w(t) = Z(\lambda,t,\tau)\theta T + \int_\tau^t \{Z\theta w_\xi + (\gamma+\lambda Q)Y\theta w\} d\xi$$

where θw satisfies (5.9). Hence multiplying (5.10) by Y we have

$$(5.44) \qquad \theta w(t) - Z(\lambda,t,\tau)\theta T = \int_\tau^t \{Y_\xi - (\alpha+\beta)Y + Z\}\theta w_\xi d\xi = 0$$

(cf. (5.30)). Hence $\theta w(t)$ is necessarily of the form $Z(\lambda,t,\tau)\theta T$ and we have proved

Theorem 5.18 The solution of Problem 5.7 given by Theorem 5.17 is unique.

As before (cf. (3.22) and Theorem 3.14) we have (cf. Carroll [8] for details and cf. also Remark 5.22).

Theorem 5.19 The unique solution of $(L+Q(t)\Lambda)w = f \in C^0(D(\Lambda))$, $w(\tau) = T \in D(\Lambda)$, $w_t(\tau) = 0$ is given (under the hypotheses of Lemma 5.11) by $w(t) = \theta^{-1}[Z(\cdot,t,\tau)\theta T(\cdot)] + \int_\tau^t \theta^{-1}Y(\cdot,t,\xi)\theta f(\xi)d\xi$ $(0 \le \tau \le t \le b)$.

As in Definition 3.15 we will say that the problem $(L+Q(t)\Lambda)w = f$, $w(\tau) = T \in D(\Lambda)$, $w_t(\tau) = 0$, and $f \in C^0(D(\Lambda))$ is uniformly well posed if w depends continuously in H on (t,τ,T,f). By hypocontinuity and Lemmas 5.8 and 5.13 we can state

Theorem 5.20 The problem $(L+Q(t)\Lambda)w = f \in C^0(D(\Lambda))$, $w(\tau) = T \in D(\Lambda)$, and $w_t(\tau) = 0$ is uniformly well posed under the hypotheses of Lemma 5.11 $(0 \le \tau \le t \le b)$.

Remark 5.21 We can generalize some of the results of Section 1.3 to the case of $L = \partial^2/\partial t^2 + (\alpha(t)+\beta(t))\,\partial/\partial t + \gamma(t)$ in place of $\partial^2/\partial t^2 + (\frac{2m+1}{t})\,\partial/\partial t$; the details are in Carroll [8]. Thus one considers for $0 \le \tau \le t \le b$ e.g., $Lw + Q(t)A_x * w = 0$, $w(\tau) = T \in S'$, and $w_t(\tau) = 0$ with Condition 3.1 on $A(y)$ fulfilled. It is in this problem that the analyticity of Z, Y, etc. in λ is utilized (cf. Lemmas 5.8, 5.9, and 5.13).

One proves that the unique solution $w \in C^2(S')$ of this general problem is given by $w(t) = Z(t,\tau) * T$ where $Z(t,\tau) = F^{-1}Z(A(y),t,\tau)$ and the problem is uniformly well posed.

Remark 5.22 In Carroll [10; 11] some semilinear versions of the singular Cauchy problem were treated by spectral methods. Let L be as above and construct, under the hypotheses of Lemma 5.11, resolvants $Z(\lambda,t,\tau)$ and $Y(\lambda,t,\tau)$ as above. Then consider in H

$$(5.45) \qquad (L+Q(t)\Lambda)w + f(t,w) = 0; \quad w(\tau) = T \in D(\Lambda); \quad w_t(\tau) = 0$$

$(0 \leq \tau \leq t \leq b)$. Following Theorem 5.19 this leads to the integral equation (where $\mathcal{Z}(t,\tau) = \theta^{-1}Z(\cdot,t,\tau)\theta$ and $\mathcal{Y}(t,\tau) = \theta^{-1}Y(\cdot,t,\tau)\theta$ belong to $L_s(H)$)

$$(5.46) \qquad w(t) = \mathcal{Z}(t,\tau)T - \int_\tau^t \mathcal{Y}(t,\xi)f(\xi,w(\xi))d\xi = \mathcal{T}w$$

Now parts of Lemma 5.16 can be somewhat improved by using estimates already established plus some stronger variations; for example one can show that $(t,\xi) \to \mathcal{Y}(t,\xi) \in C^0(L_s(H,D(\Lambda^{1/2})))$ and $t \to \mathcal{Z}(t,\tau) \in C^1(L_s(D(\Lambda^{1/2}),H)$ (cf. (5.27) for $\mathcal{Z}$ and for $\mathcal{Y}$ one can produce an estimate $|Y|^2(\lambda,t,\tau) \leq \hat{c}/\lambda$ for $\lambda > 0$ and $0 \leq \tau \leq t \leq b$, stronger than that of Lemma 5.15 - cf. also Chapter 3). One looks for fixed points of $\mathcal{T}$ using some modifications of a technique of Foiaş-Gussi-Poenaru [1; 2]. For example if we assume $\xi \to \tilde{f}(\xi,v) = \Lambda^{1/2}f(\xi,\Lambda^{-1}v) \in C^0(H)$ for $v \in C^0(H)$ with $\|\tilde{f}(\xi,v) - \tilde{f}(\xi,u)\| \leq k_1(\xi)w_1(\|v-u\|)$ where $k_1 \in L^1$ and w_1 is an

Osgood function (i.e., $\int_0^\varepsilon d\rho/w_1(\rho) = \infty$ for $\varepsilon > 0$) while $\|\tilde{f}(\xi,v)\| \leq k_2(\xi)w_2(\|v\|)$ where $k_2 \in L^1$ and w_2 is a Wintner function (i.e., $\int_\varepsilon^\infty d\rho/w_2(\rho) = \infty$ for $\varepsilon > 0$), then there exists a unique solution $w \in C^0(D(\Lambda))$ of (5.46) on $[o,b]$. Furthermore, this solution satisfies (5.45) with $w \in C^2(H)$, $w \in C^1(D(\Lambda^{1/2}))$, and $w \in C^0(D(\Lambda))$ while (5.45) is uniformly well posed. The improved estimate on Y also leads to a stronger version of Theorem 5.19 where only $f \in C^0(D(\Lambda^{1/2}))$ is required while $w \in C^2(H)$, $w \in C^1(D(\Lambda^{1/2}))$, and $w \in C^0(D(\Lambda))$. One can also add a nonlinear term to $f(t,w)$ in (5.45) which generates a compact operator in H and obtain another existence result for (5.45).

1.6 EPD equations in general spaces. We go now to a technique of Hersh [1; 2] in Banach spaces which was generalized by Carroll [18; 19; 24; 25] to more general locally convex spaces (cf. also Bragg [10], Carroll-Donaldson [20], Donaldson [1; 5], Donaldson-Hersh [6], Hersh [3]). Thus let E be a complete separated locally convex space and A a closed densely defined linear operator in E which generates a locally equicontinuous group $T(\cdot)$ in E (cf. remarks after Definition 6.3). This means that $T(t) \in L(E)$ for $t \in \mathbf{R}$, $T(t)\,T(s) = T(t+s)$, $T(o) = I$, $t \to T(t)e \in C^0(E)$ for $e \in E$, and given any (continuous) seminorm p on E there exists a (continuous) seminorm q such that $p(T(t)e) \leq q(e)$ for $|t| \leq t_0 < \infty$ (any t_0); also $\lim\,(T(t) - I)e/t = Ae$ as $t \to 0$ for $e \in D(A)$ (dense). We refer here to Komura [1] and Dembart [1] for locally equicontinuous semigroups or groups and remark that it is absolutely necessary to consider such groups in "large" spaces E in order to deal for example with growth properties of the solutions of certain differential equations (see also for example Babalola [1], Komatsu [1], Lions [7], Miyadera [1], Oucii [1], Schwartz [6], Waelbroeck [1], Yosida [1], etc. for other "general" semigroups - for strongly continuous semigroups in Banach spaces see Hille-Phillips [2]).

Let us consider now the problem

$$(6.1) \qquad w^m_{tt} + \frac{2m+1}{t} w^m_t = A^2 w^m; \quad w^m(0) = e \varepsilon D(A^2); \quad w^m_t(0) = 0$$

for $w \varepsilon C^2(E)$. Following Hersh [1] one replaces A by $\partial/\partial x$ and looks at

$$(6.2) \qquad R^m_{tt} + \frac{2m+1}{t} R^m_t = R^m_{xx}; \quad R^m(\cdot,0) = \delta; \quad R^m_t(\cdot,0) = 0$$

which is a resolvant equation as in Section 3 whose solution $\hat{R}^m = FR^m$ is given by (3.6) with $z = ty$. We recall the Sonine integral formula (4.6) and pick $p = -1/2$, which corresponds to $2p + 1 = 0 = n - 1$, as the pivotal index (note $p = \frac{n}{2} - 1$ as in Theorem 4.12). Clearly the unique solution of (6.2) with $m = -1/2$ is $R^{-1/2}(\cdot,t) = \mu_x(t) = 1/2[\delta(x+t) + \delta(x-t)]$ so that, from (4.6), for $m > -1/2$

$$(6.3) \qquad R^m(\cdot,t) = \frac{2\Gamma(m+1)}{\Gamma(1/2)\Gamma(m+\frac{1}{2})}\int_0^1 (1-\xi^2)^{m-\frac{1}{2}} \mu_x(\xi t)d\xi$$
$$= c_m \int_{-1}^1 (1-\xi^2)^{m-\frac{1}{2}} \delta(x-\xi t)d\xi$$

where $c_m = \frac{\Gamma(m+1)}{\Gamma(1/2)\Gamma(m+1/2)}$ (note that $\delta(x \pm \xi t)$ could be used in the last expression in (6.3) because of symmetry involved in μ_x - cf. (6.6)). Now one writes formally

$$(6.4) \qquad w^m(t) = \langle R^m(\cdot,t), \ T(\cdot)e\rangle$$

as the solution of (6.1) where for t fixed <, > denotes a pairing between the distribution $R^m(\cdot,t) \in E_x'$ of order zero and $T(\cdot)e \in C^0(E)$. Note here that for $e \in D(A^2)$, $T(\cdot)e \in C^2(E)$ with $\frac{d}{dx} T(x)e = AT(x)e = T(x)Ae$, and of course $t \to R^m(\cdot,t) \in C^\infty(E_x')$ from Section 1.3. If we write out (6.4) in terms of (6.3) there results

$$(6.5) \qquad w^m(t) = c_m\int_{-1}^{1} (1-\xi^2)^{m-\frac{1}{2}} \ \langle \delta(x-\xi t), T(x)e\rangle \, d\xi$$

$$= c_m\int_{-1}^{1} (1-\xi^2)^{m-\frac{1}{2}} \ T(\xi t)e d\xi$$

which coincides with the formula established by Donaldson [1] using a different technique. We can also use the formula

$$(6.6) \qquad w^m(t) = 2c_m \int_{0}^{1} (1-\xi^2)^{m-\frac{1}{2}} \cosh(A\xi t)e d\xi$$

where $\cosh A\xi t = 1/2[T(\xi t) + T(-\xi t)]$ with $\langle \mu_x(\xi t), T(x)e\rangle = \cosh (A\xi t)e$. To verify that (6.4) represents a solution of (6.1) one can of course work directly with the vector integrals (6.5) or (6.6) but we follow Hersh [1] and Carroll [18; 19; 24; 25] in treating (6.4) as a certain distribution pairing. Thus $R^m(\cdot,t)$, $R_x^m(\cdot,t)$, and $R_{xx}^m(\cdot,t) = \Delta_x R^m(\cdot,t)$ are all of order less than or equal to two in E' with supports contained in a fixed compact set $K = \{x; |x| \le x_0\}$ for $0 \le t \le b < \infty$ and we let $\hat{K} = \{x; |x| \le x_0 + \beta\}$ for any $\beta > 0$. If $S \in E'$ is of order less than or equal to two with supp $S \subset K$ we can think of $S \in C^2(\hat{K})'$ (cf. Schwartz [1]). Recall further that on $\hat{K}, C^2(E) = C^2 \hat{\otimes}_\varepsilon E$ (see e.g., Treves [1]) where the ε topology on $C^2 \otimes E$ is the topology of uniform convergence on products of equicontinuous sets in $(C^2)' \times E'$. It is easy to show (see Carroll [18] for details)

<u>Lemma 6.1</u> The composition $S \to \langle S, \vec{\phi} \rangle$: $C^2(\hat{K})' \to E$ is continuous for $\vec{\phi} \in C^2(E)$ fixed. The map $\Delta \otimes 1 : C^2(E) = C^0 \hat{\otimes}_\varepsilon E$ (defined by $(\Delta \times 1) \sum \phi_i \otimes e_i = \sum \Delta\phi_i \otimes e_i$) is continuous.

Let us indicate first the formal calculations necessary to show that (6.4) is a solution of (6.1), while establishing at the same time some recursion relations. Their validity is essentially obvious upon suitable interpretation (cf. below). Thus, from (6.4) and (3.11) transformed,

$$
\begin{aligned}
(6.7) \qquad w_t^m(t) &= \langle R_t^m(\cdot,t), T(\cdot)e\rangle = \frac{t}{2(m+1)} \langle \Delta R^{m+1}(\cdot,t), T(\cdot)e\rangle \\
&= \frac{t}{2(m+1)} \langle R^{m+1}(\cdot,t), \Delta T(\cdot)e\rangle \\
&= \frac{t}{2(m+1)} \langle R^{m+1}(\cdot,t), T(\cdot)A^2 e\rangle \\
&= \frac{t\, c_{m+1}}{2(m+1)} \int_{-1}^{1} (1-\xi^2)^{m+\frac{1}{2}} \quad T(\xi t)A^2 e d\xi \\
&= \frac{A^2 t}{2(m+1)} w^{m+1}(t)
\end{aligned}
$$

Similarly from (6.4) and (3.12) transformed

$$(6.8)\qquad w_t^m(t) = \langle R_t^m(\cdot,t),\ T(\cdot)e\rangle = \frac{2m}{t}\,[\langle R^{m-1}(\cdot,t),\ T(\cdot)e\rangle$$

$$- \langle R^m(\cdot,t),\ T(\cdot)e\rangle] = \frac{2m}{t}\,[w^{m-1}(t) - w^m(t)]$$

Then differentiating (6.7) and using (6.7) - (6.8) we obtain (6.1). The calculations under the bracket $\langle , \rangle$ can be justified using Lemma 6.1 and in order to transport Δ around in (6.7) one can think of $\langle , \rangle$ as a distribution pairing over the interior of $\hat{K}$ (cf. Carroll [18]). This proves, given A, T(t), and E as above ($e \in D(A^2)$)

<u>Theorem 6.2</u> The equivalent formulas (6.4) - (6.6) give a solution to (6.1) for $m \geq -1/2$ and the recursion relations indicated by (6.7) - (6.8) are valid.

The question of uniqueness for (6.1) had seemed to be rather more complicated than that of existence but a new theorem of Carroll [26] (Theorem 6.5) gives a satisfactory response. There are several types of theorems available (cf. Carroll [18; 24; 25; 26], Carroll-Donaldson [20], Donaldson [1], Donaldson-Goldstein [2], Hersh [1]). The first one we discuss (for completeness) is based on a variation of the classical "adjoint" method and simplifies somewhat the presentation in Carroll [18]; it is not as strong however, as Theorem 6.5.

<u>Definition 6.3</u> The space E (as above) will be called A-adapted if E' is complete and $D(A^*)$ is dense.

When E is A-adapted (A being the generator of a locally

equicontinuous group in E) it follows (cf. Komura [1]) that A^* generates a locally equicontinuous group $T^*(t)$ in E'. The requirement of completeness is a luxury which we permit ourselves for simplicity. Indeed, for the discussion of locally equicontinuous semigroups sequential completeness is sufficient (in view of the Riemann type integrals employed for example). Thus, in particular, if E is reflexive (not necessarily complete) then it is quasicomplete (cf. Schaeffer [1]), hence sequentially complete, as is the reflexive space E', and A^* will generate a locally equicontinuous group $T^*(t)$ in E' with $D(A^*)$ dense (cf. Komura [1]). We remark also that if E is bornological then E' is in fact complete (cf. Schaeffer [1]). We use completeness basically only in asserting that $C^2(E) = C^2 \hat{\otimes}_\varepsilon E$ but a version of this is probably true for E sequentially complete; the increase in detail and explanation throughout our discussion does not seem to justify the refinement however.

We consider now the resolvants $R^m(\cdot,t,\tau)$ and $S^m(\cdot,t,\tau)$ of Section 3 when $A(y) = y^2$ with $A_x = -\Delta_x = -\partial^2/\partial x^2$ (cf. also Section 4). We recall (cf. Lemma 4.3) that $R^m(\cdot,t,\tau)$ and $S^m(\cdot,t,\tau)$ belong to E_x', with suitable orders (exercise), and Lemma 3.13 holds with O_M replaced by Exp $O_M = F E'$. Thus let $w^m(t)$ satisfy (6.1) with $w^m(0) = w_t^m(0) = 0$ and consider (for an A-adapted E)

$$(6.9) \qquad u^m(t,\xi) = \langle S^m(\cdot,t,\xi),\ T^*(\cdot)e' \rangle$$

where $e' \in D((A^*)^2)$ (dense); w^m is <u>any</u> solution of (6.1), not necessarily arising from (6.4). Then from (3.17)

$$(6.10) \qquad u^m_\xi(t,\xi) = <S^m_\xi(\cdot,t,\xi),\ T^*(\cdot)e'>$$

$$= <(-R^m(\cdot,t,\xi) + \frac{2m+1}{\xi} S^m(\cdot,t,\xi)),\ T^*(\cdot)e'>$$

$$= -v^m(t,\xi) + \frac{2m+1}{\xi} u^m(t,\xi)$$

where $v^m(t,\xi) = <R^m(\cdot,t,\xi),\ T^*(\cdot)e'>$. Further, from (3.16),

$$(6.11) \qquad v^m_\xi(t,\xi) = -<\Delta S^m(\cdot,t,\xi),\ T^*(\cdot)e'> =$$

$$- <S^m(\cdot,t,\xi),\ \Delta T^*(\cdot)e'> = -(A^*)^2\ u^m(t,\xi)$$

Now take (E, E') brackets $<, >_E$ with $u^m(t,\xi)$ in (6.1), where t has been replaced by ξ and w^m is <u>any</u> solution of (6.1) with zero initial data; then integrate in ξ from 0 to t (cf. (3.18) - (3.20)). We note first that

$$(6.12) \qquad \int_0^t <w^m_{\xi\xi}(\xi),\ u^m(t,\xi)>_E d\xi = -\int_0^t <w^m_\xi(\xi),\ u^m_\xi(t,\xi)>_E d\xi$$

$$(6.13) \qquad \int_0^t <A^2 w^m(\xi),\ u^m(t,\xi)>_E d\xi = \int_0^t <w^m(\xi),\ (A^*)^2 u^m(t,\xi)>_E d\xi$$

$$= -\int_0^t <w^m(\xi),\ v^m_\xi(t,\xi)>_E d\xi = - < w^m(t),\ e'>_E$$

$$+ \int_0^t w^m_\xi(\xi),\ v^m(t,\xi)>_E d\xi$$

Consequently, using (6.10), $<w^m(t), e'>_E \equiv 0$ for any $e' \in D((A^*)^2)$ (dense) - which implies that $w^m(t) \equiv 0$ - since we will have from (6.1), (6.12), and (6.13)

$$(6.14) \qquad 0 = \int_0^t \langle w_\xi^m(\xi), -u_\xi^m(t,\xi) + \frac{2m+1}{\xi} u^m(t,\xi) - v^m(t,\xi)\rangle_E d\xi$$
$$+ \langle w^m(t), e'\rangle_E = \langle w^m(t), e'\rangle_E$$

Theorem 6.4 If E is A-adapted then the solution to (6.1), given by Theorem 6.2, is unique.

This uniqueness theorem has been included primarily to illustrate the adjoint method. If can be supplanted by the following recent result of Carroll [26] which was motivated by (and improves) a result in Carroll [18] based partially on a technique of Fattorini [1].

Theorem 6.5 For Rem > - 1/2 let w^m be any solution of (6.1) with zero initial data and let A generate a locally equicontinuous group T(x) in E as above. Then $w^m(t) \equiv 0$.

Proof: We note first that our existence-uniqueness calculations in Sections 3-4 are valid for suitable complex m (Rem ≥ - 1/2), as indicated partially in Section 5, but we will not spell out the details. Thus our demonstration is strictly justified only for real m ≥ - 1/2. We use $R^m(\cdot,t, s)$ and $S^m(\cdot, t, s)$ with $A_x = -\Delta_x = -\partial^2/\partial x^2$ as in the proof of Theorem 6.4. Define, with brackets as in (6.4),

$$(6.15) \qquad R(t,s) = \langle R^m(\cdot,t,s), T(\cdot)\, w^m(s)\rangle$$

$$(6.16) \qquad S(t,s) = \langle S^m(\cdot,t,s), T(\cdot)\, w_s^m(s)\rangle$$

Then evidently (cf. (3.16) - (3.17) and the calculations leading to Theorem 6.4)

$$(6.17)\qquad R_s(t,s) = -\langle S^m(\cdot,t,s),\, T(\cdot)\, A^2 w^m(s)\rangle + \langle R^m(\cdot,t,s),\, T(\cdot) w_s^m(s)\rangle$$

$$(6.18)\qquad S_s(t,s) = \langle -R^m(\cdot,t,s) + \frac{2m+1}{s} S^m(\cdot,t,s),\, T(\cdot) w_s^m(s)\rangle + \langle S^m(\cdot,t,s),\, T(\cdot) w_{ss}^m(s)\rangle$$

$$(6.19)\qquad (R+S)_s(t,s) = \phi_s(t,s) = 0 = \langle S^m(\cdot,t,s), T(\cdot)[w_{ss}^m + \frac{2m+1}{s} w_s^m - A^2 w^m]\rangle$$

Consequently $\phi(t,o) = \phi(t,t)$ and recalling that $S^m(\cdot,t,t) = 0$ with $R^m(\cdot,t,t) = \delta$ we have $\phi(t,o) = 0 = \phi(t,t) = w^m(t)$. QED

We sketch here in passing some relations of (4.6) and (4.10) to the Riemann-Liouville (R-L) integral (cf. Carroll [18; 19], Donaldson [1], and Rosenbloom [1]); questions of uniqueness for (6.1) in E can be treated in this context but Theorem 6.5 includes everything known. The relation of EPD equations to R-L integrals was of course known by Diaz, Weinberger, Weinstein, etc. many years ago and some facts are of obvious general interest. We recall first (cf. Riesz [1]) that the R-L integral of $f \in C^0$ or "suitable" $f \in L^1$ is defined for Re $\alpha > 0$ by

$$(6.20)\qquad (I^\alpha f)(t) = \frac{1}{\Gamma(\alpha)} \int_0^t (t-s)^{\alpha-1} f(s)\,ds$$

One can obviously deal with $f \in C^0(E)$ for example in the same manner and we will assume f is E valued in what follows. For $\mathrm{Re}\,\alpha > -p$ ($p \geq 1$ an integer) the "continuation" of I^α is given by

$$(6.21) \qquad (I^\alpha f)(t) = \sum_{k=0}^{p-1} \frac{f^{(k)}(o)t^{\alpha+k}}{\Gamma(\alpha+k+1)} + (I^{\alpha+p}f^{(p)})(t)$$

for $f \in C^p(E)$ or $f \in C^{p-1}(E)$ with $f^{(p)}(\cdot) \in L^1(E)$ "suitably". One knows that $I^\alpha I^\beta = I^{\alpha+\beta}$, except possibly when β is a negative integer $\beta = -p$, where $I^{-p}f = f^{(p)}$, in which case $f(o) = \ldots = f^{(p-1)}(o) = 0$ is required in order that $I^\alpha I^{-p} = I^{\alpha-p}$ in general, and I^0 is defined to be the identity. Evidently $(d/dt)(I^\alpha f) = I^{\alpha-1}f$ for any α. We record now some easily established facts.

Lemma 6.6 If $f \in C^{n+1}(E)$ with $-n < \mathrm{Re}\,\alpha \leq -n + 1$ and α nonintegral then

$$(6.22) \qquad (I^\alpha f)'(t) = (I^{\alpha-1}f)(t) = \frac{f(o)t^{\alpha-1}}{\Gamma(\alpha)} + (I^\alpha f')(t)$$

If $f(o) = 0$ or $\alpha = -n$ with $f \in C^{n+1}(E)$ then

$$(6.23) \qquad (I^\alpha f)' = I^\alpha f' = I^{\alpha-1}f$$

(when $\mathrm{Re}\,\alpha > 0$ (6.22) - (6.23) hold for $f \in C^1(E)$). If $f \in C^n(E)$ with $-n < \mathrm{Re}\,\alpha \leq -n + 1$ or $\alpha = -n$ (or $f \in C^0(E)$ when $\mathrm{Re}\,\alpha > 0$) then

$$(6.24) \qquad t(I^\alpha f)(t) = (I^\alpha(sf))(t) + \alpha(I^{\alpha+1}f)(t)$$

Proof: Routine computation yields (6.22) - (6.23) and (6.24) follows by an induction argument. QED

Now, referring to the Sonine integral formula (4.6) (cf. also (6.3) and (6.6)), we can write, after a change of variables,

$$R^m(\cdot,\tau^{1/2}) = c_m\tau^{-m}\int_0^\tau (\tau-s)^{m-\frac{1}{2}}s^{-\frac{1}{2}}R^{-\frac{1}{2}}(\cdot,s^{1/2})ds \tag{6.25}$$

where $c_m = \frac{\Gamma(m+1)}{\Gamma(1/2)\Gamma(m+1/2)}$ as before. Let then

$$W^m(t) = \frac{t^m w^m(t^{1/2})}{\Gamma(m+1)} \tag{6.26}$$

with $w^m(t^{1/2}) = \langle R(\cdot,t^{1/2}),T(\cdot)e\rangle$; it follows from (6.25) that for Re $m > -1/2$

$$W^m(t) = (I^{m+\frac{1}{2}}W^{-\frac{1}{2}})(t) \tag{6.27}$$

Setting $\tilde{w}^m(t) = w^m(t^{1/2})$ it is easy to show (note that $\partial/\partial t^{1/2} = 2t^{1/2}(\partial/\partial t)$)

<u>Lemma 6.7</u> Given w^m, $\tilde{w}^m$, and W^m as above, where w^m satisfies the differential equation in (6.1), one has for $W^m \in C^2(E)$

$$4t\tilde{w}^m_{tt} + 4(m+1)\tilde{w}^m_t = A^2\tilde{w}^m \tag{6.28}$$

$$4tW^m_{tt} - 4(m-1)W^m_t = A^2W^m \tag{6.29}$$

<u>Remark 6.8</u> A previous uniqueness argument (cf. Carroll [18; 19]) for not necessarily A-adapted E proceeded from w^m any solution of (6.1) with zero initial data to W^m determined by (6.26) and <u>defined</u> $W^{-1/2} = I^{-m-\frac{1}{2}} W^m$, together with the corresponding $w^{-1/2}$. Then, using Lemma 6.6, given suitable (but excessive) hypotheses of differentiability and "regularity" on W^m,

$W^{-1/2}$ was shown to satisfy (6.29) with index -1/2, and a uniqueness theorem for the corresponding $w^{-1/2}$ was established (weaker than Theorem 6.5). This led to uniqueness for w^m under suitable hypotheses. One feature of the technique involved embedding W^m in a "Weinstein complex" determined by (cf. (4.10)

$$(6.30) \qquad W^{m+\alpha-p}(t) = (\partial/\partial t)^p(I^{\alpha}W^m)(t) = (I^{\alpha-p}W^m)(t)$$

where p is an integer ($0 \le p \le \ell$) and Re $\alpha > 0$ or $\alpha = 0$ (ℓ and α are chosen so that $-1/2 = m + \alpha - \ell$ which means that if $k + \frac{1}{2} < \text{Re } m < k + \frac{3}{2}$, $k = -1, 0, 1, 2, \ldots$, then $\alpha = k + \frac{3}{2} - m$ with $\ell = k + 2$). (4.10) propagates initial values and (6.7) - (6.8) hold. $W^{m+\alpha-p}(o)$ and $W_t^{m+\alpha-p}(o)$ ($m+\alpha-p \ge \frac{1}{2}$) are critical in (6.30) and to study this one notes from (6.28) - (6.29) that W^m corresponds to some $\tilde{w}^{-m}$ (for any index m) which we write in the form

$$(6.31) \qquad W^n(t) = \hat{w}^{-n}(t^{1/2})/\Gamma(n+1)$$

for some $\hat{w}^{-n}(\cdot)$ satisfying the differential equation in (6.1) with index -n. The connection is somewhat interesting and it is sufficient to indicate this for the crucial index 1/2 (note $W_t^{1/2} = W^{-1/2}$ and in (6.26) $w^{-1/2}(o) = 0$ does not automatically insure $W^{-1/2}(o) = 0$). Thus differentiating (6.31) with index 1/2 and noting the obvious difference between $f_t(t^{1/2})$ and $f(t^{1/2})_t = (d/dt)f(t^{1/2})$ we obtain (cf. (6.26))

$$(6.32) \qquad \tfrac{1}{2}t^{-1/2}w^{1/2}(t^{1/2}) + t^{1/2}w^{1/2}(t^{1/2}) = \hat{w}^{-1/2}(t^{1/2})_t$$

whereas from (6.8) with index 1/2 we have (since $\partial/\partial t^{1/2} = 2t^{1/2}(\partial/\partial t)$)

$$(6.33)\qquad 2t^{1/2}w^{1/2}(t^{1/2})_t = t^{-1/2}[w^{-1/2}(t^{1/2}) - w^{1/2}(t^{1/2})]$$

Consequently from (6.32) - (6.33) we have

$$(6.34)\qquad w^{-1/2}(t^{1/2}) = 2t^{1/2}\hat{w}^{-1/2}(t^{1/2})_t$$

and thus given $w^{-1/2}(o) = 0$ there results

$$(6.35)\qquad \hat{w}^{-1/2}(t^{1/2})_t = \frac{1}{2}t^{-1/2}\int_0^{t^{1/2}} w_\xi^{-1/2}(\xi)d\xi$$

Therefore $\hat{w}^{-1/2}(t^{1/2})_t \to 0$ as $t \to 0$ since $w_\xi^{-1/2}(\xi) \to 0$ which means by (6.31) that $W_t^{1/2}(t) = W^{-1/2}(t) \to 0$ as $t \to 0$, as desired.

In any event we have existence and uniqueness determined by Theorems 6.2 and 6.5 and some new growth and convexity theorems will follow, with realistic examples, of a type first developed by Carroll [18; 19; 25]. Thus, referring to (6.6) - (6.7), and taking now m real so that $c_m > 0$ for $m > -1/2$ we have under the hypotheses of Theorem 6.2

Theorem 6.9 Let E be a space of functions or equivalence classes of functions and assume $e \in D(A^2)$ with $\cosh(At)A^2e \geq 0$ $(m \geq -1/2)$. Then $w^m(\cdot)$ is a nondecreasing function of t.

Proof: From (6.6) - (6.7) we have for $m \geq -1/2$

$$(6.36)\qquad w_t^m(t) = \frac{tA^2}{2(m+1)}w^{m+1}(t) = \frac{tc_{m+1}}{m+1}\int_0^1 (1-\xi^2)^{m+\frac{1}{2}}\cosh(A\xi t)A^2e\,d\xi \geq 0$$

QED

Next we recall that (from Section 4) $L_m^t = \partial^2/\partial t^2 + \frac{2m+1}{t}\partial/\partial t = (2m)^2 t^{-2(2m+1)}\, \partial^2/\partial^2(t^{-2m})$ for $m \neq 0$ while $L_0^t = t^{-2}\, \partial^2/\partial^2(\log t)$. The differential equation in (6.1) can be written as $L_m^t w^m = A^2 w^m$ so we have from (6.6) with $m > -1/2$ real

$$(6.37) \qquad L_m^t w^m(t) = 2c_m \int_0^1 (1-\xi^2)^{m-\frac{1}{2}} \cosh(A\xi t)A^2 e \, d\xi$$

Theorem 6.10 Under the hypotheses of Theorem 6.9, $w^m(\cdot)$ is a convex function of t^{-2m} for $m \neq 0$ and of $\log t$ for $m = 0$.

Proof: The result for $m = -1/2$ is obvious since $w^{-1/2}(t) = \cosh(At)e$ and the rest follows from (6.37). QED

Example 6.11 Let $E = C^0(\mathbb{R})$ with the topology of uniform convergence on compact sets. The operator $A = d/dx$ generates a locally equicontinuous group $(T(t)f)(x) = f(x+t)$ which is not equicontinuous. If $f \geq 0$ then evidently $T(t)f \geq 0$ for $t \in \mathbb{R}$ so to fulfill the hypotheses of Theorems 6.9 - 6.10 we need only find $e = f$ such that $A^2 f = f'' \geq 0$; such functions are abundant, but of course they increase as $t \to \infty$ and we do indeed want to work in "large" enough spaces E to permit such growth.

Example 6.12 Let $E = C^\infty(\mathbb{R}^3) \times C^\infty(\mathbb{R}^3)$ where $C^\infty(\mathbb{R}^3)$ has the Schwartz topology and recall the mean value operators $\mu_x(t)$ and $A_x(t)$ defined in Section 2 which we extend as even functions for t negative. Define

$$(6.38) \qquad A = \begin{pmatrix} 0 & 1 \\ \Delta & 0 \end{pmatrix} \; ; \; A^2 = \begin{pmatrix} \Delta & 0 \\ 0 & \Delta \end{pmatrix}$$

where Δ is the three-dimensional Laplacian. Then A generates a locally equicontinuous group T(t) in E acting by convolution (i.e., $T(t)\vec{f} = T(t) * \vec{f}$) determined by

$$(6.39) \qquad T(t) = \begin{pmatrix} \mu_x(t) + \frac{t^2}{3}\Delta A_x(t) & t\mu_x(t) \\ t\Delta\mu_x(t) & \mu_x(t) + \frac{t^2}{3}\Delta A_x(t) \end{pmatrix}$$

$$= \mu_x(t) + At\mu_x(t) + \frac{A^2t^2}{3} A_x(t)$$

One notes again that T(t) is not equicontinuous. From (6.39) it follows that

$$(6.40) \qquad \cosh At = \mu_x(t) + \frac{t^2}{3} A^2 A_x(t)$$

so that $\cosh At\, \vec{f} \geq 0$ provided $\vec{f} \geq 0$ and $A^2\vec{f} \geq 0$. Given $\vec{f} = \binom{f_1}{f_2}$ this means that $f_i \geq 0$ with $\Delta f_i \geq 0$. Thus in order to apply Theorems 6.9 - 6.10 we want $\vec{f}$ such that $A^2\vec{f} \geq 0$ and $A^4\vec{f} \geq 0$ (i.e., $\Delta f_i \geq 0$ and $\Delta^2 f_i \geq 0$) which is possible e.g., when $f_i(x) = \exp \sum \gamma_j^i x_j$. Again the necessity of using "large" spaces E is apparent. We note also from (6.39) that T(t) and T(-t) behave quite differently relative to the preservation of positivity.

Analogues of these growth and convexity theorems for other "canonical" singular Cauchy problems appear in Carroll [18; 19; 25] (cf. also Chapter 2).

1.7 Transmutation. The idea of transmutation of operators goes back to Delsarte and Lions (cf. Delsarte [1], Delsarte-Lions [2; 3], Lions [1; 2; 3; 4; 5]) with subsequent contributions by Carroll-Donaldson [20], Hersh [3], Thyssen [1; 2], etc. The subject has many yet unexplored ramifications and is connected with the question of related differential equations (cf. Bragg-Dettman, loc. cit., Carroll-Donaldson [20], Hersh [3]); we expect to examine this more extensively in Carroll [24]. This section therefore will be partially heuristic and the word "suitable" will be used occasionally when convenient with precise domains of definition unspecified at times.

One forumlation goes as follows. Let $D_x = \partial/\partial x$ and $D_t = \partial/\partial t$ and consider polynomial differential operators $P = P(D)$ and $Q = Q(D)$ ($D = D_x$ or D_t). One says that an operator B transmutes P into Q if (formally) $QB = BP$. B will usually be an integral operator with a function or distribution kernel and in fact one often assumes this a priori, although it is perhaps too restrictive (cf. Remark 7.3). One picks a space of functions f (the choice is very important) and considers the problem

$$P(D_x)\phi(x,t) = Q(D_t)\phi(x,t) \tag{7.1}$$

$$\phi(x,o) = f(x); \quad D_t^k\phi(x,o) = 0 \qquad x,t \geq 0 \tag{7.2}$$

where $1 \leq k \leq m - 1$ with m = order Q. In order to define B we put $\phi(o,t) = (Bf)(t)$ and writing $\psi(x,t) = P(D_x)\phi(x,t)$ there results

$$(P(D_x) - Q(D_t))\psi = P(D_x)(P(D_x) - Q(D_t))\phi = 0 \tag{7.3}$$

$$(7.4)\qquad D_t^k\psi(x,o) = P(D_x)D_t^k\phi(x,o) = \begin{cases} P(D_x)f & ; \quad k = 0 \\ 0 & ; \quad 1 \le k \le m-1 \end{cases}$$

One may wish to extend f and P(D)f appropriately for $x < 0$ when desirable or necessary. There results formally

$$(7.5)\qquad \psi(o,t) = (BP(D_x)f)(t) = P(D_x)\phi(x,t)|_{x=0}$$

$$= Q(D_t)\phi(o,t) = [Q(D_t)Bf](t)$$

and evidently it is not necessary to suppose that P(D) or Q(D) have constant coefficients. It is however necessary to suppose that (7.1) - (7.2) have a unique solution in order to well define B; the same uniqueness criterion applies to (7.3) - (7.4) but this will often be trivial if f is chosen in the right space.

<u>Theorem 7.1</u> Suppose the problem (7.1) - (7.2) (resp. (7.3) - (7.4)) has a unique solution ϕ (resp. ψ) with f in a suitable space so that (7.5) makes sense and define generically $(Bf)(t) = \phi(o,t)$. Then B transmutes P(D) into Q(D).

Let us indicate some simple applications. Let $P(D) = D^2$ and $Q(D) = D$ so that (7.1) - (7.2) become

$$(7.6)\qquad \phi_{xx} = \phi_t; \quad \phi(x,o) = f(x) \qquad x,t \geq 0$$

We prolong f as an even function and take partial Fourier transforms $x \to s$ ($Ff = \hat{f}$) to obtain $\hat{\phi}_t = -s^2\hat{\phi}$ with $\hat{\phi}(s,o) = \hat{f}(s)$. The solution is $\hat{\phi}(s,t) = \hat{f}(s)\exp(-s^2t)$ and if

$$(7.7) \qquad R(x,t) = \frac{1}{2}K(x,t) = \frac{1}{2}(\pi t)^{-1/2} \exp(-\frac{x^2}{4t})$$

it is known that $F_x R(x,t) = \exp(-s^2 t)$ (cf. Titchmarsh [2]). Hence $(R(\cdot,t) * f(\cdot))(x) = \phi(x,t)$ and recalling that f is even there results

$$(7.8) \qquad \phi(o,t) = (Bf)(t) = \int_0^\infty f(\xi)K(\xi,t)d\xi$$

(cf. Carroll-Donaldson [20]). An easy calculation shows that (7.5) will hold provided $f'(o) = 0$ and of course f, f', and f'' must have suitable growth at infinity.

Now there is no difficulty in extending the transmutation idea to vector functions f with values in a complete separated locally convex space E. For example suppose that A is a suitable (closed, densely defined) operator in E with g(A) a reasonable operator function and let $P(D)w = g(A)w$ where w takes values in E (actually in D(g(A)) since A may not be everywhere defined). If B transmutes P into Q then formally $Q(D)Bw = BP(D)w = Bg(A)w = g(A)Bw$ provided that w satisfies the conditions necessary for the transmutation and that $Bw \in D(g(A))$. In this event $u = Bw$ will satisfy $Q(D)u = g(A)u$ and there arises the general question of when this situation can prevail. If B is an integral operator of Riemann type with a function kernel then g(A) can be passed under the integral sign and this part of the question is trivial up to the point where initial or boundary values arise. As an example let us consider the case (cf. Carroll-Donaldson [20])

(7.9) $P(D)w = D^2w = A^2w; \quad w(o) = u_0; \quad w_t(o) = 0$

(7.10) $Q(D)u = Du = A^2u; \quad u(o) = u_0$

If $A^2 = \Delta$ these are of course wave and heat equations and the connecting formula is well known, but we will derive it via transmutation. First formally we use (7.8) transmuting D^2 into D to obtain

(7.11) $u = Bw = \int_0^\infty w(\xi)K(\xi,t)d\xi$

and we note that $w_t(o) = 0$, corresponding to $f'(o) = 0$, as required in (7.8). Suitable growth of w, w_t, and w_{tt} will be assumed. Further $A^2Bw = BA^2w$ for $t > 0$ while for $t = 0$ one must require $u_0 \in D(A^2)$. Consequently

Theorem 7.2 Given w a solution of (7.9) it follows that u given by (7.11) satisfies (7.10), provided $u_0 \in D(A^2)$ while w, w_t, and w_{tt} grow suitably at infinity.

Such results are not new of course, at least in more classical forms, and we refer to Bragg-Dettman, loc. cit., Carroll-Donaldson [20], Hersh [3], Lions, loc. cit., and references there.

Remark 7.3 One can relax (7.2) and study not well posed Cauchy problems (or other problems) which lead to a unique solution ϕ (cf. Carroll-Donaldson [20]); for example growth conditions on ϕ and f could be imposed (see here Carroll [24]). Lions in his extensive investigations, chooses spaces of functions f where the transmutation is an isomorphism in the sense that there is an inverse transmutation B^{-1} sending Q(D) into P(D) (i.e. $PB^{-1} = B^{-1}Q$) but this seems to be an unnecessary luxury in general, although obviously of great interest (cf. Section 4.3). The choice of spaces for f is in any event of paramount importance for the transmutation method to work. If B is an integral operator with a function or distribution kernel one can begin with this as a stipulation and then discover the differential problem which the kernel must satisfy (not necessarily a simple Cauchy problem), and we refer to Carroll [24; 25], Carroll-Donaldson [20], Hersh [3], Lions, loc. cit., etc. for further information (cf. also (7.25)).

We will conclude this section with a version of Lions transmutation method sending $P(D) = D^2$ into $L_m = D^2 + \frac{2m+1}{t} D$ (cf. also Section 4.3). Thus we look for a function $\phi(x,t)$ satisfying

$$(7.12) \qquad \phi_{xx} = \phi_{tt} + \frac{2m+1}{t}\phi_t; \quad \phi(x,o) = f(x); \quad \phi_t(x,o) = 0$$

and again we extend f to be even with $f'(o) = 0$. This is of course the one dimensional EPD equation whose solution is given by (6.3) as (cf. Theorem 4.2)

$$(7.13) \qquad \phi(x,t) = c_m \int_{-1}^{1} (1-\xi^2)^{m-\frac{1}{2}} (\delta(x-\xi t) * f(x))d\xi$$

$$(7.14) \qquad \phi(o,t) = (Bf)(t) = 2c_m t^{-2m} \int_o^t (t^2-\tau^2)^{m-\frac{1}{2}} f(\tau)d\tau$$

which agrees with Lions [1; 2; 3; 5]. Consequently

$$(7.15) \qquad u = Bw = 2c_m t^{-2m} \int_o^t (t^2-\tau^2)^{m-\frac{1}{2}} w(\tau)d\tau$$

is the vector version of (7.14) and we have

Theorem 7.4 If w satisfies (7.9) with $u_o \in D(A^2)$ then u given by (7.15) satisfies $L_m u = A^2 u$ with $u(o) = u_o$ and $u_t(o) = 0$.

Proof: Evidently $(Bw)(o) = w(o)$ and $(Bw)'(o) = 0$ (cf. Lions [1] where an explicit verification of (7.5) is also given - the condition $w_t(o)$ corresponding to $f'(o) = 0$ is crucial here as in (7.8) and (7.11)). Again A^2 can be passed under the integral sign in (7.15) as before in (7.11). QED

Remark 7.5 Changing variables in (7.15) we obtain (6.6) when $w(\tau)$ is written as $w(\tau) = \cosh A\tau u_o$.

Actually in this case there is also an inverse transmutation operator B^{-1} but we will not go into details here (see Lions, loc. cit. for an exhaustive study of the matter). In Lions [3] it is also shown how to transmute $P(D) = D^2$ into $Q(D) = M_m = D^2 + \frac{2m+1}{t} D + p(t)D + q(t)$ and vice versa, and we will sketch some of the calculations here (p and q are assumed to be continuous) (cf. also Hersh [3]). Thus let $\phi(x,t)$ be the solution of

$$(7.16) \qquad \phi_{xx} = M_m \phi = \phi_{tt} + \frac{2m+1}{t}\phi_t + p(t)\phi_t + q(t)\phi$$

$$(7.17) \qquad \phi(x,o) = f(x); \quad \phi_t(x,o) = 0 \qquad (x,t \geq 0)$$

where f is extended as an even function with $f'(o) = 0$. Then again the transmutation operator B is determined by $(Bf)(t) = \phi(o,t)$ and satisfies formally the transmutation relation $M_m B = BD^2$, while to construct B Lions uses the following ingenious "trick." He sets

$$B(\cos xs)(t) = \theta(s,t) \tag{7.18}$$

and from the transmutation relation plus the fact that $(Bf)(o) = f(o)$ we obtain

$$M_m \theta(s,t) + s^2\theta(s,t) = 0 \tag{7.19}$$

with $\theta(s,o) = 1$, while $\theta_t(s,o) = 0$ automatically (for unique solutions $\theta(s,t)$ see Section 5). Assuming now (we omit verification) that B is defined by a (distribution) kernel $b(t,x)$ with support in the region $|x| \le t$, which is legitimate here but is in part incidental to the transmutation concept, one obtains, since $f(x) = \cos xs$ is even, and $b(t,\cdot)$ will be even (verification follows from (7.25) for example)

$$\theta(s,t) = 2\int_0^t b(t,x) \cos xs \, dx = 2\int_0^\infty b(t,x) \cos xs \, dx \tag{7.20}$$

from which follows, by Fourier inversion in a distribution sense,

$$b(t,x) = \frac{1}{\pi}\int_0^\infty \theta(s,t) \cos xs \, ds \tag{7.21}$$

<u>Theorem 7.6</u> The transmutation of D^2 into M_m is determined by an operator B with kernel defined by (7.21), where θ is known.

Further properties are developed by Lions [3; 5] and a general transmutation theory will be developed in Carroll [24]. Hersh's technique (cf. Hersh [3]) is based on special functions and kernels and leads formally to the same result as that of Theorem 7.6. Let us sketch this for completeness. Thus let $w(t,\lambda)$ be the solution of

$$(7.22) \qquad M_m w + \lambda^2 w = 0; \quad w(o,\lambda) = 1; \qquad w_t(o,\lambda) = 0$$

while u satisfies (cf. (7.10))

$$(7.23) \qquad D^2 u + \lambda^2 u = 0; \qquad u(o,\lambda) = 1; \quad u_t(o,\lambda) = 0$$

THen $u(t,\lambda) = \cos \lambda t$ and Hersh proposes a formula

$$(7.24) \qquad w(t,\lambda) = \int_{-\infty}^{\infty} b(t,x)u(x,\lambda)dx = \int_{-\infty}^{\infty} b(t,x) \cos \lambda x \, dx$$

Since (7.22) holds we have, provided $b(t,\cdot)$ has compact support with reasonable smoothness,

$$(7.25) \qquad M_m b(t,x) = b_{xx}(t,x)$$

This shows symmetry of $b(t,x)$ in x and (7.24) becomes (cf. (7.20))

$$(7.26) \qquad w(t,\lambda) = 2 \int_{0}^{\infty} b(t,x) \cos \lambda x \, dx$$

Again (7.25) can be solved, as can (7.19), by the methods of Section 1.5. Defining now B by $(Bf)(t) = \int_{-\infty}^{\infty} b(t,x) f(x) \, dx$ we have formally from (7.25) $M_m B = BD^2$. The idea then is to replace λ^2 by suitable operator functions $-g(A)$.

Chapter 2

Canonical Sequences of Singular Cauchy Problems

2.1 The rank one situation. In this chapter we will develop the group theoretic version of the EPD equations studied in Chapter 1. This leads to many new classes of equations and results parallel to those of Chapter 1 as well as to some interesting new situations. Moreover it exhibits the main results of Chapter 1 in their natural group theoretic context. The ideas of spherical symmetry, radial mean values and Laplacians, etc. inherent in EPD theory have natural counterparts in terms of geodesic coordinates and one can obtain recursion relations, Sonine formulas, etc. group theoretically. The results are based on Carroll [21; 22] in the semisimple case and were anticipated in part by earlier work of Carroll [18; 19], Carroll-Silver [15; 16; 17] and Silver [1] for some semisimple and Euclidean cases. The group theory is "routine" at the present time and relies heavily on Helgason's work (cf. Helgason [1; 2; 3; 4; 5; 6; 7; 8; 9; 10]) but one must of course refer to basic material of Harish-Chandra (cf. Warner [1; 2] for a summary) as well as lecture notes by Varadarajan, Ranga Rao, etc.); other specific references to Bargmann [1], Bargmann-Wigner [2], Bhanu-Munti [1], Carroll [27; 28], Coifman-Weiss [1], Ehrenpreis-Mautner [1], Flensted-Jensen [1], Furstenberg [1], Gangolli [1], Gelbart [1], Gelfand et al. [1; 2; 3; 4], Godement [1], Hermann

[1], Jacquet-Langlands [1], Jehle-Parke [1], Kamber-Tondeur [1], Karpelevič [1], Knapp-Stein [1], Kostant [1], Kunze-Stein [1; 2], Lyubarskij [1], Maurin [1], McKerrell [1], Miller [1; 2], Naimark [1], Pukansky [1], Rühl [1], Sally [1; 2], Simms [1], Smoke [1], Stein [1], Takahashi [1], Talman [1], Tinkham [1], Vilenkin [1], Wallach [1], Wigner [1], etc., as well as the fundamental work of E. Cartan and H. Weyl, are to be taken for granted, even if not mentioned explicitly. The basic Lie theory is developed somewhat concisely in this section; for a rather more leisurely treatment we refer to Bourbaki [5], Carroll [23], Hausner-Schwartz [1], Helgason [1; 2; 3], Hochschild [1], Jacobson [1], Loos [1], Serre [1; 2], Tondeur [1], Želobenko [1], etc.

We will start out with the full machinery for the rank one semisimple case, following Carroll [21; 22], and later will give extremely detailed examples for special cases. This avoids some repetition and presents a "clean" theory immediately; the reader unfamiliar with Lie theory might look at the examples first where many details and definitions are covered. We deliberately omit the treatment of invariant differential operators acting in sheaves or in sections of vector bundles even though this is one of the more important subjects in modern work (some references are mentioned above). The preliminary material will be expository and specific theorems will not be proved here. The Euclidean group cases have been covered in Carroll-Silver [15; 16; 17] and especially Silver [1] so that we will only give a few remarks later about this at the end of the chapter; the basic results are in any event included in Chapter 1. Thus let G be a real connected noncompact semisimple Lie group with finite center and K a maximal compact subgroup so that V = G/K is a symmetric space of noncompact type. Let $\tilde{g} = \tilde{k} + p$ be a Cartan decomposition, $a \subset p$ a maximal abelian subspace, and we will suppose until further notice that dim a = rank V = 1.

One sets A = exp a, K = exp $\tilde{k}$, and N = exp $\tilde{n}$ where $\tilde{n} = \sum g_\lambda$ for $\lambda > 0$ where the g_λ are the standard root spaces corresponding to positive roots α and possibly 2α in the rank one case. One sets $\rho = \frac{1}{2}\sum m_\lambda \lambda$ for $\lambda > 0$ where $m_\lambda = \dim g_\lambda$ and we pick an element $H_0 \in a$ such that $\alpha(H_0) = 1$. Thus $\rho = (\frac{1}{2} m_\alpha + m_{2\alpha})\alpha$ and we can identify a Weyl chamber as a connected component $a_+ \subset a' \subset a$ with $(0,\infty)$ in writing $\alpha(tH_0) = t$ where $\mu \in \mathbb{R}$ corresponds to $\mu \in a^*$ by $\mu(tH_0) = \mu t$. The Iwasawa decomposition of G is G = KAN which we write in the form g = k(g) exp H(g) n(g) where the notation $a_t = \exp tH_0$ is used. Let M (resp. M') denote the centralizer (resp. normalizer) of A in K so that the Weyl group is $W = M'/M$ and the maximal boundary of V is B = K/M (thus M = $\{k \in K;\ \mathrm{Ad}kH = H$ for $H \in a\}$ and $M' = \{k \in K;\ \mathrm{Ad}k a \subset a\}$ - see the examples for specific details). There are natural polar coordinates in a dense submanifold of V arising from the decomposition $G = K\bar{A}_+K$ ($A_+ = \exp a_+$) provided by the diffeomorphism $(kM,a) \to kaK : B \times A_+ \to V$ (one could also work with the decomposition G = KAK). Thus the polar coordinates of $\pi(g)$ =

$\pi(k_1ak_2) \varepsilon V$ are (k_1M,a) where $\pi : G \to V$ is the natural map. Now given $v = gK \varepsilon V$ and $b = kM \varepsilon B$ one writes $A(v,b) = -H(g^{-1}k)$ and the Fourier transform of $f \varepsilon L^2(V)$ is defined by $Ff = \tilde{f}$ where (using Warner's notation)

$$(1.1) \qquad \tilde{f}(\mu,b) = \int_V f(v) \exp (i\mu+\rho)A(v,b)dv$$

for $\mu \varepsilon a^*$ and $b \varepsilon B$. All measures are suitably normalized in this treatment. This sets up an isometric isomorphism $f \leftrightarrow \tilde{f}$ between $L^2(V)$ and the space $L^2(a_+^* \times B)$ (cf. here below for a_+^*) with inversion formula

$$(1.2) \qquad f(v) = \frac{1}{2} \int_{a^*\times B} \tilde{f}(\mu,b)[\exp(-i\mu+\rho)A(v,b)]|c(\mu)|^{-2}d\mu db$$

Here the 1/2 comes from the order of the Weyl group (which is two) and $c(\mu)$ is the standard Harish-Chandra function. For a general expression of $c(\mu)$ we can write (cf. also (3.14)) $c(\mu) = I(i\mu)/I(\rho)$ where (cf. Gindikin-Karpelevic [1], Helgason [3; 8], Warner [1; 2])

$$I(\nu) = \prod_{\alpha>0} \beta\left(\frac{1}{2} m_\alpha, \frac{1}{4} m_{\alpha/2} + \frac{(\nu,\alpha)}{(\alpha,\alpha)}\right)$$

Here $\beta(x,y) = \Gamma(x)\Gamma(y)/\Gamma(x+y)$ is the Beta function and one defines (ν,α) in this context in terms of the Killing form $B(\cdot,\cdot)$ as $B(H_\nu,H_\alpha)$ where for $\lambda \varepsilon a_C^*$, $H_\lambda \varepsilon a_C$ is determined by $\lambda(h) = B(H_\lambda,H)$ for $H \varepsilon a$ (note here that a_C^* is the space of $\mathbf{R}$ linear maps of a into $\mathbb{C}$ while a_C is the complexification of a which is formally the set of all sums $\lambda + i\mu$ for $\lambda,\mu \varepsilon a$). Then

a_+^* is the preimage of a_+ under the map $\lambda \to H_\lambda$ and is a Weyl chamber in a^*. Specific examples of $c(\mu)$ will be written down later. Now $a^*/W \sim a_+^*$ (recall $W = \{1,s\}$ where $sH = -H$ for $H \in a$ or equivalently $sa_t = a_{-t} = a_t^{-1}$) and one can write

$$(1.3) \qquad L^2(V) = \int_{a^*/W} H_\mu \; |c(\mu)|^{-2} d\mu$$

where, for $\phi \in L^2(B)$,

$$(1.4) \qquad H_\mu = \{\hat{\phi}_\mu(v) = \int_B [\exp\,(-i\mu+\rho)A(v,b)]\phi(b)db\}$$

The quasiregular representation L of G on $L^2(V)$, defined by $L(g)f(v) = f(g^{-1}v)$ decomposes in the form

$$(1.5) \qquad L = \int_{a^*/W} L_\mu |c(\mu)|^{-2} \, d\mu$$

where L_μ acts in H_μ by the same rule as L. L_μ is in fact irreducible and unitary and is equivalent to the so-called class one principal series representation induced from the parabolic subgroup MAN by means of the character man $\to a^{i\mu} = \exp i\mu \log a$.

We recall here also the definition of the mean value of a function ϕ over the orbit of $g\pi(h) = gu$ under the isotropy subgroup $I_v = gKg^{-1}$ at $v = \pi(g)$. Thus writing $M^h\phi = M^u\phi$ we have (recalling that $\pi(h) = u$)

$$(1.6) \qquad (M^u\phi)(v) = \int_K \phi(gk\pi(h))dk = \Phi(u,v)$$

The so-called Darboux equation is

$$(1.7) \qquad D_u\Phi = D_v\Phi = M^u(D\phi)(v)$$

(cf. Helgason [1]) when $D \in D(G/K)$. The symbol $D(G/K)$ denotes the left invariant differential operators on G/K which are defined as follows. If $\psi : G \to G$ is a diffeomorphism and f is a function on G one sets $f^{\psi}(g) = (f \circ \psi^{-1})(g)$ while if D is a linear differential operator we write $D^{\psi} : f \to (Df^{\psi^{-1}})^{\psi}$. Thus $(D^{\psi}f)(h) = (D_x(f(\psi(x))) \circ \psi^{-1})(h) = D_x f(\psi(x))|_{x=\psi^{-1}(h)}$ and in particular we have $(D^{\tau_g}f)(h) = D_x f(gx)|_{x=g^{-1}h}$ where $\tau_g v = gv$ on G or on V = G/K. Writing $\sigma_k g = gk$ for $g \in G$ and $k \in K$, the space of differential operators on G such that $D^{\tau_g} = D$ and $D^{\sigma_k} = D$ is denoted by $D(G/K)$; this is identical with the space of linear differential operators on G/K (called left invariant) such that $D^{\tau_g} = D$ (the condition $D^{\sigma_k} = D$ is automatic on G/K - proof obvious). Now the zonal spherical functions on G are defined by the formula ($\mu \in a^*$)

$$(1.8) \qquad \tilde{\phi}_{\mu}(g) = \int_K \exp(i\mu-\rho)H(gk)dk$$

and we can write by invariant integration $\tilde{\phi}_{\mu}(g) = \phi_{\mu}(gK)$ (actually the $\tilde{\phi}_{\mu}(g)$ are K biinvariant in the sense that $\tilde{\phi}_{\mu}(g) = \tilde{\phi}_{\mu}(\hat{k}g\tilde{k}))$ and K is unimodular - cf. Helgason [1], Maurin [1], Nachbin [2], Wallach [1], or Weil [1] - and thus $\int_K f(\hat{k}k\tilde{k})dk = \int_K f(k^{-1})dk = \int_K f(k)dk$. It is known further that $\tilde{\phi}_{\mu}(g) = \tilde{\phi}_{-\mu}(g^{-1})$ (cf. Harish-Chandra [2] or Warner [1; 2]) while the $\phi_{\mu} \in C^{\infty}(G/K)$, and are characterized by their eigenvalues $\lambda_D = \lambda_D(\phi_{\mu})$ for $D \in D(G/K)$, with $\phi_{\mu}(\pi(e)) = 1$, plus the biinvariance of $\tilde{\phi}_{\mu}$. We now demonstrate a lemma which has some interesting consequences.

Lemma 1.1 The Fourier transform of $M^h = M^u \in E'(V)$ ($u = \pi(h)$) is given by $FM^h = \tilde{\phi}_\mu(h)$ and if $h = \hat{k}a\tilde{k}$ with $a \in A_+$ then $(M^u f)(v) = (M^a f)(v)$ so that $(M^u f)(v)$ depends only on the radial component a in the polar decomposition $(\hat{k}M, a)$ of $u = \pi(h)$.

Proof: The action of M^h or M^u as a distribution in $E'(V)$ is determined by (cf. (1.6)) $\langle M^h, \phi\rangle = (M^h\phi)(\omega)$ where $\omega = \pi(e)$ (e being the identity in G). Thus in (1.6) take $\phi = \exp\,(i\mu + \rho)\,A(\omega, b)$ where $b = \hat{k}M$. Since $A(ekhK, \hat{k}M) = -H((kh)^{-1}\hat{k})$ one has by the unimodularity of K (cf. (1.8) and comments thereafter)

$$(1.9)\qquad (M^h\phi)(\omega) = \int_K \exp\,[-(i\mu+\rho)H(h^{-1}k^{-1}\hat{k})dk$$

$$= \int_K \exp\,[-(i\mu+\rho)H(h^{-1}k\hat{k})dk = \int_K e^{-(i\mu+\rho)H(h^{-1}k)}dk$$

$$= \tilde{\phi}_{-\mu}(h^{-1}) = \tilde{\phi}_\mu(h) = FM^h$$

One notes here that $M^h = M^u$ works on functions in V such as $\exp\,(i\mu+\rho)A(v,b) = \phi$ and $\langle M^h, \phi\rangle$ is precisely (1.1) with f replaced by $M^u = M^h$ and $\phi = \exp\,(i\mu+\rho)A(\omega,b)$; no integration over V is involved since $v = \omega$ in ϕ, only a distribution evaluation is of concern here. Finally we note that with $h = \hat{k}a\tilde{k}$ as above there follows from (1.6)

$$(1.10)\qquad (M^h f)(v) = \int_K f(gk\hat{k}aK)dk = \int_K f(gkaK)dk = (M^a f)(v) \qquad \text{QED}$$

This leads to a symmetric space version of theorem of Zalcman [1], generalizing an old formula of Pizzetti; here G/K is of arbitrary rank and we refer to Helgason [1; 2] or Warner [1; 2] for general information on higher rank situations.

Theorem 1.2 (Pizzetti-Zalcman). Suppose the operators Δ_k with eigenvalues λ_k generate D(G/K) as an algebra; then locally

(1.11) $\quad (M^u f)(v) = (\phi_\mu(u, \Delta_k)f)(v)$

where $\phi_\mu(u) = 1 + \sum_{n=1}^{\infty} P_n(u,\lambda_k(\phi_\mu))$ is expressed in terms of its eigenvalues as $\phi_\mu(u,\lambda_k)$.

Zalcman [1], in an $\mathbb{R}^n$ context, has generalized this kind of theorem considerably.

Proof: One notes from Helgason [1] that $(M^h\phi_\lambda)(v) = \int_K \phi_\lambda(gk\pi(h))dk = \int_K \tilde{\phi}_\lambda(gkh)dk = \tilde{\phi}_\lambda(g)\tilde{\phi}_\lambda(h)$ and the local expression $(M^u f)(v) = ([1 + \sum_{n=1}^{\infty} P_n(u,\Delta_k)]f)(v)$ holds. Consequently applying this local expression to ϕ_λ we obtain $\tilde{\phi}_\lambda(h) = 1 + \sum_{n=1}^{\infty} P_n(u,\lambda_k(\phi_\lambda)) = \phi_\lambda(u)$ and (1.11) follows. One should remark also that the polynomials P_n are without constant terms.

QED

2.2 Resolvants. The objects of interest in a generalized EPD theory are the radial components of a basis for the H_μ spaces of (1.4), multiplied by a suitable weight function, the result of which we will denote by $\hat{R}^m(t,\mu)$ (cf. Carroll [21; 22], Carroll-Silver [15; 16; 17], Silver [1]) and we mention that m can denote a multiindex here. First we remark that V is endowed with the Riemannian structure induced by the Killing form B(•,•) (but the Riemannian structure does not play an important role here, e.g., $\frac{1}{2}B(\cdot,\cdot)$ could serve - see Example 3.2) and for rank V = 1, D(G/K) is generated by a single Laplacian Δ, determined by the standard Casimir operator C in the enveloping algebra of $\tilde{g}$ (cf. Remark 3.15 about C). We look at the radial component Δ_R of Δ, passing this from the coordinate t in $a_t \in A$ to $\pi(A)$ in an obvious manner, and setting $M_t = M^{a_t}$ with $\hat{R}^o(t,\mu) = FM_t = \tilde{\phi}_\mu(a_t)$ one obtains an eigenvalue equation (cf. Helgason [5])

$$(2.1)\qquad [D_t^2 + (m_\alpha + m_{2\alpha})\coth t\, D_t + m_{2\alpha}\tanh t\, D_t]\tilde{\phi}_\mu$$

$$+ [\mu^2 + (\tfrac{1}{2}m_\alpha + m_{2\alpha})^2]\tilde{\phi}_\mu = 0$$

The solution of (2.1), analytic at t = 0, is e.g.,

$$(2.2)\qquad \hat{R}^o(t,\mu) = \hat{\phi}_\mu(\exp tH_o) = F(\delta,\beta,\gamma, - sh^2 t)$$

where evidently $\hat{R}^o(o,\mu) = 1$ and $\hat{R}_t^o(o,\mu) = 0$ ($\delta = (m_\alpha + 2m_{2\alpha} + 2i\mu)/4$, $\beta = (m_\alpha + 2m_{2\alpha} - 2i\mu)/4$, and $\gamma = (m_\alpha + m_{2\alpha} + 1)/2$). The general EPD situation involves embedding $\hat{R}^o$ in a "canonical" sequence of "resolvants" $\hat{R}^m(t,\mu)$, for m > 0 a positive integer or a multi-index, such that the resolvant initial conditions

$$(2.3)\qquad \hat{R}^m(o,\mu) = 1; \qquad \hat{R}_t^m(o,\mu) = 0$$

are satisfied. There will also be "canonical" recursion relations between the $\hat{R}^m$ of indices differing by ±1 or ±2 which will arise group theoretically from considering a full set of basis elements in the H_μ spaces.

Thus we must first determine a basis for $L^2(B)$ and this is well known (thanks are due here to R. Ranga Rao for some helpful information). We let $\{\pi_\tau, V_\tau\}$ with $\dim V_\tau = d_\tau$ be a complete set of inequivalent irreducible unitary representations of K and let $V_\tau^M \subset V_\tau$ be the set of elements fixed by M. One knows by a result of Kostant [1] that $\dim V_\tau^M = 1$ or 0 in the rank one case (cf. also Helgason [5]) and for the set $\tau \in T$ where $\dim V_\tau^M = 1$ we let w_1^τ be a basis vector for V_τ^M with $w_i^\tau (1 \leq i \leq d_\tau)$ an orthonormal basis for V_τ under a scalar product $< \, , \, >_\tau$. Then for example the collection of functions $(\tau \in T) kM \to \langle w_i^\tau, \pi_\tau(k) w_1^\tau \rangle_\tau$ is known to be a basis for $L^2(B)$ and we define (cf. (1.4) with $v = a_t K$ and $b = kM$)

$$(2.4) \qquad \hat{\phi}^j_{\mu,\tau}(a_t K) = \int_B e^{(i\mu-\rho)H(a_t^{-1}k)} \langle w_j^\tau, \pi_\tau(k) w_1^\tau \rangle_\tau \, db$$

$$= \int_K e^{(i\mu-\rho)H(a_t^{-1}k)} B^\tau \pi_\tau(k)^{-1} w_j^\tau \, dk$$

$$= E_{1,\tau}(B^\tau : -i\mu : a_t^{-1}) w_j^\tau$$

where $B^\tau \in \mathrm{Hom}_M(V_\tau, \mathbb{C})$ is determined by the rule $B^\tau w_s^\tau = \delta_{1,s}$ (Kronecker symbol) so that $B^\tau \pi_\tau(k)^{-1} w_j^\tau = \langle w_j^\tau, \pi_\tau(k) w_1^\tau \rangle_\tau = \langle \pi_\tau(k)^{-1} w_j^\tau, w_1^\tau \rangle_\tau$ (note that $\int_K \phi(k) dk = \int_B (\int_M \phi(km) dm) db$ and

$H(g^{-1}km) = H(g^{-1}k)$ with $\pi_\tau(km) = \pi_\tau(k)\pi_\tau(m))$. The $E_{1,\tau}$ above are Eisenstein integrals as defined in Wallach [1] and the $\hat{\phi}^j_{\mu,\tau}(a_tK)$ are the radial components of basis elements in H_μ (cf. below). It is possible to obtain an explicit evaluation of these functions, using results of Helgason [5; 11] as follows. One defines (cf. Helgason [5])

$$(2.5)\qquad f^\lambda_{\tau,j}(x) = d_\tau^{1/2} \int_K e^{(-i\lambda-\rho)H(g^{-1}k)} <w_j^\tau, \pi_\tau(k)w_1^\tau>_\tau \, dk$$

for $x = gK$. Then it is proved in Helgason [5; 11] (to whom we gratefully acknowledge some conversation on the matter) that

$$(2.6)\qquad f^\lambda_{\tau,j}(\hat{k}aK) = d_\tau^{1/2} < w_j^\tau, \pi_\tau(\hat{k})w_1^\tau >_\tau \Psi_{\lambda,\tau}(aK)$$

where $\Psi_{\lambda,\tau}(x)$ is defined by

$$(2.7)\qquad \Psi_{\lambda,\tau}(x) = \int_K e^{(-i\lambda-\rho)H(g^{-1}k)} < w_1^\tau, \pi_\tau(k)w_1^\tau >_\tau \, dk$$

Thus recalling the polar decomposition $(\hat{k}M,a) \to \hat{k}aK$ we have

<u>Theorem 2.1</u> General basis elements in H_μ are

$$(2.8)\qquad \hat{\phi}^j_{\mu,\tau}(\hat{k}a_tK) = \int_K e^{(i\mu-\rho)H(a_t^{-1}\hat{k}^{-1}k)} <w_j^\tau, \pi_\tau(k)w_1^\tau>_\tau \, dk$$

$$= d_\tau^{-1/2} f^{-\mu}_{\tau,j}(\hat{k}a_tK) = <w_j^\tau, \pi_\tau(\hat{k})w_1^\tau>_\tau \Psi_{-\mu,\tau}(a_tK)$$

In particular one has $\hat{\phi}^j_{\mu,\tau}(a_tK) = \hat{\phi}^1_{\mu,\tau}(a_tK) = \Psi_{-\mu,\tau}(a_tK)$ since $\delta_{1,j} = <w_j^\tau, \pi_\tau(1)w_1^\tau>_\tau$ (this "collapse" was observed in special cases by Silver [1]).

Now let $\hat{K}_0 = \{\pi_\tau, V_\tau\}$ for $\tau \in T$ (i.e., dim $V_\tau^M = 1$) and following Helgason [5] we use a parametrization due to Johnson and

Wallach (see Helgason [5] for references and cf. also Kostant [1]). If $m_{2\alpha} = 0$, $\hat{K}_0 = \{(p,q)\}$ with $p \varepsilon Z_+$ and $q = 0$; if $m_{2\alpha} = 1$, $\hat{K}_0 = \{(p,q)\}$ with $(p,q) \varepsilon Z_+ \times Z$ where $p \pm q \varepsilon 2Z_+$; if $m_{2\alpha} = 3$ or 7, $\hat{K}_0 = \{(p,q)\}$ with $(p,q) \varepsilon Z_+ \times Z_+$ where $p \pm q \varepsilon 2Z_+$. One sets $\ell = (i\mu+\rho)(H_0)$ and then (cf. Helgason [5; 11])

<u>Theorem 2.2</u> The radial components of basis elements in H_μ are given by

$$(2.9) \qquad \Psi_{-\mu,\tau}(a_tK) =$$

$$= c_{-\mu,\tau} \mathrm{th}^p t \ \mathrm{ch}^{-\ell} t F\left(\frac{\ell+p+q}{2}, \frac{\ell+p-q+1-m_{2\alpha}}{2}, p+\frac{m_\alpha+m_{2\alpha}+1}{2}, \mathrm{th}^2 t\right)$$

$$(2.10) \qquad c_{-\mu,\tau} = \frac{\Gamma\left(\frac{\bar{\ell}+p+q}{2}\right)\Gamma\left(\frac{\bar{\ell}+p-q+1-m_{2\alpha}}{2}\right)\Gamma\left(\frac{m_\alpha+m_{2\alpha}+1}{2}\right)}{\Gamma\left(\frac{\bar{\ell}}{2}\right)\Gamma\left(\frac{\bar{\ell}+1-m_{2\alpha}}{2}\right)\Gamma\left(p+\frac{m_\alpha+m_{2\alpha}+1}{2}\right)}$$

where th = tanh, ch = cosh, and F is the standard hypergeometric function.

Note here that our ℓ is the negative of the ℓ in Helgason [5; 11] where $\ell = (i\lambda-\rho)(H_0) = (-i\mu-\rho)(H_0)$ (cf. here our notation in (1.1)) and recall that $\alpha(H_0) = 1$ with $a_t = \exp tH_0$. If one now sets $d_{2\alpha} = -4q(q+m_{2\alpha}-1)$ and $d_\alpha = -p(p+m_\alpha+m_{2\alpha}-1) + q(q+m_{2\alpha}-1)$ it is shown in Helgason [5] that $\Psi(t) = \Psi_{-\mu,\tau}(a_tK)$ also satisfies the differential equation

$$\Psi_{tt} + (m_\alpha + m_{2\alpha})\coth t\,\Psi_t + m_{2\alpha}\tanh t\,\Psi_t + [d_\alpha sh^{-2}t + d_{2\alpha}sh^{-2}2t + \rho(H_o)^2 + \mu(H_o)^2]\Psi = 0 \tag{2.11}$$

where sh = sinh, and we have used the identity coth 2t = $\frac{1}{2}$(coth t + tanh t). Recalling that $\rho \sim (\frac{1}{2}m_\alpha + m_{2\alpha})$ one sees that the trivial representation $1 \in T$, corresponding to $p = q = 0$, gives rise to (2.1), so that (since $c_{-\mu,1} = 1$) we should have $\Psi_{-\mu,1}(a_tK) = \tilde{\phi}(a_t) = ch^{-\ell}t\, F(\ell/2, \frac{\ell+1-m_{2\alpha}}{2}, \frac{m_\alpha+m_{2\alpha}+1}{2}, th^2t)$. This is borne out by the Kummer relation $F(a,b,c,z) = (1-z)^{-a}F(a,c-b,c,z/z-1)$ with $z = -sh^2t$, $a = \delta = \ell/2$, $b = \beta$, and $c = \gamma$, so that $(1-z)^{-a} = ch^{-\ell}t$ and $c - b = \gamma - \beta = (m_\alpha+2+2i\mu)/4 = (\ell+1-m_{2\alpha})/2$; hence $\Psi_{-\mu,1}(a_tK) = \tilde{\phi}(a_t)$ as indicated.

2.3 Examples with $m_{2\alpha} = 0$. Now to construct resolvants $\hat{R}^m(t,\mu) = \hat{R}^{p,q}(t,\mu)$ from the $\Psi_{-\mu,\tau}(a_tK)$ one multiplies the $\Psi_{-\mu,\tau}$ by a suitable factor in order to obtain the resolvant initial conditions (2.3) and to produce $\tilde{\phi}_\mu(a_t)$ when $p = q = 0$. These requirements are not alone sufficient to produce the "canonical" resolvants since one needs to incorporate certain group theoretic recursion relations into the theory which serve to "split" the second order singular differential equation for the $\hat{R}^{p,q}$ arising from (2.11) into a composition of two first order equations (cf. here Infeld-Hull [1]). Even then we remark that the resolvants will not be unique since one can always multiply $\hat{R}^m(t,\mu)$ by a function $\psi_m \in C^2$ such that $\psi_m(o) = 1$, $\psi_m'(o) = 0$, $\psi_0(t) \equiv 1$. This will simply give a different second order singular differential equation for the new resolvant and different splitting recursion relations but the resolvant initial conditions will remain valid and the reduction to $\tilde{\phi}_\mu(a_t)$ for m = 0 is unaltered. Thus we will choose the simplest form of resolvant fulfilling the stipulations imposed while referring to these resolvants and equations as canonical.

Example 3.1 Take the case where $m_\alpha = 1$, $m_{2\alpha} = 0$, $d_\alpha = -p^2$, $d_{2\alpha} = 0$, and $\rho = 1/2$. This corresponds to $G = SL(2,\mathbb{R})$ with $K = SO(2)$ and the resolvants are given in Carroll [18; 19; 22], Carroll-Silver [15; 16; 17], Silver [1] as

$$(3.1)\qquad \hat{R}^m(t,\mu) = \left(\frac{\zeta+1}{2}\right)^{i\mu-\frac{1}{2}-m} F\left(\frac{1}{2}-i\mu, m+\frac{1}{2}-i\mu, m+1, \frac{\zeta-1}{\zeta+1}\right)$$

$$= \frac{2^m\Gamma(m+1)}{(\zeta^2-1)^{m/2}} P^{-m}_{i\mu-\frac{1}{2}}(\zeta)$$

where $\zeta = \mathrm{ch}t$ and P^{-m}_ν denotes the standard associated Legendre function of the first kind. Now we recall a formula (cf. Snow [1], p. 18)

$$(3.2)\qquad F(\alpha,\alpha+1/2,\nu, z) = \left(\frac{1+\sqrt{1-z}}{2}\right)^{-2\alpha} F\left(2\alpha, 2\alpha+1-\nu, \nu, \frac{1-\sqrt{1-z}}{1+\sqrt{1-z}}\right)$$

Setting $z = \mathrm{th}^2t$ in (3.2) and noting that $1 - \mathrm{th}^2t = \mathrm{sech}^2t$ we have $\sqrt{1-z} = \mathrm{sech}\, t$ and $1 - \sqrt{1-z}/(1+\sqrt{1-z}) = \zeta-1/\zeta+1$ with $\zeta = \mathrm{ch}t$ as above. Now one observes that $\hat{R}^m(t,\mu)$ in (3.1) is symmetric in μ (and t) and this is exhibited for example in the form of the resolvant (3.1) when one uses the formula (cf. Robin [1])

$$(3.3) \qquad P^{-m}_{i\mu-\frac{1}{2}}(\mathrm{ch}t) = \frac{2^{-m}\mathrm{sh}^m t}{\Gamma(m+1)} F\left(\frac{m+\frac{1}{2}+i\mu}{2}, \frac{m+\frac{1}{2}-i\mu}{2}, m+1, -\mathrm{sh}^2 t\right)$$

Hence (3.1) can be written (recall that $F(a,b,c,z) = F(b,a,c,z)$)

$$(3.4) \qquad \hat{R}^m(t,\mu) = F\left(\frac{m+\frac{1}{2}+i\mu}{2}, \frac{m+\frac{1}{2}-i\mu}{2}, m+1, -\mathrm{sh}^2 t\right)$$

Now, returning to (3.2) and (2.9) - (2.10), we write for $\tau \sim (p,o)$ with $\alpha = i\mu/2$, $\beta = 1/2(p+1/2)$, and $\ell = i\mu + 1/2$

$$(3.5) \qquad \Psi_{-\mu,\tau}(a_t K) = c_{-\mu,\tau}\mathrm{th}^p t\ \mathrm{ch}^{-\ell} t\ F(\alpha+\beta, \alpha+\beta+\tfrac{1}{2}, p+1, \mathrm{th}^2 t)$$

$$= c_{-\mu,\tau}\mathrm{th}^p t\ \mathrm{ch}^{-\ell} t\left(\frac{1+\mathrm{sech}\ t}{2}\right)^{-2(\alpha+\beta)}$$

$$\cdot\ F\left(2(\alpha+\beta), 2(\alpha+\beta)-p, p+1, \frac{\zeta-1}{\zeta+1}\right)$$

$$= c_{-\mu,\tau}\mathrm{sh}^p t\left(\frac{1+\zeta}{2}\right)^{-\ell-p} F\left(i\mu+\frac{1}{2}+p, i\mu+\frac{1}{2}, p+1, \frac{\zeta-1}{\zeta+1}\right)$$

Rewriting the first equation in (3.1) with μ replaced by $-\mu$ in the right hand side we obtain

$$(3.6) \qquad \hat{R}^m(t,\mu) = \left(\frac{\zeta+1}{2}\right)^{-\ell-m} F\left(\frac{1}{2}+i\mu, \frac{1}{2}+i\mu+m, m+1, \frac{\zeta-1}{\zeta+1}\right)$$

Then identifying p with m we can say from (3.5) - (3.6)

$$(3.7) \qquad \hat{R}^p(t,\mu) = c^{-1}_{-\mu,\tau}\ \mathrm{sh}^{-p} t\ \Psi_{-\mu,\tau}(a_t K)$$

For completeness we check the differential equation satisfied by the R^p of (3.7), given that (2.11) holds. An elementary calculation yields then

$$(3.8)\qquad \hat{R}^p_{tt} + (2p+1)\coth t\, \hat{R}^p_t + [p(p+1) + \mu^2 + \tfrac{1}{4}]\, \hat{R}^p = 0$$

which agrees with previous calculations (cf. Carroll [21; 22], Carroll-Silver [15; 16; 17], Silver [1]). The canonical recursion relations associated to the $\hat{R}^p$ of (3.1), (3.4), (3.6), or (3.7) can of course be read off from known formulas for the associated Legendre functions for example but they can also be obtained group theoretically (cf. Example 3.5 - Theorem 3.8) by using a full set of basis elements in the H_μ spaces. For now we simply list them in the form

$$(3.9)\qquad \hat{R}^p_t = \frac{-\text{sh}t}{2p+2}\,[p(p+1) + \mu^2 + \tfrac{1}{4}]\hat{R}^{p+1};$$

$$\hat{R}^p_t + 2p\coth t\, \hat{R}^p = 2p\,\text{csch}\, t\, \hat{R}^{p-1}$$

Evidently the composition of these two relations yields (3.8) and this is the sense in which we speak of "splitting" the resolvant equation (3.8).

<u>Example 3.2</u> We give now explicit matrix details to clarify Example 3.1. (The Euclidian case can also be treated group theoretically as in Carroll-Silver [15; 16; 17] and especially Silver [1] and a simple example is worked out at the end of this chapter; the results of course agree with those of Chapter 1.)

Thus let $G = SL(2,\mathbb{R})$ and $K = SO(2)$ with Lie algebras $\tilde{g} = s\ell(2,\mathbb{R})$ and $\tilde{k} = so(2)$. We recall that G is connected and semisimple, consisting of real 2 x 2 matrices of determinant one, while the matrices in the compact subgroup K are orthogonal; $\tilde{g}$ consists of real 2 x 2 matrices of trace zero and $\tilde{k} \subset \tilde{g}$ is composed of skew symmetric matrices. We write $V = G/K$ and set

$$(3.10) \qquad X = \frac{1}{2}\begin{pmatrix} 0 & 1 \\ 1 & 0 \end{pmatrix} \; ; \; Y = \frac{1}{2}\begin{pmatrix} 1 & 0 \\ 0 & -1 \end{pmatrix} \; ; \; Z = \frac{1}{2}\begin{pmatrix} 0 & 1 \\ -1 & 0 \end{pmatrix}$$

so that $\tilde{k} = \mathbb{R}\, Z$ and we write $p = \{\mathbb{R}X + \mathbb{R}Y\}$ for the subspace of $\tilde{g}$ spanned by X and Y. One has a Cartan decomposition $\tilde{g} = \tilde{k} + p$ (of vector spaces) with $[\tilde{k},\tilde{k}] \subset \tilde{k}$, $[p,p] \subset \tilde{k}$, $[\tilde{k},p] \subset p$ and the Cartan involution $\theta : \xi + \eta \to \xi - \eta$ ($\xi \in \tilde{k}$, $\eta \in p$) is a Lie algebra automorphism of $\tilde{g}$. Recall here that if $P = \exp p$ then $G = PK$ is the standard polar decomposition of $g \in G$ into a product of a positive definite and an orthogonal matrix. The Killing form $B(\xi,\eta) = \text{trace ad}\xi \;\; \text{ad}\eta$ ($=4$ trace $\xi\eta$) is negative definite on $\tilde{k}$ and positive definite on p (with $B(\tilde{k},p) = 0$). One checks easily that X and Y form an orthonormal basis in p for the scalar product $((\xi,\eta)) = \frac{1}{2}B(\xi,\eta)$; we repeat that the 1/2 factor is of no particular significance here in constructing resolvants, etc. and is used mainly to be consistent with some previous work and with the exposition in Helgason [1].

Now setting $X_\alpha = X + Z$, $X_{-\alpha} = X - Z$, and $H_\alpha = 2Y$ we have a standard (or "canonical") triple (cf. Serre [1])

(3.11) $X_\alpha = \begin{pmatrix} 0 & 1 \\ 0 & 0 \end{pmatrix}$; $X_{-\alpha} = \begin{pmatrix} 0 & 0 \\ 1 & 0 \end{pmatrix}$

$$[H_\alpha, X_\alpha] = 2X_\alpha \; ; \quad [H_\alpha, X_{-\alpha}] = -2X_{-\alpha} \; ;$$
$$[X_\alpha, X_{-\alpha}] = H_\alpha$$

where the root subspaces $g_\alpha = \mathbb{R}X_\alpha$ and $g_{-\alpha} = \mathbb{R}X_{-\alpha}$ are characterized by the rule $g_\lambda = \{\xi \in \tilde{g};\ \text{ad}\, H\xi = \lambda(H)\xi \text{ for all } H \in a = \mathbb{R}Y\}$, while the map $\alpha : tY \to t$ determines then an element (called a root) in the dual a^* (note that $\alpha(H_\alpha) = 2$ and $\mathbb{R} \sim a^* = \mathbb{R}\alpha$). Here a is a maximal abelian subspace of p and in this case $a = \tilde{a}$ is also a Cartan subalgebra of $\tilde{g}$. Set now $\tilde{n} = g_\alpha = \{\sum g_\alpha, \alpha > 0\}$, $N = \exp \tilde{n}$, and $A = \exp \tilde{a}$. The Iwasawa decomposition of G can be written $(0 \le \theta < 4\pi)$

(3.13) $g = k_\theta a_t n_\xi = \exp \theta Z \exp t\, Y \exp \xi X_\alpha$

$$= \begin{pmatrix} \cos\theta/2 & \sin\theta/2 \\ -\sin\theta/2 & \cos\theta/2 \end{pmatrix} \begin{pmatrix} e^{t/2} & 0 \\ 0 & e^{-t/2} \end{pmatrix} \begin{pmatrix} 1 & \xi \\ 0 & 1 \end{pmatrix}$$

which we express as before in the form $g = k(g) \exp H(g) n(g)$.

Next we set $M = \{\pm \begin{pmatrix} 1 & 0 \\ 0 & 1 \end{pmatrix}\}$ for the centralizer of A in K and $M' = M \cup \{\pm \begin{pmatrix} 0 & 1 \\ -1 & 0 \end{pmatrix}\}$ will be the normalizer of A in K (thus $M = \{k \in K;\ \text{Ad}kH = H \text{ for } H \in a$ and $M' = \{k \in K;\ \text{Ad}k\, \tilde{a} \subset \tilde{a}\}$). Again $W = M'/M$ has order two and $B = K/M$ is essentially K with the angle variation cut in half. We recall that for $v \doteq gK \in V$ and $b = kM \in B$ $A(v,b) = -H(g^{-1}k)$ and here $\rho = 1/2 \sum m_\lambda \lambda\ (\lambda > 0)$ equals

$\frac{1}{2}\alpha$. Thus $\rho(tY) = t/2$ and for $\mu \in \mathbf{R}$, $\mu \in \tilde{a}^*$ is determined again by the rule $\mu(tY) = t\mu(Y) = \mu t$ (such μ of course exhaust $\tilde{a}^*$ and $\alpha \sim \mu = 1$). The Fourier transform Ff of $f \in L^2(V)$ is given by (1.1) and a_+ is defined as in Section 1, as is the scalar product $(\lambda,\nu) = B(H_\lambda,H_\nu)$. When $\mu \in \tilde{a}^*$ is characterized as above then since $B(Y,Y) = 2$ it follows that $H_\mu = \frac{1}{2}\mu Y$ and obviously a_+^* can thus be identified with (o,∞). Evidently for μ, $\nu \in \tilde{a}^*$ one has $(\mu,\nu) = \frac{1}{2}\mu\nu$

<u>Example 3.3</u> A geometrical realization of the present situation can be described in terms of G acting on the upper half plane Im $z > 0$ by the rule $g \cdot z = (az+b)/(cz+d)$ for $g = \begin{pmatrix} a & b \\ c & d \end{pmatrix} \in G$. Then K is the isotropy subgroup of G at the point $z = i$ (i.e., $K \cdot i = i$) and the upper half plane can be identified with G/K under the map $gK \to g \cdot i$ (cf. Helgason [1]). In geodesic polar coordinates (τ,θ) at i the Riemannian structure is described by $ds^2=d\tau^2 + \sinh^2\tau d\theta^2$ where we use the metric tensor determined by $g_V(\xi,\eta) = \frac{1}{2}B(\xi,\eta)$ for ξ, $\eta \in p$. If $\omega = \pi(e)$ (which corresponds to i under our identification since $\pi(e) = K$) then, denoting left translation in G by τ_g, the Riemannian structure is determined by $g_V((\tau_g)_*\xi,(\tau_g)_*\eta) = g_\omega(\xi,\eta)$ (for the notation $(\tau_g)_*$ see e.g. Kobayashi-Nomizu [1; 2]). The geodesics on V through $\omega = \pi(e)$ are of the form $\gamma : s \to (\exp s\xi)\omega$ and if $\xi = \alpha X + \beta Y \in p$ the square τ^2 of the geodesic distance from w to $(\exp \xi)\omega = \pi(\exp \xi)$ is equal to $\alpha^2 + \beta^2$ (since X and Y form an orthonormal basis for p under the scalar product $((\xi,\eta)) = 1/2\ B(\xi,\eta)$) and

the polar angle is determined by $\tan\theta = \beta/\alpha$. Many calculations can easily be made using standard formulas and one can describe the action $g \cdot i$ for example in general (relevant material for the computations appears in Gangolli [1], Helgason [1; 2; 3], and Kobayashi-Nomizu [1; 2]). We will spare the reader the results of our computations since they are essentially routine. Let us however remark that the volume element dv in (1.1) can be written in this example as $dv = sh\tau\, d\tau\, d\theta$ (up to a normalization factor).

Example 3.4 First we refer back to the formula for $c(\mu)$ given after (1.2) and mention that in general (cf. Helgason [2; 3], Warner [2])

$$(3.14)\qquad c(\mu) = \int_{\overline{N}} e^{-(i\mu+\rho)H(\overline{n})} d\overline{n}$$

where $\overline{n} = \theta\tilde{n} = g_{-\alpha}$ here, $\overline{N} = \exp \overline{n}$, and $d\overline{n}$ is normalized by $c(-i\rho) = 1$. Here θ is the Cartan involution $\theta : \xi + \eta \to \xi - \eta$ for $\xi \in \tilde{k}$ and $\eta \in p$ and one observes that it is an involutive automorphism of $\tilde{g}$ such that $(\xi,\eta) \to -B(\xi,\theta\eta)$ is strictly positive definite on $\tilde{g} \times \tilde{g}$. In terms of Beta functions we obtain

$$(3.15)\qquad c(\mu) = \frac{\beta(1/2, \frac{(i\mu,\alpha)}{(\alpha,\alpha)})}{\beta(1/2, \frac{(\rho,\alpha)}{(\alpha,\alpha)})}$$

Since $(\alpha,\alpha) = 1/2$, $(i\mu,\alpha) = i\mu$, and $(\rho,\alpha) = 1/4$ there results (cf. Bhanu Murti [1])

$$(3.16)\qquad |c(\mu)|^{-2} = \left|\frac{\Gamma(\frac{1}{2}+i\mu)\Gamma(\frac{1}{2})}{\Gamma(i\mu)}\right|^2 = \pi\mu\tanh\pi\mu$$

Thus we can recover $f(v)$ from $\tilde{f}(\mu,b)$ using (3.16) and (1.2).

<u>Example 3.5</u> We now want to show how the resolvants (3.1) can be obtained by a cumbersome and "classical" method in order to further confirm (3.7) as a legitimate resolvant (the equivalence of (3.1) and (3.7) has already been established). The technique has a certain interest and moreover we will obtain a full basis for H_μ while showing how the recursion relations (3.9) can be obtained group theoretically. Some of the more tedious calculations will be omitted. We remark that Theorem 2.1 does not seem convenient here to obtain a full set of basis elements in H_μ since the unitary irreducible representations of $K = SO(2)$ are one dimensional of the form $e^{in\phi}$ (see Vilenkin [1]). Thus first we can remark that $a_t k_\theta = \hat{k}\,\hat{a}\,\hat{n}$ where $\hat{a} = a_s$ with

$$(3.17) \qquad e^s = (\mathrm{ch}\ t + \mathrm{sh}\ t \cos\theta)$$

(cf. Helgason [1] and Warner [1; 2]). Thus $\rho H(a_t k_\theta) = \rho(sY) = s/2$ and for $\mu(sY) = \mu s$ one has

$$(3.18) \qquad e^{(i\mu-\rho)H(a_t k_\theta)} = e^{(i\mu-\frac{1}{2})s} = (\mathrm{ch}\ t + \mathrm{sh}\ t\cos\theta)^{i\mu-\frac{1}{2}}$$

Now, given basis vectors $\phi^m(b) = \exp im(\theta-\pi)$ for example in $L^2(B)$ (recall that $B \sim K$ with the angle variation, in our notation of (3.13), cut from $(o, 4\pi)$ to $(o, 2\pi)$), we get basis vectors in H_μ by means of (1.4) (the factor of $\exp(-im\pi)$ is inserted for convenience in calculation). In particular we obtain "radial objects" when $v = a_t K$ in (1.4) and since $A(gK, k_\theta M) = -H(g^{-1}k_\theta)$ we consider $H(a_t^{-1}k_\theta) = s'Y$ in (1.4), where s' is determined as in (3.17) but with t replaced by -t. Thus the radial part of the basis vactors in H_μ will be

$$\hat{\phi}^m_\mu(a_tK) = \frac{e^{-im\pi}}{2\pi}\int_0^{2\pi} e^{(i\mu-\rho)s'Y} e^{im\theta} d\theta \tag{3.19}$$

$$= \frac{e^{-im\pi}}{2\pi}\int_{-\pi}^{\pi} (\text{ch } t - \text{sh } t \cos\theta)^{i\mu-\frac{1}{2}} e^{im\theta} d\theta$$

$$= \frac{1}{2\pi}\int_0^{2\pi} (\text{ch}t + \text{sh}t \cos\eta)^{i\mu-\frac{1}{2}} e^{im\eta}\, d\eta$$

$$= P^{i\mu-\frac{1}{2}}_{m,o}(\text{ch } t)$$

(see Vilenkin [1]). In order now to produce resolvants $\hat{R}^m(t,\mu)$ as in Section 2.2 from $P^{i\mu - 1/2}_{m,o}(\text{ch } t)$ of (3.19) one must first consider the growth of these functions as $t \to 0$ and introduce suitable weight factors in order to satisfy the resolvant initial conditions (2.3) (note that $\partial/\partial t = (z^2-1)^{1/2}\, \partial/\partial z$). Let us recall in this direction that (cf. Vilenkin [1] and Robin [1])

$$P^\ell_{o,m}(z) = \frac{\Gamma(\ell+m+1)\Gamma(\ell-m+1)}{\Gamma^2(\ell+1)} P^\ell_{m,o}(z) \tag{3.20}$$

$$= \frac{\Gamma(\ell-m+1)}{\Gamma(\ell+1)} P^m_\ell(z)$$

where P^m_ℓ is the standard associated Legendre function of the first kind. Then we can exhibit the resolvants in the form

$$(3.21)\qquad \hat{R}^m(t,\mu) = \frac{2^m\Gamma(m+1)\Gamma(\ell-m+1)}{\Gamma(\ell+m+1)(z^2-1)^{m/2}} P_\ell^m(z)$$

$$= \left(\frac{z+1}{2}\right)^{\ell-m} F\left(-\ell, m-\ell, m+1, \frac{z-1}{z+1}\right)$$

where $z = \mathrm{ch}\, t$, $\ell = -1/2 + i\mu$, $m \geq 0$ is an integer, and F denotes a standard hypergeometric function (cf. Carroll-Silver [15; 16; 17] and Carroll [18; 19]). Recalling that $\Gamma(\ell+m+1)P_\ell^{-m}(z) = \Gamma(\ell-m+1)P_\ell^m(z)$ for $m \geq 0$ an integer, we remark in passing that, for $m \in \mathbb{C}$ not a negative integer, resolvants can be determined by the formula

$$(3.22)\qquad \hat{R}^m(t,\mu) = \frac{2^m\Gamma(m+1)}{(z^2-1)^{m/2}} P_\ell^{-m}(z)$$

(which reduces to (3.21) when m is an integer) and this should be taken as the generic expression for $\hat{R}^m$ in this case (cf. (3.1)). Evidently the $\hat{R}^m$ of (3.21) satisfy (2.3).

We recall the next well known formulas

$$(3.23)\qquad (z^2-1)^{1/2}\frac{dP_\ell^m}{dz} + \frac{mz}{(z^2-1)^{1/2}} P_\ell^m = (\ell+m)(\ell-m+1)P_\ell^{m-1}$$

$$(3.24)\qquad (z^2-1)^{1/2}\frac{dP_\ell^m}{dz} - \frac{mz}{(z^2-1)^{1/2}} P_\ell^m = P_\ell^{m+1}$$

(cf. Vilenkin [1] and Robin [1]) and these, with (3.22), lead to (3.9), where p is replaced by m (note that $\ell(\ell+1) = -(\frac{1}{4} + \mu^2)$); consequently (3.8) will follow as before.

The most interesting fact however about (3.9) is that these recursion relations have a group theoretic significance (cf.

Carroll-Silver [15; 16; 17]) and the variable t in (3.8) - (3.9) may be identified with the geodesic distance τ of Example 3.3. For purposes of calculation it is convenient to consider

$$(3.25) \qquad X_+ = -X + iY; \quad X_- = -X - iY; \quad H = -2iZ$$

and to note that (H, X_+, X_-) form a "canonical" triple relative to the root space decomposition $\tilde{g}_{\mathbb{C}} = \tilde{k}_{\mathbb{C}} + \mathbb{C}X_+ + \mathbb{C}X_-$, $\tilde{k}_{\mathbb{C}} = \mathbb{C}H$, in the sense that

$$(3.26) \qquad [H,X_+] = 2X_+; \quad [H,X_-] = -2X_-; \quad [X_+,X_-] = H$$

(one notes the difference here between (3.12) complexified as $\tilde{g}_{\mathbb{C}} = \mathbb{C}H_\alpha + \mathbb{C}X_\alpha + \mathbb{C}X_{-\alpha}$ and the decomposition above relative to (H,X_+,X_-) based on (3.26) - such decompositions occur relative to any Cartan subalgebra such as $\mathbb{C}H$ or $\mathbb{C}Y$ as indicated in Serre [1] for example). It will be useful to exploit the isomorphism Q between $SL(2,\mathbb{R})$ and the group SU(1,1) of unimodular quasiunitary matrices given by $Q(g) = \hat{g} = m^{-1} g\, m$ where $m = \begin{pmatrix} 1 & i \\ i & 1 \end{pmatrix}$. Thus $g = \begin{pmatrix} \alpha & \beta \\ \gamma & \delta \end{pmatrix} \to \hat{g} = \begin{pmatrix} a & b \\ \bar{b} & \bar{a} \end{pmatrix}$ with $a = 1/2[\alpha+\delta+i(\beta-\gamma)]$ and $b = 1/2[\beta+\gamma+i(\alpha-\delta)]$ while $\det g = \det \hat{g} = 1$ (cf. Vilenkin [1]). There is a natural parametrization of SU(1,1) in terms of generalized Euler angles (ϕ,τ,ψ) so that any $\hat{g} \in SU(1,1)$ can be written in the form

$$(3.27) \qquad \hat{g} = \begin{pmatrix} \operatorname{ch} \tau/2\; e^{i(\frac{\phi+\psi}{2})} & \operatorname{sh} \tau/2\; e^{i(\frac{\phi-\psi}{2})} \\ \operatorname{sh} \tau/2\; e^{-i(\frac{\phi-\psi}{2})} & \operatorname{ch} \tau/2\; e^{-i(\frac{\phi+\psi}{2})} \end{pmatrix}$$

where $0 \le \phi < 2\pi$, $0 \le \tau < \infty$, and $-2\pi \le \psi < 2\pi$; this will be expressed as $\hat{g} = (\phi,\tau,\psi)$. In particular, setting $w_1(s) = \exp sX$, $w_2(s) = \exp sY$, and $w_3(s) = \exp sZ$, we have $Qw_1(s) = (0,s,0)$, $Qw_2(s) = (\pi/2, s, -\pi/2)$, and $Qw_3(s) = (s,0,0) = (0,0,s)$. We note that $(\phi,\tau,\psi)^{-1} = (\pi-\psi,\tau,-\pi-\phi)$ and $(\phi,0,0)(0,\tau,0)(0,0,\psi) = (\phi,\tau,\psi)$ while if $\hat{g}_1 = (0,\tau_1,0)$ and $\hat{g}_2 = (\phi_2,\tau_2,0)$ then $\hat{g}_1\hat{g}_2 = (\phi,\tau,\psi)$ with

(3.28)
$$\tan\phi = \frac{\text{sh}\,\tau_2\sin\phi_2}{\text{ch}\,\tau_1\,\text{sh}\,\tau_2\cos\phi_2 + \text{sh}\,\tau_1\,\text{ch}\,\tau_2};$$

$$\text{ch}\,\tau = \text{ch}\,\tau_1\,\text{ch}\,\tau_2 + \text{sh}\,\tau_1\,\text{sh}\,\tau_2\cos\phi_2;$$

$$\tan\psi = \frac{\text{sh}\,\tau_1\sin\phi_2}{\text{sh}\,\tau_1\,\text{ch}\,\tau_2\cos\phi_2 + \text{ch}\,\tau_1\,\text{sh}\,\tau_2}$$

Now going to the notation of Example 3.3 with $\hat{p} = \exp(\alpha X + \beta Y)$, $\tau^2 = \alpha^2 + \beta^2$, and $\tan\theta = \beta/\alpha$ an easy calculation shows that $Q(\hat{p}) = (\theta,\tau,-\theta)$. Consequently the geodesic polar coordinates of $\pi(\hat{p}) = \hat{p}K$ can be read off directly from the Euler angles of $Q(\hat{p})$.

Now the representation L_μ of G on H_μ (cf. (1.5)) induces a representation, which we again call L_μ, of $\tilde{g}$ on dense subspaces $W_\mu \in H_\mu$ of differentiable functions by the rule

(3.29)
$$L_\mu(\xi)\,f(v) = \frac{d}{ds}L_\mu(\exp s\xi)f(v)\Big|_{s=0}$$

Recall here that $L_\mu(h)f(\pi(g)) = f(h^{-1}\pi(g)) = f(\pi(h^{-1}g))$ and to apply (3.29) for $\xi = X$, Y, or Z one wants then to differentiate $f(\pi(g_i(s)))$ with respect to s where $g_i(s) = w_i(-s)g = w_i^{-1}(s)g$

(and $v = \pi(g)$). We work in the geodesic polar coordinates $(\tau(s), \theta(s))$ of $\pi(g(s)) = \pi(\hat{p}(s)k(s)) = \pi(\hat{p}(s))$ and consider therefore $Q(g(s)) = Q(\hat{p}(s))Q(k(s)) = (\theta(s), \tau(s), -\theta(s))(o,o,\phi(s)) = (\theta(s), \tau(s), \phi(s)-\theta(s)) = (\theta(s),\tau(s),\psi(s))$. In particular $Q(g_1(s)) = Q(\exp(-sX))Q(g)=(o,-s,o)(\theta,\tau,o)(o,o,u-\theta)$ where $Q(g) = Q(p)Q(k) = (\theta,\tau,-\theta)(o,o,u)$. Then one uses (3.28) to compute $(o,-s,o)(\theta,\tau,o)$ with $\phi = \theta(s)$, $\tau = \tau(s)$, $\tau_1 = -s$, $\tau_2 = \tau$, and $\phi_2 = \theta$ (evidently $\psi(s)$ is of no interest here). Similarly $Q(g_2(s)) = (\pi/2,-s,-\pi/2)(\theta,\tau,-\theta)(o,o,u) = (\pi/2,o,o)(o,-s,o)(\theta-\pi/2,\tau,o)(o,o,u-\theta)$ and (3.28) can be applied to the middle two terms. Finally $Q(g_3(s)) = (-s,o,o)(\theta,\tau,o)(o,o,u-\theta)$ which lends itself immediately to calculation. We write $H_+ = L_\mu(X_+)$, $H_- = L_\mu(X_-)$, and $H_3 = \frac{1}{2}L_\mu(H)$ and using the relation $d/ds\, f(\tau(s),\theta(s))\big|_{s=0} = f_\tau\, \tau'(o) + f_\theta\, \theta'(o)$ a routine calculation yields (cf. Carroll-Silver [16])

<u>Proposition 3.6</u> In geodesic polar coordinates (τ,θ) on V one has

(3.30) $\quad H_+ = e^{-i\theta}\, [-i \coth \tau\, \partial/\partial\theta + \partial/\partial\tau]$

(3.31) $\quad H_- = e^{i\theta}\, [i \coth \tau\, \partial/\partial\theta + \partial/\partial\tau]$

(3.32) $\quad H_3 = i\, \partial/\partial\theta$

From general known facts about irreducible unitary principal series representations of G, or SU(1,1), and their complexifications (see e.g. Miller [1; 2] or Vilenkin [1]) - other representations contribute nothing new in this context - it is natural now to look for "canonical" dense differentiable basis vectors in H_μ of the form $f_m^\mu(\tau,\theta)$ satisfying

$$(3.33)\qquad H_+f_m^\mu = (m-\ell)f_{m+1}^\mu \quad ; \quad H_-f_m^\mu = -(m+\ell)f_{m-1}^\mu ;$$

$$H_3f_m^\mu = mf_m^\mu$$

where $\ell = -1/2 + i\mu(\mu > 0)$ and $m = 0, \pm 1, \pm 2, \ldots$. (It seems excessive here to go into the general representation theory of $SL(2, \mathbb{R})$, or $SU(1,1)$; the references cited here - and previously at the beginning of the chapter - are adequate and accessible.) It is then easy to verify, using (3.30) - (3.32), that the following proposition holds.

<u>Proposition 3.7</u> Canonical basis vectors in H_μ can be taken in the form

$$(3.34)\qquad f_m^\mu(\tau,\theta) = (-1)^m \exp(-im\theta)\ P_{0,m}^\ell(\text{ch }\tau)$$

where $\ell = -1/2 + i\mu(\mu > 0)$ and $m = 0, \pm 1, \pm 2, \ldots$.

In view of (3.20) and the definition (3.21) (or (3.22)) of $\hat{R}^m$ we can write for $m \geq 0$, $z = \text{ch }\tau$, and $\ell = -1/2 + i\mu$

$$(3.35)\qquad \hat{R}^m(\tau,\mu) = \frac{2^m\Gamma(m+1)\Gamma(\ell+1)}{\Gamma(\ell+m+1)(z^2-1)^{m/2}}(-1)^m e^{im\theta}\ f_m^\mu(\tau,\theta)$$

Using (3.30) - (3.31) one can then easily prove (cf. Carroll-Silver [16])

<u>Theorem 3.8</u> The "canonical" relations (3.33) for H_+ and H_- are equivalent to the recursion relations (3.9) for $\hat{R}^m$. In particular the variable t in Example 3.1 can be identified with the geodesic distance τ, from $\omega = K$ to $k_\psi a_t K$.

Proof: The recursion relations (3.9) (and hence (3.8)) follow immediately from (3.30) - (3.31) and (3.34) - (3.35). This suggests of course the identification of τ and t but one can give a computational proof also. We write $k_1 a k_2 = \hat{p} k_3$ so that $k_1 a = \hat{p} k_4$ and for $\hat{p} = \exp(\alpha X + \beta Y)$ we obtain from Example 3.3 the distance $\tau^2 = \alpha^2 + \beta^2$ between $\omega = \pi(e) = K$ and $\hat{p}K$ together with angle measurements $\cos\theta = \alpha/\tau$ and $\sin\theta = \beta/\tau$. Set now $k_1 = k_\psi$, $a = a_t$, and $k_4 = k_\phi$, using the notation of (3.13); there result a number of equations connecting the variables $(\tau, t, \psi, \phi, \alpha, \beta)$, the solution of which involves the identification of t with τ. Thus the radial variable t in the natural polar expression $(k_\psi M, a_t)$ for $u = \pi(h) = k_\psi a_t K$ is in fact the geodesic radius τ of $\pi(h) = \hat{p}K$. The actual calculations are somewhat tedious but completely routine so we will omit them. QED

We note now that the second equation in (3.9) can be written in the form $(\text{sh}^{2m} t\, \hat{R}^m)' = 2m\,\text{sh}^{2m-1} t\, \hat{R}^{m-1}$ which yields, in view of (2.3),

$$\text{(3.36)} \qquad \text{sh}^{2m} t\, \hat{R}^m(t,\mu) = 2m \int_0^t \text{sh}^{2m-1}\eta \;\; \hat{R}^{m-1}(\eta,\mu)\, d\eta$$

One rewrites the integrand in $(3.36)_m$ in terms of $(3.36)_{m-1}$ and integrates out the η variable; iterating this procedure we obtain

Theorem 3.9 If $m \geq k \geq 1$ are integers the "Sonine" formula

$$(3.37) \qquad sh^{2m}t\ \hat{R}^m(t,\mu) = \frac{2^k\Gamma(m+1)}{\Gamma(m-k+1)\Gamma(k)} \int_0^t (cht-ch\eta)^{k-1} \cdot (sh\eta)^{2m-2k+1}\ \hat{R}^{m-k}(\eta,\mu)d\eta$$

is a direct consequence of (3.9). In particular for $m = k$ we have

$$(3.38) \qquad sh^{2m}t\ \hat{R}^m(t,\mu) = 2^m m \int_0^t (cht-ch\eta)^{m-1} sh\eta\ \hat{R}^0(\eta,\mu)d\eta$$

Remark 3.10 Referring back to the formula for the zonal spherical functions (1.8) and using the notation of (3.17) - (3.18) we see that (cf. (3.19))

$$(3.39) \qquad \tilde{\phi}_\mu(a_t) = \frac{1}{2\pi}\int_0^{2\pi} (cht + sht\cos\theta)^{i\mu-\frac{1}{2}} d\theta = P_{i\mu-\frac{1}{2}}(cht)$$

where $P_{i\mu-\frac{1}{2}}(cht) = P_{0,0}^{i\mu-\frac{1}{2}}(cht)$ is the standard Legendre function. We recall also by Lemma 1.1 that $FM^{a_t} = FM_t = \tilde{\phi}_\mu(a_t)$, which equals $\hat{R}^0(t,\mu)$.

Using (3.38) we can define $R^m(t,\cdot) = F^{-1}\hat{R}^m(t,\mu) \in E'(V)$ and integrating in $E'(V)$ (cf. Carroll [14]) we have for $m \geq 1$ an integer,

$$(3.40) \qquad sh^{2m}t\ R^m(t,\cdot) = 2^m m \int_0^t (cht-ch\eta)^{m-1} sh\eta\ M_\eta d\eta$$

We recall that in the present situation $D(G/K)$ is generated by the Laplace-Beltrami operator Δ which arises from the Casimir operator C in the enveloping algebra $E(\tilde{g})$ (here Δ corresponds to 2C due to our choice of Riemiannian structure determined by $1/2B(\cdot,\cdot)$ and $C = 1/2(X^2+Y^2-Z^2)$ - see Remark 3.15 below on Casimir operators). In $E(\tilde{g}_{\mathbb{C}})$ we have then also $2C = (\frac{1}{2}H)^2 + \frac{1}{2}(X_+X_- + X_-X_+)$ and in geodesic polar coordinates (cf. Proposition 3.6)

$$(3.41) \qquad \Delta = H_3^2 + \frac{1}{2}(H_+H_- + H_-H_+) = \partial^2/\partial\tau^2 + \coth\tau\ \partial/\partial\tau$$
$$+ \operatorname{csch}^2\tau\ \partial^2/\partial\theta^2$$

We note here that an alternate expression for Δ is given by $\Delta = H_+H_- + H_3^2 - H_3$ (cf. Carroll-Silver [15; 16; 17] and Silver [1]).

<u>Theorem 3.11</u> For v fixed $(M^uf)(v)$ depends only on the geodesic radius $\tau = t$ of $u = \pi(h) = k_\psi a_t K$ and can be identified with the geodesic mean value $M(v,t,f)$.

Proof: This can be proved by tedious calculation based on the coordinates introduced in previous examples but it is simply a consequence of Lemma 1.1 and Theorem 3.8. Indeed, setting $g = \hat{k}_1\hat{a}\hat{k}_2$ with $h = k_1a_tk_2$ we have $v = \pi(g) = \hat{k}_1\hat{a}K$ and $u = \pi(h) = k_1a_tK$ with

$$(3.42) \qquad (M^uf)(v) = (M^{a_t}f)(v) = (M_tf)(v)$$
$$= \int_K f(\hat{k}_1\hat{a}\hat{k}_2k\pi(a_t))dk = \int_K f(\hat{k}_1\hat{a}k\pi(a_t))dk$$

We note that the distance from $v = \hat{k}_1\hat{a}K$ to $\hat{k}_1\hat{a}ka_tK$ is the same as the distance from $\omega = \pi(e) = K$ to $ka_tK = (\exp \xi)\,\omega = \hat{p}\omega$, which as indicated in Theorem 3.8, will simply be $\tau = t$. Hence as k varies the points $\hat{k}_1\hat{a}k\pi(a_t)$ describe a geodesic "circle" around $v = \hat{k}_1\hat{a}\omega$ of radius $\tau = t$; (3.42) is, upon suitable normalization already provided, the definition of $M(v,t,f)$. QED

Remark 3.12 The validity of the Darboux equation (1.7) for geodesic mean values $M(v,t,f)$ and $D = \Delta$ in spaces of constant negative curvature, harmonic spaces, etc. has been discussed, using purely geometrical arguments by Günther [1], Olevskij [1], and Willmore [1] (cf. also Fusaro [1], Günther [2; 3], Weinstein [12], and Remark 5.6).

We can now define, for $m \geq 1$ an integer, a composition (or convolution) of $R^m(t,\cdot) \in E'(V)$ with a function $f(\cdot)$ on V by means of (3.40). Thus for $v = \pi(g)$ we write

$$(3.43)\qquad (M_t \# f)(v) = \langle M_t(\cdot), f(g\cdot)\rangle = (M_t f(g\cdot))(w) = (M_t f)(v)$$

where $\langle\ ,\ \rangle$ denotes a distribution pairing as in Lemma 1.1 (cf. Helgason [2] for a similar notation). Then, setting $(R^m(t,\cdot) \# f(\cdot))(v) = u^m(t,v)$, we have from (3.40) a Sonine formula

$$(3.44)\qquad sh^{2m}t\ u^m(t,v) = 2^m m \int_0^t (cht - ch\eta)^{m-1} sh\eta (M_\eta f)(v)\,d\eta$$

Since $F\Delta T = \ell(\ell+1)FT$ with $\ell = -1/2 + i\mu$ when $T \in E'(V)$ (cf. Ehrenpreis-Mautner [1]) there follows from (3.8), (3.9), (3.43) and the definition of $u^m = u^m(t,v) = u^m(t,v,f)$

Theorem 3.13 For $m \geq 1$ an integer and $f \in C^2(V)$ the function $u^m(t,v,f) = u^m(t,v) = (R^m(t,\cdot) \# f(\cdot))(v)$ satisfies

(3.45) $$u_t^m = \frac{\text{sh } t}{2(m+1)} [\Delta - m(m+1)]u^{m+1}$$

(3.46) $$u_t^m + 2m \coth t \, u^m = 2m \text{ csch } t \, u^{m-1}$$

(3.47) $$u_{tt}^m + (2m+1) \coth t \, u_t^m + m(m+1)u^m = \Delta u^m$$

(3.48) $$u^m(o,v) = f(v); \quad u_t^m(o,v) = 0$$

where (3.45), (3.47), and (3.48) hold for $m \geq 0$.

Proof: The notation here directs that $(\Delta M_t \# f)(v) = \Delta_v(M_t f)(v) = (M_t \Delta f)(v)$ (see e.g. (1.7) and recall also that $\Delta = \delta d + d\delta$ in the notation of de Rham [1] is formally symmetric) while (3.48) is immediate from the definitions and (3.45) for example. One can also prove Theorem 3.13 directly from the definition (3.44) and the Darboux equation (1.7) without using the Fourier transformation (cf. Carroll-Silver [15; 16; 17]). Indeed (3.46) follows immediately from (3.44) by differentiation while (3.45) and (3.47) result then by induction using (1.7) and the radial Laplacian $D_u = \Delta = \partial^2/\partial t^2 + \coth t \, \partial/\partial t$ (cf. (3.41)) for $m = 0$ together with (3.46) (see Carroll-Silver [16] for details).

QED

Definition 3.14 The sequence of singular Cauchy problems

(3.47) - (3.48) for $m \geq 0$ an integer will be called a canonical sequence. The canonical resolvant sequence corresponding to (3.8) under inverse Fourier transformation is (cf. also (2.3))

$$R^m_{tt} + (2m+1)\coth t\, R^m_t + m(m+1)R^m = \Delta R^m \tag{3.49}$$

$$R^m(o,\cdot) = \delta \; ; \qquad R^m_t(o,\cdot) = 0 \tag{3.50}$$

where δ is a Dirac measure at $\omega = \pi(e)$. Similarly from the equations (3.9) we have

$$R^m_t = \frac{\operatorname{sh} t}{2(m+1)} [\Delta - m(m+1)]R^{m+1} \tag{3.51}$$

$$R^m_t + 2m \coth t\, R^m = 2m \operatorname{csch} t\, R^{m-1} \tag{3.52}$$

Remark 3.15 We recall at this point a few facts about Casimir operators for a semisimple Lie algebra $\tilde{g}$ (cf. Warner [1]). If $X_1, \ldots, X_n$ is any basis for $\tilde{g}_{\mathbb{C}}$ let $g_{ij} = B(X_i, X_j)$ where $B(\cdot,\cdot)$ denotes the Killing form and let g^{ij} be the elements of the matrix (g^{ij}) inverse to (g_{ij}). The Casimir operator is then defined as

$$C = \sum g^{ij} X_i X_j \tag{3.53}$$

This operator C is independent of the basis $\{X_i\}$ and lies in the center Z of the enveloping algebra $E(\tilde{g}_{\mathbb{C}})$. Another formulation can be based on a Cartan decomposition $\tilde{g} = \tilde{k} + p$ and a θ stable Cartan subalgebra $\tilde{j}$ of $\tilde{g}$ (i.e., $\tilde{j} = \tilde{j} \cap \tilde{k} + \tilde{j} \cap p = j_{\tilde{k}} + j_p$); such $\tilde{j}$ always exist. Let a basis H_i $(1 \leq i \leq m)$ for $j_{\tilde{k}}$ and a basis $H_j (m+1 \le j \le \ell)$ for j_p be chosen so that $B(H_i,H_j) = -\delta_{ij}$ for $1 \le i, j \le m$ and $B(H_i,H_j) = \delta_{ij}$ for $m + 1 \le i, j \le \ell$. Pick root vectors X_α in $\tilde{g}_{\mathbb{C}}$ so that $B(X_\alpha,X_{-\alpha}) = 1$ (where $[X_\alpha,X_{-\alpha}] = H_\alpha$). Then

$$(3.54) \qquad C = -\sum_1^m H_i^2 + \sum_{m+1}^{\ell} H_j^2 + \sum_{\alpha>0} (X_\alpha X_{-\alpha} + X_{-\alpha}X_\alpha)$$

$$= -\sum_1^m H_i^2 + \sum_{m+1}^{\ell} H_j^2 + \sum_{\alpha>0} H_\alpha + 2\sum_{\alpha>0} X_{-\alpha}X_\alpha$$

When G has finite center with K a maximal compact subgroup then the operator in D(G/K) determined by C is the Laplace-Beltrami operator for the Riemannian structure induced by the Killing form $B(\cdot,\cdot)$ (see Helgason [1] and cf. (3.41)).

<u>Example 3.16</u> We consider now the case where $G = SO_0(3,1) = SH(4)$ is the connected component of the identity in the Lorentz group $L = SO(3,1)$ and $K = SO(3) \times SO(1) \simeq SO(3)$. The resulting Lobačevskij space $V = G/K$ has dimension 3 and rank 1. In $\mathbb{R}^4$ G consists of so called proper Lorentz transformations which do not reverse the time direction. Thus set $[x,x] = -r^2(x,x) = -x_0^2 + x_1^2 + x_2^2 + x_3^2$ and think of x_0 as time. Then G corresponds to 4 x 4 matrices of determinant one leaving $[x,x]$ invariant and preserving the sign of x_0. In particular if Ω_4 denotes the interior of the positive light cone $r^2(x,x) > 0$, $x_0 > 0$, then $G : \Omega_4 \to \Omega_4$. The Lobačevskij space V can be identified with the points of the "pseudosphere" $r^2(x,x) = 1$, $x_0 > 0$ (which lies within Ω_4). If coordinates

$$(3.55)\qquad x_1 = r\ \text{sh}\ \theta_3 \sin\theta_2 \sin\theta_1; \quad x_3 = r\ \text{sh}\ \theta_3 \cos\theta_2;$$

$$x_2 = r\ \text{sh}\ \theta_3 \sin\theta_2 \cos\theta_1; \quad x_0 = r\ \text{ch}\ \theta_3$$

with $0 \leq \theta_1 < 2\pi$, $0 \leq \theta_2 < \pi$, $0 \leq \theta_3 < \infty$, and $0 \leq r < \infty$ are prescribed in Ω_4 then coordinates $\theta_i (i = 1,2,3)$ can be used on V and the invariant measure dv on V is given by $dv = \text{sh}^2\theta_3 \sin\theta_2 \Pi d\theta_i$ $(i = 1,2,3)$. We will now follow Takahashi [1] in notation because of his more "canonical" formulation but refer also to Vilenkin [1] for many interesting geometrical insights.

Thus as a basis of the Lie algebra $\tilde{g}$ of G we take

$$(3.56)\qquad Y_1 = \begin{pmatrix} 0 & 1 & 0 & 0 \\ 1 & 0 & 0 & 0 \\ 0 & 0 & 0 & 0 \\ 0 & 0 & 0 & 0 \end{pmatrix}; \quad Y_2 = \begin{pmatrix} 0 & 0 & 1 & 0 \\ 0 & 0 & 0 & 0 \\ 1 & 0 & 0 & 0 \\ 0 & 0 & 0 & 0 \end{pmatrix};$$

$$Y_3 = \begin{pmatrix} 0 & 0 & 0 & 1 \\ 0 & 0 & 0 & 0 \\ 0 & 0 & 0 & 0 \\ 1 & 0 & 0 & 0 \end{pmatrix}; \quad X_{12} = \begin{pmatrix} 0 & 0 & 0 & 0 \\ 0 & 0 & 1 & 0 \\ 0 & -1 & 0 & 0 \\ 0 & 0 & 0 & 0 \end{pmatrix};$$

$$X_{13} \begin{pmatrix} 0 & 0 & 0 & 0 \\ 0 & 0 & 0 & 1 \\ 0 & 0 & 0 & 0 \\ 0 & -1 & 0 & 0 \end{pmatrix}; \quad X_{23} \begin{pmatrix} 0 & 0 & 0 & 0 \\ 0 & 0 & 0 & 0 \\ 0 & 0 & 0 & 1 \\ 0 & 0 & -1 & 0 \end{pmatrix}$$

Note that if E_{pq} denotes the matrix with 1 in the (p,q) position (p^{th} row and q^{th} column) with zeros elsewhere then $Y_p = E_{op} + E_{po}$ (p = 1,2,3) and $X_{pq} = E_{pq} - E_{qp}$ (p<q; p,q = 1,2,3). Then setting $\tilde{k} = \mathbb{R}X_{12} + \mathbb{R}X_{13} + \mathbb{R}X_{23}$ and $p = \mathbb{R}Y_1 + \mathbb{R}Y_2 + \mathbb{R}Y_3$ we have a Cartan decomposition $\tilde{g} = \tilde{k} + p$ and $K = \exp \tilde{k}$. Take now $a = \mathbb{R}Y_1$ as a maximal abelian subspace of p and set $X_p = Y_p + X_{1p}$ (p = 2,3) so that

$$(3.57) \qquad X_2 = \begin{pmatrix} 0 & 0 & 1 & 0 \\ 0 & 0 & 1 & 0 \\ 1 & -1 & 0 & 0 \\ 0 & 0 & 0 & 0 \end{pmatrix} ; \qquad X_3 = \begin{pmatrix} 0 & 0 & 0 & 1 \\ 0 & 0 & 0 & 1 \\ 0 & 0 & 0 & 0 \\ 1 & -1 & 0 & 0 \end{pmatrix}$$

Then writing $n = \mathbb{R}X_2 + \mathbb{R}X_3$ an Iwasawa decomposition of $\tilde{g}$ is given by $\tilde{g} = \tilde{k} + a + \tilde{n}$ and writing A = exp a with $N = \exp \tilde{n}$ one has G = KAN. We will use the notation $a_t = \exp tY_1$ so that

$$(3.58) \qquad a_t = \begin{pmatrix} \text{ch } t & \text{sh } t & 0 & 0 \\ \text{sh } t & \text{ch } t & 0 & 0 \\ 0 & 0 & 1 & 0 \\ 0 & 0 & 0 & 1 \end{pmatrix}$$

One can take $Y_1 = H_0$ in the general rank one picture with $\alpha(H_0) = 1$ since $[Y_1,X_2] = X_2$ and $[Y_1,X_3] = X_3$. Thus α, which is the only positive root, has multiplicity two and $\rho = \frac{1}{2}m_\alpha\alpha = \alpha$ with $\rho(tY_1) = t$. We recall that $M = \{k\varepsilon k;\ \text{Ad}kH = H \text{ for } H \varepsilon a\}$ and that $M' = \{k\varepsilon k;\ \text{Ad}ka \subset a\}$. Since $\exp \text{Ad}kH = k \exp Hk^{-1}$ we see that $AdkH = H$ for $H \varepsilon a$ implies $k \exp Hk^{-1} = \exp H$ while $AdkH \varepsilon a$ for $H \varepsilon a$ implies $k \exp Hk^{-1} = \exp H'$ for $H' \varepsilon a$. It is immediate that $M = \{\exp \theta X_{23}\}$ consists of matrices

$$(3.59) \qquad m = \begin{pmatrix} 1 & 0 & \\ 0 & 1 & O \\ & O & \hat{m} \end{pmatrix} \quad ; \quad \hat{m} \varepsilon SO(2)$$

whereas $M' = M \cup \{\exp \pi X_{12}, \exp \pi X_{13}\}$, where $\exp \pi X_{1p}$ $(p = 2,3)$ sends a_t to a_{-t}. Thus $W = M'/M$ has the properties indicated before in the general rank one situation and $B = K/M$ can be identified with the two sphere $S^2 \subset \mathbb{R}^3$. This identification can be spelled out in at least two ways and we will adopt the one leading to the simplest integral expressions later. Thus, following Takahashi [1], we write $u_\theta = \exp \theta X_{12}$ and $v_\phi = \exp \phi X_{23}$ for $0 \leq \theta \leq \pi$ and $-\pi < \phi \leq \pi$ (in particular $v_\phi \varepsilon M$). Then any $k \varepsilon K$ can be written in the form $k = m\, u_\theta v_\phi$ for $m \varepsilon M$ so that if $m = v_\psi$ then $k_{11} = \cos\theta$, $k_{12} = \sin\theta\cos\phi$, $k_{13} = \sin\theta\sin\phi$, $k_{21} = -\sin\theta\cos\psi$, and $k_{31} = \sin\theta\sin\psi$. One can pass the map $k \to x = (k_{11}, k_{21}, k_{31}) : k \to \mathbb{R}^3$ to quotients to obtain a map $K/M \to S^2$ where $k \to v_\psi u_\theta M$ and $x \to (\psi,\theta)$; here $-\pi < \psi < \pi$ and $0 < \theta < \pi$ on the domain of uniqueness. The invariant measure on S^2 corresponding to this identification is given by $ds = (4\pi)^{-1} \sin\theta\, d\theta d\psi$ and the functions

$$(3.60) \qquad Z_{n,m}(\theta,\psi) = c_{m,n}\, P_n^m(\cos\theta) e^{im\psi}$$

with $c_{m,n} = [(2n+1)(n-m)!/(n+m)!]^{1/2}$, $-n \le m \le n$, and $n = 0,1,2, \ldots$, form a complete orthonormal system in $L^2(S^2)$ (cf. Talman [1] and Vilenkin [1]).

We can now give explicit formulas for a canonical set of radial objects in H_μ (cf. (1.4)). Thus let $v = a_t K$ and $b = kM$ with $k = v_\psi u_\theta$ and consider (cf. (3.19))

$$(3.61) \qquad \hat{\phi}_\mu^{m,n}(a_t K) = \frac{1}{4\pi}\int_{-\pi}^{\pi}\int_0^{\pi} e^{(i\mu-1)H(a_t^{-1}k)} Z_{n,m}(\theta,\psi) \sin\theta \, d\theta d\psi$$

It follows from general formulas in Takahashi [1] that $\exp H(a_t^{-1} v_\psi u_\theta) = \operatorname{ch} t - \operatorname{sh} t \cos\theta$ and therefore (3.61) becomes

$$(3.62) \qquad \hat{\phi}_\mu^{m,n}(a_t K) = \hat{\phi}_\mu^{0,n}(a_t K) = Z_\mu^n(t)$$

$$= \frac{c_{0,n}}{2}\int_0^{\pi} (\operatorname{ch} t - \operatorname{sh} t \cos\theta)^{i\mu-1} P_n(\cos\theta) \sin\theta \, d\theta$$

since $\int_{-\pi}^{\pi} e^{im\psi} d\psi = 0$ for $m \neq 0$ and $P_n^0(\cos\theta) = P_n(\cos\theta)$. We note that (3.62) reduces to $\hat{R}^0(t,\mu) = FM_t = \tilde{\phi}_\mu(a_t)$ when $n = 0$ since $c_{0,0} = 1$ and $P_0(\cos\theta) = 1$ whereas, setting $k = v_\psi u_\theta v_\phi$ and recalling that $\tilde{\phi}_\mu$ is an even function of t, the expression for $\hat{R}^0(t,\mu)$ can be written in the form (cf. Takahashi [1])

$$(3.63) \qquad \hat{R}^0(t,\mu) = \int_K e^{(i\mu-1)H(a_t^{-1}k)} dk$$

$$= \frac{1}{8\pi^2}\int_{-\pi}^{\pi}\int_{-\pi}^{\pi}\int_0^{\pi} (\operatorname{ch}t - \operatorname{sh}t \cos\theta)^{i\mu-1} \sin\theta \, d\theta d\psi d\phi$$

$$= \frac{1}{2}\int_0^{\pi}(\text{cht-sht }\cos\theta)^{i\mu-1}\ \sin\theta d\theta = Z_\mu^0(t)$$

$$= \frac{1}{2i\mu \text{sh } t}[(\text{cht} + \text{sht})^{i\mu} - (\text{cht} - \text{sht})^{i\mu}]$$

$$= \sin\mu t/\mu \text{sht}$$

To evaluate (3.62) in general one has recourse to various formulas for special functions and we are grateful here to R. G. Langebartel for indicating a general formula for such integrals in terms of Meijer G functions (cf. Luke [1]). The details are somewhat complicated so we simply state the result here. Thus, since $c_{0,n} = (2n+1)^{1/2}$,

$$(3.64) \qquad Z_\mu^n(t) = \left(\frac{(2n+1)\pi}{2}\right)^{1/2} \frac{\Gamma(n+1-i\mu)}{\Gamma(1-i\mu)}\ \text{sh}^{-\frac{1}{2}}t\ P_{i\mu-\frac{1}{2}}^{-n-\frac{1}{2}}(\text{ch } t)$$

We note that for n = 0 this formula yields (cf. Magnus-Oberhettinger-Soni [1])

$$(3.65) \qquad Z_\mu^0(t) = \sqrt{\frac{\pi}{2}}\ \text{sh}^{-\frac{1}{2}}t\ P_{i\mu-\frac{1}{2}}^{-\frac{1}{2}}(\text{ch } t)$$

$$= \left(\frac{1}{2i\mu}\right)\ \text{sh}^{-1}t[(\text{ch } t + \text{sh } t)^{i\mu} - (\text{ch } t + \text{sh } t)^{-i\mu}]$$

$$= \left(\frac{1}{2i\mu}\right)\ \text{sh}^{-1}t(e^{i\mu t} - e^{-i\mu t})$$

in accordance with (3.63). Further let us remark that (cf. Robin [1])

$$(3.66)\qquad P^{-\frac{1}{2}}_{i\mu-\frac{1}{2}}(\text{ch } t) = \sqrt{\frac{2}{\pi}}\,\text{sh}^{1/2}t\; F\left(\frac{1+i\mu}{2}, \frac{1-i\mu}{2}, \frac{3}{2}, -\text{sh}^2 t\right)$$

and hence from (3.65) one has $Z^0_\mu(t) = F(\frac{1+i\mu}{2}), \frac{1-i\mu}{2}, \frac{3}{2}, -\text{sh}^2 t)$ as required by (2.2).

To obtain resolvants $\hat{R}^n(t,\mu)$ for n = 1, 2, . . . we now multiply the $Z^n_\mu(t)$ of (3.64) by an appropriate weight factor. Since one has for example

$$(3.67)\qquad P^{-n-\frac{1}{2}}_{i\mu-\frac{1}{2}}(\text{ch } t) = \frac{2^{-n-\frac{1}{2}}\text{sh}(t)^{n+\frac{1}{2}}}{\Gamma(n+\frac{3}{2})}\, F\left(\frac{n+1+i\mu}{2}, \frac{n+1-i\mu}{2}, n+\frac{3}{2}, -\text{sh}^2 t\right)$$

(cf. Robin [1]), in order to satisfy the resolvant initial conditions (2.3), it is natural to take (cf. (3.21) - (3.22)

$$\begin{aligned}(3.68)\qquad \hat{R}^n(t,\mu) &= F\left(\frac{n+1+i\mu}{2}, \frac{n+1-i\mu}{2}, n+\frac{3}{2}, -\text{sh}^2 t\right)\\ &= 2^{n+\frac{1}{2}}\,\Gamma(n+\tfrac{3}{2})\,\text{sh}(t)^{-n-\frac{1}{2}} P^{-n-\frac{1}{2}}_{i\mu-\frac{1}{2}}(\text{ch } t)\\ &= 2^{n+\frac{1}{2}}\left(\frac{(2n+1)\pi}{2}\right)^{-1/2} \frac{\Gamma(1-i\mu)\Gamma(n+\frac{3}{2})}{\Gamma(n+1-i\mu)}\,\text{sh}^{-n}t\; Z^n_\mu(t)\end{aligned}$$

We now use (3.23) - (3.24) to obtain recursion relations for the $\hat{R}^n$. Some routine calculation yields then

<u>Theorem 3.17</u> The resolvants $\hat{R}^n$ for the case of three dimensional Lobačevskij space are defined by (3.68) and satisfy (ν = $i\mu-1$)

$$(3.69)\qquad \hat{R}^n_t + (2n+1)\,\coth t\, \hat{R}^n = (2n+1)\,\text{csch } t\, \hat{R}^{n-1}$$

(3.70) $\hat{R}^n_t = \frac{[\nu(\nu+2) - n(n+2)]}{2n+3} \text{sh } t\, \hat{R}^{n+1}$

(3.71) $\hat{R}^n_{tt} + (2n+2) \coth t\, \hat{R}^n_t + n(n+2)\hat{R}^n = \nu(\nu+2)\hat{R}^n$

These equations are in agreement with Silver [1] where they are obtained in a different manner and the resolvants are expressed somewhat differently in terms of the functions $P^{\alpha}_{\beta,\gamma}$ of Vilenkin [1]. One can now proceed as before to determine resolvants R^n and canonical sequences of singular Cauchy problems. Similarly one can expand the matrix theory to deal with higher dimensional Lobačevskij spaces (cf. Silver [1]) but we will not spell this out here (cf. Section 4). The question of deducing the recursion relations (3.69) - (3.70) from group theoretic information about the irreducible unitary representations of G in the H_μ will be deferred for the moment.

Remark 3.18 Going back to the general format of Sections 2.1 - 2.2, Example 3.16 corresponds to the case $m_\alpha = 2$, $m_{2\alpha} = 0$, $\rho = 1$, $d_\alpha = -p(p+1)$, $d_{2\alpha} = 0$, and $n = p$. Then using the Kummer relation indicated at the end of Section 2.2 with $z = \text{th}^2 t$ now, we have $z/z-1 = -\text{sh}^2 t$ and for $\tau \sim (p,o)$

(3.72 $\Psi_{-\mu,\tau}(a_t K)$

$$= c_{-\mu,\tau} \text{th}^p t\, \text{ch}^{-i\mu-1} t\, F\left(\frac{i\mu+1+p}{2}, \frac{i\mu+2+p}{2}, p+\frac{3}{2}, \text{th}^2 t\right)$$

$$= c_{-\mu,\tau} \text{sh}^p t\, F\left(\frac{p+1+i\mu}{2}, \frac{p+1-i\mu}{2}, p+\frac{3}{2}, -\text{sh}^2 t\right)$$

Hence we can write from (3.68)

$$\hat{R}^p(t,\mu) = c^{-1}_{-\mu,\tau} \, sh^{-p} t \, \Psi_{-\mu,\tau}(a_t K) \tag{3.73}$$

exactly as in (3.7). Putting this in (2.11) yields (3.71) and the recurrence relations follow as before from (3.23) - (3.24).

2.4 Expressions for general resolvants. Let first m_α = m and $m_{2\alpha} = 0$ so that $\rho = m/2$, $\ell = i\mu + \frac{m}{2}$, $d_\alpha = -p(p+m-1)$, and $d_{2\alpha} = 0$, while $\tau \sim (p,o)$.

<u>Theorem 4.1</u> Resolvants for the case $m_\alpha = m$ and $m_{2\alpha} = 0$ are given by

$$\begin{aligned} \hat{R}^p(t,\mu) &= c^{-1}_{-\mu,\tau} \, sh^{-p} t \, \Psi_{-\mu,\tau}(a_t K) \\ &= ch^{-p-\ell} t \, F\left(\frac{\ell+p}{2}, \frac{\ell+p+1}{2}, p+\frac{m+1}{2}, th^2 t\right) \\ &= \frac{\Gamma(p+\frac{m}{2}+\frac{1}{2}) 2^{p+\frac{m}{2}-\frac{1}{2}}}{sh^{p+\frac{m}{2}-\frac{1}{2}} t} \, P^{-p-\frac{m}{2}+\frac{1}{2}}_{i\mu-\frac{1}{2}} (ch\, t) \end{aligned} \tag{4.1}$$

These satisfy the resolvant initial conditions as well as the differential equation and splitting recursion relations below.

$$\hat{R}^p_{tt} + (2p+m) \coth t \, \hat{R}^p_t + [p(p+m) + \mu^2 + (\tfrac{m}{2})^2]\hat{R}^p = 0 \tag{4.2}$$

$$\hat{R}^p_t = \frac{-sh\, t}{2p+m+1} [p(p+m) + \mu^2 + (\tfrac{m}{2})^2] \, \hat{R}^{p+1} \tag{4.3}$$

$$\hat{R}^p_t + (2p+m-1) \coth t \, \hat{R}^p = (2p+m-1) csch\, t \, \hat{R}^{p-1} \tag{4.4}$$

Proof: Equation (4.1) follows from definitions (cf. (3.7) and (3.73) for motivation), (2.9), the Kummer relation $F(a,b,c,z)=(1-z)^{-a}F(a,c-b,c,z/z-1)$, and an extended version of (3.67); the recursion formulas (4.3) - (4.4) follow from the known relations

$$(4.5)\qquad (z^2-1)^{1/2}\frac{d}{dz}P_b^a + \frac{az}{(z^2-1)^{1/2}}P_b^a = (a+b)(b-a+1)P_b^{a-1}$$

$$(4.6)\qquad (z^2-1)^{1/2}\frac{d}{dz}P_b^a - \frac{az}{(z^2-1)^{1/2}}P_b^a = P_b^{a+1}$$

(cf. Robin [1] for details). Finally (4.2) results from (2.11) or (4.3) - (4.4). QED

In the situation now when $m_{2\alpha} = 1$ the situation becomes somewhat different. We recall that $\tau \sim (p,q)$ with $(p,q) \varepsilon Z_+ \times Z$ and $p \pm q \varepsilon 2Z_+$. We take $m_\alpha = m$ so that $d_\alpha = -p(p+m) + q^2$ and $d_{2\alpha} = -4q^2$ with $\rho = \frac{m}{2} + 1$. For the resolvants we use (3.7) again (cf. (3.73) also) with (2.9) to obtain

$$(4.7)\qquad \hat{R}^{p,q}(t,\mu) = c^{-1}_{-\mu,\tau}\, sh^{-p}t\, \Psi_{-\mu,\tau}(a_tK)$$

$$= ch^{-p-2x}F(x+\tfrac{p+q}{2},\ x+\tfrac{p-q}{2},\ y,\ th^2t)$$

where $x = \frac{1}{2}(i\mu + \frac{m}{2} + 1) = \frac{\ell}{2}$ and $y = p + \frac{m}{2} + 1$. An elementary computation now yields the resolvant equation from (2.11) in the form

$$(4.8)\qquad \hat{R}^{p,q}_{tt} + [(2p+m+1)\coth t + tht]\hat{R}^{p,q}_t$$

$$+ [p(p+m+2) + \mu^2 + (\tfrac{m}{2}+1)^2 + q^2 sech^2 t]\hat{R}^{p,q} = 0$$

Theorem 4.2 For the case $m_{2\alpha} = 1$ with $m_\alpha = m$ resolvants are given by (4.7), satisfying the resolvant initial conditions (2.3) and (4.8). There are various splitting recursion relations which we list below.

$$\hat{R}_t^{p,q} = \left[\frac{2(x+\frac{p+q}{2})(x+\frac{p-q}{2})}{y} - p - 2x\right] \text{tht } \hat{R}^{p,q} \tag{4.9}$$

$$+ \frac{2(x+\frac{p+q}{2})(x+\frac{p-q}{2})(y-x-\frac{p+q}{2})(x+\frac{p-q}{2}-y)}{y^2(y+1)} \text{sh}^2 t \text{ tht } \hat{R}^{p+2,q}$$

$$\hat{R}_t^{p,q} = 2(y-1)\coth t \text{ sech}^2 t \hat{R}^{p-2,q} \tag{4.10}$$

$$+ \left[2(1-6)\coth t + \left\{\frac{2(x+\frac{p+q}{2}-1)(y-x-\frac{p-q}{2}-1)}{y-2} - q\right\}\text{th } t\right]$$

$$\cdot \hat{R}^{p,q}$$

$$\hat{R}_t^{p,q} = q \text{ th } t \hat{R}^{p,q} + \frac{2}{y}(x+\frac{p+q}{2})(x+\frac{p-q}{2}-y)\text{sh } t \hat{R}^{p+1,q+1} \tag{4.11}$$

$$\hat{R}_t^{p,q} = -q \text{ th } t \hat{R}^{p,q} - 2(y-1)\coth t\hat{R}^{p,q} \tag{4.12}$$

$$+ 2(y-1)\text{csch } t \hat{R}^{p-1,q-1}$$

$$\hat{R}_t^{p,q} = -q \text{ th } t \hat{R}^{p,q} - \frac{2}{y}(x+\frac{p-q}{2})(y-x-\frac{p+q}{2})\text{sh } t \hat{R}^{p+1,q-1} \tag{4.13}$$

$$\hat{R}_t^{p,q} = q\text{th } t \hat{R}^{p,q} - 2(y-1)\coth t \hat{R}^{p,q} \tag{4.14}$$

$$+ 2(y-1)\text{csch } t \hat{R}^{p-1,q+1}$$

$$(4.15) \qquad \hat{R}_t^{p,q} = q \text{ th } t\, \hat{R}^{p,q} + \frac{2 \coth t}{q+1}\left(x + \frac{p+q}{2}\right)$$

$$\cdot \left(x + \frac{p-q}{2} - y\right)\left(\hat{R}^{p,q} - \hat{R}^{p,q+2}\right)$$

$$(4.16) \qquad \hat{R}_t^{p,q} = -q \text{ th } t\, \hat{R}^{p,q} + \frac{2 \coth t}{q-1}\left(x + \frac{p-q}{2}\right)$$

$$\cdot \left(y - x - \frac{p+q}{2}\right)\left(\hat{R}^{p,q} - \hat{R}^{p,q-2}\right)$$

Proof: The recursion relations are derived using the formula $\frac{d}{dz} F(a,b,c,z) = (ab/c)F_+ = (ab/c)F(a+1,b+1,c+1,z)$ and various contiguity relations for hypergeometric functions (cf. Magnus-Oberhettinger-Soni [1]. Thus to obtain (4.9) - (4.10) one uses the easily derived formulas

$$(4.17) \qquad (1-z)F_+ = F + \frac{(c-a)(b-c)}{c(c+1)} zF(a+1,b+1,c+2,z)$$

$$(4.18) \qquad abz(1-z)F_+ = c(c-1)F(a-1,b-1,c-2,z)$$

$$+ c\left[\frac{(a-1)(c-b-1)}{(c-2)} z - (c-bz-1)\right]F$$

For (4.11) - (4.12) one uses the formulas

$$(4.19) \qquad b(1-z)F_+ = cF + (b-c)F(a+1,b,c+1,z)$$

$$(4.20) \qquad abz(1-z)F_+ = c(c-1)F(a-1,b,c-1,z) - c(c-bz-1)F$$

whereas for (4.13) - (4.14) we utilize

$$(4.21) \qquad abz(1-z)F_+ = c(c-1)F(a,b-1,c-1,z) - c(c-az-1)F$$

$$(4.22)\qquad a(1-z)F_+ = cF - (c-a)F(a,b+1,c+1,z)$$

and finally for (4.15) - (4.16) one has

$$(4.23)\qquad bz(1-z)F_+ = c\left(z+\frac{(b-c)}{a-b+1}\right)F - \frac{c(b-c)}{a-b+1}F(a+1,b-1,c,z)$$

$$(4.24)\qquad abz(1-z)F_+ = \frac{-bc(c-a)}{a-b-1}F(a-1,b+1,c,z)$$

$$+ c\left[\frac{(c-a)(a-1)}{a-b-1} - (c-bz-a)\right]F$$

Let us also remark briefly about the splitting phenomenon. Thus for example (4.19) yields

$$(4.25)\qquad F' = \frac{a}{c(1-z)}[cF + (b-c)F(a+1,b,c+1,z)]$$

whereas (4.20), after an index change, gives

$$(4.26)\qquad F'(a+1,b,c+1,z) = \frac{1}{z(1-z)}[cF - (c-bz)F(a+1,b,c+1,z)]$$

Now differentiate (4.25), insert (4.26), and then use (4.25) again to eliminate $F(a+1,b,c+1,z)$. Multiplying by $z(1-z)$ one obtains then the hypergeometric equation $z(1-z)F'' + [c - (a+b+1)z]F' - abF = 0$. It is easy to show that if the hypergeometric equation splits in this manner then so does (4.8) under the composition of (4.11) - (4.12), for example. QED

For completeness we will write down some formulas for the cases $m_{2\alpha} = 3$ or 7 but will omit the recursion relations since the pattern is exactly as above. Thus for $m_{2\alpha} = 3$ we set $m_\alpha = m$ and recall that $(p,q) \in Z_+ \times Z_+$ with $p \pm q \in 2Z_+$. One has

$\rho = \frac{m}{2} + 3$ and $\ell = i\mu + \rho$ with $d_{2\alpha} = -4q(q+2)$ and $d_\alpha = -p(p+m+2) + q(q+2)$. We use (3.7) and (3.73) again to define $\hat{R}^{p,q}$ and from (2.9) this yields

$$(4.27)\qquad \hat{R}^{p,q}(t,\mu) = c^{-1}_{-\mu,\tau} sh^{-p} t\ \Psi_{-\mu,\tau}(a_t K)$$

$$= ch^{-p-\ell}\ t\ F\left(\frac{\ell+p+q}{2}, \frac{\ell+p-q-2}{2}, p+\frac{m}{2}+2, th^2 t\right)$$

$$= ch^{-p-2x} t\ F\left(x+\frac{p+q}{2}, x+\frac{p-q}{2}-1, y, th^2 t\right)$$

where $x = \ell/2 = \frac{1}{2}(i\mu+\frac{m}{2}+3)$ and $y = p + \frac{m}{2} + 2$. The differential equation arising from (2.11) is then

$$(4.28)\qquad \hat{R}^{p,q}_{tt} + [(2p+m+3)\coth t + 3\ th\ t]\hat{R}^{p,q}_t$$

$$= [p(p+m+6) + \mu^2 + (\tfrac{m}{2}+3)^2 + q(q+2)sech^2 t]\hat{R}^{p,q} = 0$$

For $m_{2\alpha} = 7$ one has $\rho = \frac{m}{2} + 7$ and setting $m_\alpha = m$ it follows that $d_\alpha = -p(p+m+6) + q(q+6)$ with $d_{2\alpha} = -4q(q+6)$. In this case from (3.7) and (2.9) again

$$(4.29)\qquad \hat{R}^{p,q}(t,\mu) = ch^{-p-\ell}\ t\ F\left(\frac{\ell+p+q}{2}, \frac{\ell+p-q-6}{2}, p+\frac{m}{2}+4, th^2 t\right)$$

and from (2.11) the differential equation for $\hat{R}^{p,q}$ is

$$(4.30)\qquad \hat{R}^{p,q}_{tt} + [(2p+m+7)\coth t + 7th\ t]\hat{R}^{p,q}_t$$

$$+ [p(p+m+14) + \mu^2 + (\tfrac{m}{2}+7)^2 + q(q+6)sech^2 t]\hat{R}^{p,q} = 0$$

<u>Theorem 4.3</u> For $m_{2\alpha} = 3$ (resp. 7) the resolvants are given by (4.27) (resp. (4.29)) and satisfy (4.28) (resp. 4.30)).

These results completely solve the rank 1 case and the connection of the Fourier theory to the associated singular partial differential equations has been indicated already (cf. Theorem 3.13 and Chapter 1).

2.5 The Euclidean case plus generalizations. We follow here Carroll-Silver [15; 16; 17] and will indicate results only for one simple Euclidean case; a complete exposition appears in Silver [1]. Thus let $G = \mathbb{R}^2 \times_\gamma SO(2)$ and $K = SO(2)$ where $(\vec{x},\alpha)(\vec{y},\beta) = (\vec{x}+\gamma(\alpha)\vec{y},\alpha+\beta)$ is a semidirect product with

$$(5.1) \qquad \gamma(\alpha)\vec{y} = \begin{pmatrix} \cos\alpha & -\sin\alpha \\ \sin\alpha & \cos\alpha \end{pmatrix} \begin{pmatrix} y_1 \\ y_2 \end{pmatrix}$$

(cf. Helgason [1], Miller [1; 2], and Vilenkin [1]) for background information). Thus $G = M(2)$ is the group of orientation and distance preserving motions of $\mathbb{R}^2$ and is a split extension of $\mathbb{R}^2$ (with additive structure) by K so that $\mathbb{R}^2$ and K are subgroups of G with K compact and $\mathbb{R}^2$ normal (cf. Rotman [1]). Elements $g = (\vec{x},\alpha)$ can be represented in the form

$$(5.2) \qquad g = \begin{pmatrix} \cos\alpha & -\sin\alpha & x_1 \\ \sin\alpha & \cos\alpha & x_2 \\ 0 & 0 & 1 \end{pmatrix}$$

and multiplication is faithful. As generators of the Lie algebra $\tilde{g}$ of G we take

$$(5.3)\qquad a_1 = \begin{pmatrix} 0 & 0 & 1 \\ 0 & 0 & 0 \\ 0 & 0 & 0 \end{pmatrix};\ a_2 \begin{pmatrix} 0 & 0 & 0 \\ 0 & 0 & 1 \\ 0 & 0 & 0 \end{pmatrix};\ a_3 = \begin{pmatrix} 0 & -1 & 0 \\ 1 & 0 & 0 \\ 0 & 0 & 0 \end{pmatrix}$$

with multiplication table

$$(5.4)\qquad [a_1,a_2] = 0; \qquad [a_2,a_3] = a_1; \qquad [a_3,a_1] = a_2$$

Thus $\tilde{g}$ is solvable (but not nilpotent, nor semisimple). Now set $V = G/K$ and since $(\vec{x},\alpha)(\vec{o},\beta) = (\vec{x},\alpha+\beta)$ there is an obvious global analytic diffeomorphism $V \to \mathbb{R}^2$. One defines as before $L(g)f(\vec{x}) = f(g^{-1}\vec{x}) = f(\gamma(-\alpha)(\vec{x}-\vec{y}))$ where $g = (\vec{y},\alpha)$ and $g^{-1} = (\gamma(-\alpha)(-\vec{y}),-\alpha)$. One thinks here of $\vec{x}$ as $\pi(h) = \pi(\vec{x},\beta) = \pi((\vec{x},0)(0,\beta))$ so that $g^{-1}\vec{x} = g^{-1}\pi(h) = \pi(g^{-1}h) = \pi(\gamma(-\alpha)(-\vec{y}) + \gamma(-\alpha)\vec{x},\beta-\alpha) = \gamma(-\alpha)(\vec{x}-\vec{y})$. If f is differentiable then L induces a representation of $\tilde{g}$ as in (3.29). Writing $A_i = L(a_i)$ we have first, for $\vec{x} = \binom{x_1}{x_2} = \pi(\vec{x},\beta) = (x_1,x_2)$ (for simplicity of notation), $\exp(-ta_1)\vec{x} = (x_1-t,x_2)$, $\exp(-ta_2)\vec{x} = (x_1,x_2-t)$, and $\exp(-ta_3)\vec{x} = \gamma(-t)\vec{x}$. Consequently we obtain $A_1 = -\partial/\partial x_1$, $A_2 = -\partial/\partial x_2$, and $A_3 = x_2\,\partial/\partial x_1 - x_1\,\partial/\partial x_2$. Writing $H_+ = A_1 + iA_2$, $H_- = A_1 - iA_2$, and $H = 2iA_3$ it follows from (5.4) that (note the contrast here with (3.12) and (3.26))

$$(5.5)\qquad [H,H_+] = 2H_+; \qquad [H,H_-] = -2H_-; \qquad [H_+,H_-] = 0$$

The Riemannian structure on V is described in (geodesic) polar coordinates by $ds^2 = dt^2 + t^2 d\theta^2$ (cf. Helgason [1]) and the geometry is "trivial." There follows immediately (cf. Vilenkin

[1])

Lemma 5.1 In polar coordinates (t,θ) on V

(5.6) $$H_+ = -e^{i\theta}[\partial/\partial t + \frac{i}{t}\,\partial/\partial\theta]; \quad H = -2i\,\partial/\partial\theta;$$

$$H_- = -e^{-i\theta}[\partial/\partial t - \frac{i}{t}\,\partial/\partial\theta]$$

(5.7) $$\Delta = H_+H_- = \partial^2/\partial t^2 + \frac{1}{t}\,\partial/\partial t + \frac{1}{t^2}\,\partial^2/\partial\theta^2$$

As before (cf. Proposition 3.7) we work with (dense) differentiable basis "vectors" f^μ_m in Hilbert spaces H_μ where G provides a unitary irreducible representation L_μ and these are characterized by the conditions (cf. Miller [1; 2], Vilenkin [1])

(5.8) $$H_+f^\mu_m = i\mu f^\mu_{m+1}; \quad H_-f^\mu_m = i\mu f^\mu_{m-1}; \quad H_3f^\mu_m = mf^\mu_m$$

where $H_3 = 1/2\ H$ and μ is real (see Carroll-Silver [15; 16; 17] and Silver [1] for further details). Here we can write $L = \int_0^\infty L_\mu\ \mu d\mu$ (cf. (1.5)), but emphasize that the semisimple theory does not apply. There results (cf. Vilenkin [1])

Theroem 5.2 Canonical basis vectors in H_μ can be written in the form

(5.9) $$f^\mu_m = (-i)^m \exp\,(im\theta)J_m(\mu t)$$

Proof: Take $f^\mu_m = e^{im\theta}w^\mu_m(t)$ which will assure the third requirement in (5.8). Then the w^μ_m must satisfy

(5.10) $$\frac{d^2}{dt^2} w_m^\mu + \frac{1}{t}\frac{d}{dt} w_m^\mu - \frac{m^2}{t^2} w_m^\mu = -\mu^2 w_m^\mu$$

(cf. (5.7) - (5.8)). The solutions $J_m(\mu t)$ are chosen for finiteness conditions and the factor $(-i)^m$ makes valid the recursion relations indicated in (5.8) (cf. Vilenkin [1]); the resolvants age given by $\hat{R}^m(\mu,t) = (i)^m 2^m \Gamma(m+1)(\mu t)^{-m} w_m^\mu(t)$ and satisfy (1.3.11) - (1.3.12) with $A(y)$ replaced by μ^2 (resp. y by μ) - similarly (1.3.4) - (1.3.5) hold for $\hat{R}^m$ with $\hat{T} = 1$. QED

Remark 5.3 It is easy to show that the mean value as defined by (1.6) coincides in this case with the mean value $\mu_X(t)$ (cf. formula (1.2.1) - and see Carroll-Silver [16] for details). Thus the results of Chapter 1 may be carried over to this group theoretic situation, which is equivalent.

We conclude this chapter with some "generalizations" of the growth and convexity theorems (1.4.12) and (1.4.16) in the special case $V = SL(2,\mathbb{R})/SO(2)$ (cf. Theorem 3.13 and Carroll-Silver [15; 16; 17]). First it is clear that if $f \geq 0$ then, referring to (3.43), $(M_t \# f)(v) = (M_t f)(v) \geq 0$ for $m \geq 0$ an integer (here we assume $f \in C^2(V)$ for convenience later). Hence $u^m(t,v,f) = u^m(t,v) \geq 0$ when $f \geq 0$ by (3.44) for $m > 0$ an integer. Now write $\Delta_m = \Delta - m(m+1)$ so that by (3.45) we have (recall $M_t = M^{a_t}$)

Theorem 5.4 Let $\Delta_m f \geq 0$, $m \geq 0$ an integer; then $u^m(t,v)$ is monotone nondecreasing in t for $t \geq 0$.

Proof: $\Delta M_\eta f = M_\eta(\Delta f)$ by (1.7) and Lemma 1.1. QED

Now let ψ be any function such that $d\psi/dt = \mathrm{csch}^{2m+1} t$ so that $d/d\psi = \mathrm{sh}^{2m+1} t\, d/dt$ (cf. Weinstein [12]). Then (3.47) can be written

(5.11) $$\partial^2/\partial\psi^2\, u^m(t,v,f) = \mathrm{sh}^{4m+2} t\, u^m(t,v,\Delta_m f)$$

Consequently there follows

Theorem 5.5 If $\Delta_m f \geq 0$ then $u^m(t,v,f)$ is a convex function of ψ.

Remark 5.6 Working in a harmonic space H^m(cf. Ruse-Walker-Willmore [1]), Fusaro [1] proves ($M = M(v,t,f)$ as in Theorem 3.11)

(5.12) $$M_{tt} + \left(\frac{(m-1)}{t} + \log' g(t)^{1/2}\right)M_t = \Delta M$$

where $g = \det(g_{ij})$, g_{ij} denoting the metric tensor, and g depends on the geodesic distance t alone (Δ denotes the Laplace-Beltrami operator). If $\Delta f \geq 0$, M will be nondecreasing in t and a convex function of ψ where $\psi'(t) = t^{1-m}/g(t)^{1/2}$. Weinstein [12] works in spaces of constant negative curvature $-\alpha^2$ (which are harmonic) and proves similar theorems for

(5.13) $$M_{tt} + \alpha(m-1)\coth(\alpha t)M_t = \Delta M$$

We refer to Helgason [4] for such "Darboux" equations in a group context; the extension to "canonical sequences" is due to Carroll

[21; 22], Carroll-Silver [15; 16; 17], and Silver [1].

Chapter 3

Degenerate Equations with Operator Coefficients

3.1 Introduction. In the preceding chapters we were concerned with Cauchy problems for evolution equations with operator coefficients that were permitted to become infinite or singular. We turn now to the dual situation in which the operator coefficients are permitted to degenerate in some sense. The examples to follow will illustrate some typical partial differential equations that occur in the initial boundary value problems to which our abstract results will apply.

Example 1.1 Various diffusion and fluid flow models lead to the partial differential equation

$$(1.1)\qquad D_t\{m_1(x,t)u(x,t) - D_x(m_2(x,t)D_x u)\} - D_x^2 u = f(x,t)$$

with nonnegative real coefficients. This equation can be elliptic ($m_1 = m_2 = 0$), parabolic ($m_1 > 0$, $m_2 = 0$), or of pseudoparabolic or Sobolev type ($m_2 > 0$), and the type may change with position = x and time = t. The equation (1.1) with $m_2 > 0$ was proposed in 1926 by Milne [1]. Similar equations have been proposed for numerous other applications by Barenblat-Zheltov-Kochina [1], Benjamin [1], Benjamin-Bona-Mahoney [2], Buckmaster-Nachman-Ting [1], Chen-Gurtin [1], Coleman-Noll [1], Huilgol [1], Lighthill [1], Peregrine [1; 2], Taylor [1], and Ting [2]. This equation appears below in Examples 3.5, 4.5, 4.6, 4.7, 5.10,

5.13, 5.14, 5.15, 6.21 and 7.6.

Example 1.2 A partial differential equation analogous to (1.1) but containing a second order time derivative was introduced in 1885 by Poincaré. Similar equations of the form

$$(1.2) \qquad D_t^2 \{c_1 u(x,t) - c_2 \Delta_n u(x,t)\} - \{a_1 \Delta_{n-1} u(x,t) + a_2 D_n^2 u(x,t)\} = f(x,t)$$

with nonnegative coefficients have been proposed; here Δ_n is the Laplacian in $x \in \mathbb{R}^n$ and Δ_{n-1} denotes the first n-1 terms of Δ_n. Applications of (1.2) are given by Boussinesq [1], Lighthill [1], Love [1], and Sobolev [1]. Equations of the form (1.2) are considered below as Examples 3.9, 4.8 and 6.22 (cf., Theorem 3.8).

Remark 1.3 The name *Sobolev equation* has been used extensively to designate partial differential equations or, more generally, evolution equations with a nontrivial operator acting on the highest order time derivative. This operator is usually--not always--elliptic in the space variable.

Example 1.4 Certain applications lead to problems with standard evolution or stationary equations in a region G of $\mathbb{R}^n$ but with a constraint in the form of a partial differential equation on a lower dimensional submanifold (e.g., the boundary) of G. Thus one may seek a solution of

$$D_t u(x,t) - \Delta_n u(x,t) = f(x,t), \qquad x \in G,$$

(1.3)

$$D_t u(s,t) + \partial u/\partial N = \mathrm{div}(a(x)\ \mathrm{grad}\ u(s,t)), \quad s \in S$$

where S is a n-1 dimensional submanifold, N denotes a normal, and the divergence and gradient are given in local coordinates on S. Such a problem was given by Cannon-Meyer [1] to describe a diffusion process which was "singular" on S (cf. Example 6.21).

Example 1.5 Problems similar to the above but of second order in time occur in the form

$$-\Delta_n u(x,t) = f(x,t), \qquad x \in G,$$

(1.4)

$$D_t^2 u(s,t) + \partial u/\partial N = 0, \qquad s \in \partial G$$

to describe gravity waves (cf. Whitham [1]). One can view this as a degenerate problem in which the coefficient of u_{tt} is the boundary trace operator. It can be handled as the preceding (cf. Showalter [5]) or directly as by Friedman-Shinbrot [3] or Lions [10], Ch. I.11.

Example 1.6 Degenerate parabolic equations arise in various forms other than (1.1). Problems from mathematical genetics lead to

(1.5) $$tD_t u(x,t) - x(1-x)D_x^2 u(x,t) = f(x,t), \quad 0 < x < 1,\ t > 0$$

where the leading operator degenerates at t = 0 and the second degenerates on the boundary of the interval [0,1] (cf. Brezis-Rosenkratz-Singer [2], Friedman-Schuss [1], Kimura-Ohta [1], Levikson-Schuss [1], and Schuss [1; 2] for (1.5) and related problems). Also, (1.5) is considered below in Example 5.15. Degeneracies can occur in parabolic problems as the result of nonlinearities. This is the situation for the equation

$$(1.6) \qquad D_t u(x,t) - D_x(|u(x,t)|^m D_x u(x,t)) = 0$$

of flow through porous media (cf. Aronson [1] and Example 6.23).

Example 1.7 Wave equations occur with degeneracies, typically either in the form of (1.2) with $c_1(x,t) \geq 0$ and $c_2 \equiv 0$ or in the form of (1.1.4). The latter situation will be discussed in Section 3.2 below.

Each of the preceding examples is a realization in an appropriate function space of one of the abstract Sobolev equations

$$(1.7) \qquad \frac{d}{dt}(M(u)) + L(u) = f$$

$$(1.8) \qquad \frac{d^2}{dt^2}(C(u)) + \frac{d}{dt}(B(u)) + A(u) = f$$

for certain choices of the operators. These two equations will be considered in various forms in this chapter. In Section 3.2 we use spectral and energy methods to discuss (1.8) when C is the identity and B and A are (possibly degenerate) operator polynomials involving a closed densely defined self adjoint operator. This covers the situation of (1.1.4). The equation (1.7) is called *strongly regular* when M is invertible and $M^{-1}L$ is continuous on some space. Similarly, (1.8) is strongly regular when $C^{-1}B$ and $C^{-1}A$ are continuous. The strongly regular case is studied in Section 3.3; the operators are permitted to be time-dependent and nonlinear. We call the equations *weakly regular* when their leading operators are only invertible. We shall give well-posedness results in Section 3.4 for linear weakly regular equations by means of the classical generation theory for linear semigroups in Hilbert space. Section 3.5 treats degenerate linear time-dependent equations by the energy methods of Lions [5]. Applications include Examples 1.1, 1.4, 1.6, second order evolution equations of parabolic type (cf. Theorem 5.9), and many others (cf. Showalter [2]). Related nonlinear problems are discussed in Sections 3.6 and 3.7 by methods of monotone nonlinear operators. These techniques also allow the treatment of related variational inequalities (cf. Theorems 3.12 and 6.6). Each section contains a list of references to related work. The partition of these references is inexact, so one should check all sections for completeness.

3.2 The Cauchy problem by spectral techniques. Referring to Remarks 1.1.3 and 1.1.4 we will deal first with an abstract version of the degenerate Cauchy problem in the hyperbolic region with data given on the parabolic line, following Carroll-Wang [12] (cf. also for example Berezin [1], Bers [1], Carroll [7; 9; 34], Conti [3], Frank'l [1], Krasnov [1; 2; 3], Lacomblex [1],

Protter [2], Walker [1; 2; 3], Wang [1; 2] as well as other previous citations in Remark 1.1.4). Thus consider ($'$ denotes d/dt)

$$(2.1) \qquad u'' + \Lambda^{\alpha}S(t)u' + \Lambda^{\beta}R(t)u + \Lambda q(\Sigma)u = f$$

where Λ is a closed densely defined self adjoint operator in a separable Hilbert space H with $(\Lambda h,h) \geq c\|h\|^2$, $c > 0$, $\alpha,\beta \geq 0$, $\Sigma = \Lambda^{-1} \in L(H)$, $q(\Sigma) = a(t) + B(t)\Sigma$ where a(t) vanishes as $t \to 0$ ($B(t) \in L(H)$, $S(t) \in L(H)$, and $R(t) \in L(H)$) while f is "suitable" (see below). It is assumed that all operators commute and we seek $u(\cdot) \in C^2(H)$ satisfying (2.1) with $u(o) = u'(o) = 0$ (other initial conditions can also be treated but we omit this here - cf. however (2.32)). We will use first the technique of Banach algebras and spectral methods developed in Carroll [4; 6; 8; 10; 11; 14; 29; 30; 31], Carroll-Wang [12], and Carroll-Neuwirth [32; 33]; cf. also references there to Arens [1], Arens-Calderón [1], Dixmier [1], Foias [3], Gelfand-Raikov-Šilov [8], Lions [5; 6], Rickart [1], and Waelbroeck [2]. We obtain results "similar" to those of Krasnov [1; 2; 3] and Protter [2] but in a more general operator theoretical frameword; moreover in our development $a(\cdot)$ need not be monotone and some new features arise (see Remarks 2.10 - 2.12 and cf. also Theorems 2.16 and 2.17).

Let us assume for illustrative purposes that $S(t) = Ss(t)$, $R(t) = Rr(t)$, and $B(t) = Bb(t)$ where B, R, S, and Σ commute in L(H) and are normal; hence by a result of Fuglede [1] (Σ,B,R,S, B^*,R^*,S^*) are a commuting family and we denote by A the uniformly closed * algebra generated by this family and the identity I. Note that for any $h \in H$, $\Lambda B\Sigma h = \Lambda\Sigma Bh = Bh$ automatically for example, and the commutativity of say Λ with B means that for $h \in D(\Lambda)$, $Bh \in D(\Lambda)$ with $\Lambda Bh = B\Lambda h$. It will be assumed that b, r, and s belong to $C^0[o,T]$ and $a \in C^1[o,T]$ where $T < \infty$.

Remark 2.1 A few remarks about finitely generated commutative Banach algebras A with identity will perhaps be useful in what follows (cf. Carroll [14] and references there). Let $\phi : A \to \mathbb{C}$ be a continuous homomorphism so that ker ϕ is a (closed) maximal ideal in A with $A/\ker\phi \simeq \mathbb{C}$ and one identifies ϕ with ker ϕ. The set Φ_A (called the carrier space) of maximal ideals can then be identified with a weakly closed subset of the unit ball in A' and we write $\phi(a) = \hat{a}(\phi)$ for $a \in A$. Since $\hat{I}(\phi) = 1$, if A is a commutative * algebra in L(H) with generators a_0, $a_1, \ldots, a_n, a_1^*, \ldots, a_n^*$, and I ($a_0$ self adjoint), one can show that Φ_A is homeomorphic to the (compact) joint spectrum $\sigma_A = \{\alpha(\phi)\} = \{(\hat{a}_0(\phi), \hat{a}_1(\phi), \ldots, \hat{a}_n^*(\phi))\} \subset \mathbb{C}^{2n+1}$ where $\hat{a}_k^*(\phi) = \overline{a_k(\phi)}$ and $\hat{a}_0(\phi)$ is real. The functions $\hat{a}_k$ will then be continuous on $\Phi_A \sim \sigma_A$ and A will be isometrically isomorphic to $C(\Phi_A)$ where $C(\Phi_A)$ has the uniform norm. In our situation above we have $\sigma_A \subset \mathbb{C}^7$ and we associate the complex variables $(z_0, \ldots, z_6)$ to $(\hat{\Sigma}(\phi), \hat{B}(\phi), \ldots, \hat{S}^*(\phi))$ with $\lambda = 1/z_0 \geq c$ and $|z_k| \leq c_1 = \max(\|B\|, \|R\|, \|S\|)$ for $k \geq 1$. One notes that $z_0 \to 0$ corresponds to $\lambda \to \infty$ and we will call the map $\Gamma : \hat{a} \to a : C(\Phi_A) = C(\sigma_A) \to A$ the Gelfand map. A further notion of use here (cf. also

Section 1.5) is that of decomposing H by an isometric isomorphism $\theta : H \to \mathcal{H} = \int_{\sigma_A}^{\oplus} \mathcal{H}(z)d\nu$ which diagonalizes A and Γ may be extended, as an isometric isomorphism, for example to all bounded Baire functions $B(\sigma_A)$ with values in the von Neumann algebra A'' (here $A'' \subset L(H)$ is the bicommutant of A and is the closure of A in say $L_s(H)$); we will use σ_A for convenience in measure theoretic arguments instead of Φ_A. The so called basic measure ν arises naturally (cf. Dixmier [1] or Maurin [1; 3]) and we will not give details here. A ν-measurable family of Hilbert spaces $\mathcal{H}(z)$ on σ_A consists of a collection of functions $z \to h(z) \in \mathcal{H}(z)$ such that there is a vector subspace $F \subset \Pi\mathcal{H}(z)$ with $z \to \|h(z)\|_{\mathcal{H}(z)}$ measurable $(h \in F)$, $(h(\cdot), g(\cdot))_{\mathcal{H}(z)}$ measurable for all $h \in F$ implies $g \in F$, and there is a fundamental sequence $h_n \in F$ such that the closed subspace of $\mathcal{H}(z)$ generated by $h_n(z)$ is $\mathcal{H}(z)$. Now under the present circumstances $\mathcal{H} = L^2_\nu(\sigma_A, \mathcal{H}(z))$ is a Hilbert space with norm $[\int_{\sigma_A} \|h(z)\|^2 d\nu]^{1/2} = \|h\|_{\mathcal{H}}$ and a Lebesque dominated convergence theorem will be valid. Diagonalizable operators $G \in L(\mathcal{H})$ are defined in the obvious manner (cf. Section 1.5) with $\mathcal{G} = \theta^{-1}G\theta \in L(H)$ where $G \sim G(z) \in L(\mathcal{H}(z))$, all arguments being carried out in $L_s(H)$ for example.

Now in connection with (2.1) we consider the equation (under our assumptions $S(t) = s(t)S$, etc.), obtained by applying θ to (2.1)

$$\hat{u}'' + \lambda^\alpha z_3 s(t)\hat{u}' + \lambda^\beta z_2 r(t)\hat{u} + \lambda[a(t)+z_0 z_1 b(t)]\hat{u} = 0 \tag{2.2}$$

where $\hat{u} = \theta u$ and for example $\theta S \theta^{-1} \sim z_3$. We write $z = (z_1, \ldots, z_6)$ and by Coddington-Levinson [1] or Dieudonné [1] (cf. also Chapter 1) there exist unique solutions $Z(t,\tau,z,\lambda)$ and $Y(t,\tau,z,\lambda)$ of (2.2) such that $Z(\tau,\tau,z,\lambda) = 1$, $Z_t(\tau,\tau,z,\lambda) = 0$, $Y(\tau,\tau,z,\lambda) = 0$, and $Y_t(\tau,\tau,z,\lambda) = 1$ for $0 \le \tau \le t \le T < \infty$, $|z_k| \le c_1$, and $c \le \lambda < \infty$. The functions Z and Y will be continuous in (t,τ,z,λ) in this region and we need only check their behavior as $\lambda \to \infty$. To this end we construct a "Green's" matrix as in Sections 1.3 and 1.5 (cf. (1.3.13)) in the form

$$G_A(t,\tau,z,\lambda) = \begin{pmatrix} Z(t,\tau,z,\lambda) & \lambda^{1/2} Y(t,\tau,z,\lambda) \\ \lambda^{-1/2} Z_t(t,\tau,z,\lambda) & Y_t(t,\tau,z,\lambda) \end{pmatrix} \tag{2.3}$$

which will satisfy the equation

$$\frac{\partial}{\partial t} G_A(t,\tau,z,\lambda) + \lambda^{1/2} H_A(t,z,\lambda) G_A(t,\tau,z,\lambda) = 0 \tag{2.4}$$

$$H_A(t,z,\lambda) = \begin{pmatrix} 0 & -1 \\ a(t)+z_0 z_1 b(t)+\lambda^{\beta-1} z_2 r(t) & \lambda^{\alpha-\frac{1}{2}} z_3 s(t) \end{pmatrix} \tag{2.5}$$

As in Chapter 1 it follows that (cf. (1.3.15))

$$\frac{\partial}{\partial \tau} G_A(t,\tau,z,\lambda) - \lambda^{1/2} G_A(t,\tau,z,\lambda) H_A(\tau,z,\lambda) = 0 \tag{2.6}$$

Setting $\hat{u}_1 = \hat{u}$ and $\hat{u}_2 = \lambda^{-1/2} \hat{u}'$ with $\vec{u} = \begin{pmatrix} \hat{u}_1 \\ \hat{u}_2 \end{pmatrix}$ we obtain formally from $\vec{u}_t + \lambda^{1/2} H_A(t,z,\lambda) \vec{u} = \vec{f}$ the solution of (2.1) in the form

$$(2.7) \qquad u(t,\tau) = \int_\tau^t \mathcal{Y}(t,\xi)f(\xi)d\xi$$

where $\mathcal{Y}(t,\xi) = \theta^{-1}Y(t,\xi,z,\lambda)\theta$, $\vec{f} = \binom{0}{\lambda^{-1/2}\hat{f}}$, $\vec{u}(0) = 0$, and $\hat{f} = \theta f$.

In order to check the behavior of Y we replace t by ξ in (2.2) for Y and multiply by $\overline{Y}_\xi(\xi,\tau,z,\lambda)$; upon taking real parts and assuming now that a(t) is real valued one obtains

$$(2.8) \qquad \frac{d}{d\xi}|Y_\xi|^2 + 2\text{Re}(\lambda^\alpha z_3 s(\xi))|Y_\xi|^2 + 2\text{Re}(\lambda^\beta z_2 r(\xi)Y\overline{Y}_\xi)$$
$$+ \lambda a(\xi)\frac{d}{d\xi}|Y|^2 + 2\text{Re}(z_1 b(\xi)Y\overline{Y}_\xi) = 0$$

Now note that $|r\lambda^\beta Y\overline{Y}_\xi| \leq \frac{1}{2}(|r|^2\lambda^{2\beta}|Y|^2 + |Y_\xi|^2)$ so that, upon integration from τ to t,

$$(2.9) \qquad |Y_t|^2 - 1 + \int_\tau^t 2\text{Re}(\lambda^\alpha z_3 s(\xi))|Y_\xi|^2 d\xi + \lambda a(t)|Y|^2$$
$$- \lambda\int_\tau^t a'(\xi)|Y|^2 d\xi \leq \int_\tau^t |z_1|(|b(\xi)|^2|Y|^2+|Y_\xi|^2)d\xi$$
$$+ \int_\tau^t |z_2|(|r(\xi)|^2\lambda^{2\beta}|Y|^2+|Y_\xi|^2)d\xi$$

Assume now that $\text{Re}(z_3 s(\xi)) \geq 0$ if $\alpha > 0$ with $2\beta \leq 1$ and take $\lambda \geq 1$ so that $\lambda^{2\beta} \leq \lambda$ (recall $\lambda \geq c$ and if $c < 1$ a further argument will apply - see below). If $\alpha = 0$ then the λ^α term can be incorporated into the right hand side of (2.9) or (2.10) in an obvious manner. We have then (since $|z_k| \leq c_1$ for $k \geq 1$)

(2.10) $$|Y_t|^2 + \lambda a(t)|Y|^2 \leq 1 + 2c_1 \int_\tau^t |Y_\xi|^2 d\xi + \lambda \int_\tau^t P(\xi)|Y|^2 d\xi$$

(2.11) $$P(\xi) = a'(\xi) + c_1(|r(\xi)|^2 + \frac{1}{\lambda}|b(\xi)|^2)$$

Adding now $2c_1 \int_\tau^t \lambda a(\xi)|Y|^2 d\xi$ to the right hand side of (2.10) and using Gronwall's lemma (Lemma 1.5.10) one obtains, setting $E(t,\xi) = \exp 2c_1(t-\xi)$,

(2.12) $$|Y_t|^2 + \lambda a(t)|Y|^2 \leq E(t,\tau) + \int_\tau^t \lambda P(\xi)E(t,\xi)|Y|^2 d\xi$$

The following lemma now gives a somewhat sharper estimate on $|Y|^2$ than can be obtained by a direct application of the Gronwall lemma; the proof however is a simple variation.

<u>Lemma 2.2</u> Given (2.12) with $P \geq 0$ and $a > 0$ it follows that for $0 < \tau \leq t \leq T < \infty$ and $\lambda > 1$

(2.13) $$\lambda a(t)|Y|^2 \leq E(t,\tau) \exp \left(\int_\tau^t \frac{P(\sigma)}{a(\sigma)} d\sigma\right)$$

Proof: We first omit the term $|Y_t|^2$ on the left hand side of (2.12) and set $X(t,\tau) = \int_\tau^t \lambda P(\xi)E(t,\xi)|Y|^2 d\xi$ so that $X_t(t,\tau) = \lambda P(t)|Y|^2 + \int_\tau^t \lambda P(\xi)E_t(t,\xi)|Y|^2 d\xi = \lambda P(t)|Y|^2 + 2c_1 X(t,\tau)$. Then multiplying (2.12) (with $|Y_t|^2$ deleted) by $P(t)$ and using the last relation for X_t we obtain

(2.14) $$a(X_t - 2c_1 X) \leq PE + PX$$

Thus defining $F(t,\tau) = \exp\,(-\int_\tau^t [\frac{P}{a} + 2c_1]d\xi)$ for $\tau > 0$ we have from (2.14), $(FX)_t \leq (P/a)EF$. But $E(t,\tau)F(t,\tau) = \exp(-\int_\tau^t [\frac{P}{a}]d\xi)$ and hence $(FX)_t \leq -(\exp(-\int_\tau^t [\frac{P}{a}]d\xi))_t$ from which follows

$$(2.15)\qquad F(t,\tau)X(t,\tau) \leq 1 - \exp(-\int_\tau^t [\frac{P}{a}]d\xi)$$

since $F(\tau,\tau)X(\tau,\tau) = 0$. This may be written

$$(2.16)\qquad X(t,\tau) + E(t,\tau) \leq E(t,\tau)\,\exp(\int_\tau^t [\frac{P}{a}]d\xi)$$

which proves the lemma. QED

We observe from (2.11) that $P/a = a'/a + (\frac{c_1}{a})[|r|^2 + \frac{1}{\lambda}|b|^2]$ so that

$$(2.17)\qquad \exp\,(\int_\tau^t [\frac{P}{a}]d\xi) = \frac{a(t)}{a(\tau)}\,\exp\,(\int_\tau^t c_1[\frac{|r|^2}{a} + \frac{|b|^2}{\lambda a}]d\xi)$$

If $c < 1$ we can carry through the estimates with $|r|^2$ replaced by $|r|^2\lambda^{2\beta-1}$ when $\lambda < 1$ (recall $2\beta \leq 1$) and from Lemma 2.2 we obtain, using (2.17),

<u>Lemma 2.3</u> Given (2.12) with $P \geq 0$ and $a(\tau) > 0$ for $\tau > 0$ we have for $2\beta \leq 1$

$$(2.18)\qquad a(\tau)|Y|^2 \leq \frac{c_2}{\lambda}\exp(\tilde{c}_1 \int_\tau^t [\frac{|r|^2}{a} + \frac{|b|^2}{\lambda a}]d\xi)$$

where $\tilde{c}_1 = c_1 \max\,(1, c^{2\beta-1})$ and $c_2 = \exp 2c_1T$.

Define now $\phi(t,\tau) = \exp \tilde{c}_1 \int_\tau^t \frac{|r|^2}{a}d\xi$ and $\psi(t,\tau,\lambda) = \exp \frac{\tilde{c}_1}{\lambda}\int_\tau^t \frac{|b|^2}{a}\,d\xi$ and as $\tau \to 0$ these functions may of course become infinite. Noting that $\phi(t,\tau) \le \phi(T,\tau)$ and $\psi(t,\tau,\lambda) \le \psi(T,\tau,c)$ we can state

<u>Corollary 2.4</u> Let $\phi(\tau) = \phi^{-1}(T,\tau)$ and $\psi(\tau) = \psi^{-1}(T,\tau,c)$; then

(2.19) $$\phi(\tau)\psi(\tau)a(\tau)|Y|^2 \le c_2/\lambda$$

where $Y = Y(t,\tau,z,\lambda)$ is the unique solution of (2.2) with $P \ge 0$, $a(\tau) > 0$ for $\tau > 0$, $2\beta \le 1$, $\lambda \ge c$, and $\mathrm{Re}(z_3 s(\xi)) \ge 0$ satisfying $Y(\tau,\tau,z,\lambda) = 0$ with $Y_t(\tau,\tau,z,\lambda) = 1$.

We set now $W(t,\tau,z,\lambda) = Q(\tau)\, Y(t,\tau,z,\lambda)$ where $Q = (\phi\psi a)^{1/2}$ and observe that the estimate (2.19) holds for $\tau = 0$ also while W is continuous in (t,τ,z,λ) ($Q(\tau) \to 0$ as $\tau \to 0$). If $h \in H$ then then (cf. Remark 2.1) $(t,\tau) \to \mathcal{W}(t,\tau) = \theta^{-1}W\theta$: $h \to \Lambda^{1/2}\mathcal{W}h$ is continuous since $\int_{\sigma_A} ||\lambda^{1/2}W\,\theta h||^2 d\nu$ will converge appropriately by Lebesque dominated convergence for example (in fact $\lambda^{1/2}W \in B(\sigma_A)$ can be demonstrated, so $\Gamma(\lambda^{1/2}W) \in A''$ but this will not be needed - note here that $\lambda^{1/2}$ and W also commute on σ_A). One can now prove

<u>Theorem 2.5</u> Under the assumptions of Corollary 2.4 (and Lemma 2.3) we have, with $\gamma = \max(1/2,\alpha)$ and $0 \le \tau \le t \le T < \infty$,

(2.20) $$\begin{aligned} &(t,\tau) \to \mathcal{W}(t,\tau) \in C^0(L_s(H,D(\Lambda^{1/2})));\\ &t \to \mathcal{W}(t,\tau) \in C^1(L_s(H));\\ &t \to \mathcal{W}(t,\tau) \in C^2(L_s(D(\Lambda^{\gamma}),H)) \end{aligned}$$

Proof: We need only check the bounds since the rest will follow from the Lebesque dominated convergence theorem. From (2.12) and (2.16) - (2.17) one has

$$(2.21) \qquad |Y_t|^2 \leq \frac{a(t)}{a(\tau)} E(t,\tau)\phi(t,\tau)\psi(t,\tau,\lambda)$$

so that $|W_t|^2 = |QY_t|^2 \leq c_3$ where $c_3 = c_2 \max a(t)$ (max on $[o,T]$); actually $(t,\tau) \to \mathcal{W}_t(t,\tau) \in C^o(L_s(H))$ since $Q(\tau) \to 0$ as $\tau \to 0$ and this is useful later. For the last estimate we go back to (2.2) to obtain the inequality (recall that $2\beta \leq 1$)

$$(2.2) \qquad |QY_{tt}| \leq c_4\lambda^{\alpha} + c_5\lambda^{1/2}$$

and the theorem follows. We again observe that $(t,\tau) \to \mathcal{W}_{tt}(t,\tau) \in C^o(L_s(D(\Lambda^{\gamma}),H))$. QED

Now we consider (2.7) and will show that it represents a solution of the Cauchy problem (2.1) when $f \in C^o(D(\Lambda^{\gamma}))$. Thus setting $h(\xi) = f(\xi)/Q(\xi)$ (2.7) may be written in the form $u(t) = \int_0^t \mathcal{W}(t,\xi)h(\xi)d\xi$ and we can consider this as a Riemann type (vector valued) integral (in order to carry the closed operators Λ^{α}, Λ^{β}, and Λ under the integral signs). The following computations are then justified by Theorem 2.5 and remarks in its proof. First $u(o) = 0$ and

$$(2.23) \qquad u_t(t) = \mathcal{W}(t,t)h(t) + \int_0^t \mathcal{W}_t(t,\xi)h(\xi)d\xi$$

$$= \int_0^t \mathcal{W}_t(t,\xi)h(\xi)d\xi$$

(2.24) $$u_{tt}(t) = f(t) + \int_0^t \omega_{tt}(t,\xi)h(\xi)d\xi$$

Theorem 2.6 Assume $a \in C^1[o,T]$, $a(t) > 0$ real for $t > 0$, $a(o) = 0$, $b, r, s \in C^0[o,T]$, $2\beta \leq 1$, $\gamma = \max(\alpha,1/2)$, $Re(z_3 s(t)) \geq 0$, $P \geq 0$, and $f/Q \in C^0(D(\Lambda^\gamma))$. Then there exists a solution of (2.1) given by (2.7) with $u(o) = u_t(o) = 0$, $u \in C^2(H)$, $u \in C^1(D(\Lambda^\gamma))$, and $u \in C^0(D(\Lambda^{\gamma+1/2}))$.

For uniqueness we will give several results. First from (2.6) one obtains (cf. 1.3.17)

(2.25) $$Y_\tau = -Z + \lambda^\alpha z_3 s(\tau)Y$$

and some bounds for Z must be established. Duplicating our estimates leading to (2.9) we have

(2.26) $$|Z_t|^2 + \int_\tau^t 2Re(\lambda^\alpha z_3 s(\xi))|Z_\xi|^2 d\xi + \lambda a(t)|z|^2 - \lambda a(\tau) - \lambda\int_\tau^t a'(\xi)|Z|^2 d\xi \leq \int_\tau^t |Z_1|(|b|^2|Z|^2 + |Z_\xi|^2)d\xi + \int_\tau^t |z_2|(|r|^2\lambda^{2\beta}|Z|^2 + |Z_\xi|^2)d\xi$$

and under the same assumptions and procedures as before it follows that (cf. (2.9) - (2.12))

(2.27) $$\lambda a(t)|Z|^2 + |Z_t|^2 \leq \lambda a(\tau) + 2c_1 \int_\tau^t |Z_\xi|^2 d\xi + \lambda\int_\tau^t P(\xi)|Z|^2 d\xi$$

$$(2.28)\qquad |Z_t|^2 + \lambda a(t)|Z|^2 \leq \lambda a(\tau)E(t,\tau) + \int_\tau^t \lambda P(\xi)E(t,\xi)|Z|^2 d\xi$$

Then by a version of Lemmas 2.2 and 2.3 one obtains

$$(2.29)\qquad |Z|^2 \leq E(t,\tau)\phi(t,\tau)\psi(t,\tau,\lambda)$$

Consequently we have proved

Lemma 2.7 Under the hypotheses of Corollary 2.4, $\phi(\tau)\psi(\tau)|Z|^2 \leq c_2$.

We set $q = (\phi\psi)^{1/2}$ with $\mathcal{V}(t,\tau) = \theta^{-1}q(\tau)Z(t,\tau,z,\lambda)\theta$ so that by Lebesque dominated convergence again $\mathcal{V}(t,\tau) \in L(H)$ (and $\mathcal{V}(t,\tau) \in A''$ can be shown but again this is not needed). Next we observe from (2.6) again that

$$(2.30)\qquad Z_\tau = [\lambda a(\tau) + z_1 b(\tau) + \lambda^\beta z_2 r(\tau)]Y$$

so that $|Q(\tau)Z_\tau| \leq c_7\lambda^{1/2}$ while from (2.25) $|Y_\tau| \leq c_6\lambda^{\alpha-\frac{1}{2}}$ for $\alpha \leq 1/2$ with $|Y_\tau| \leq \tilde{c}_6$ for $\alpha \geq 1/2$. The case $\alpha \leq 1/2$ is essentially trivial and will not be discussed further. Thus using Lebesque dominated convergence again we can state that for $\tau > 0$, $\tau \to \mathcal{Y}(t,\tau) = \theta^{-1}Y\theta \in C^1(L_s(D(\Lambda^{\alpha-\frac{1}{2}}),H))$ while $\tau \to \mathcal{Z}(t,\tau) = \theta^{-1}Z\theta \in C^1(L_s(D(\Lambda^{1/2}),H))$. Therefore if u is a solution of (2.1) with data prescribed at $\tau > 0$ we operate on (2.1) with $\mathcal{Y}(t,\xi)$ (changing t to ξ in (2.1)) and integrate to obtain for $u \in C^0(D(\Lambda^{\gamma+\frac{1}{2}}))$, $u \in C^1(D(\Lambda^\gamma))$, and $u \in C^2(H)$

$$(2.31)\qquad y(t,\xi)u_\xi\Big|_\tau^t - \int_\tau^t [y_\xi - y\Lambda^\alpha Ss(\xi)]u_\xi d\xi$$

$$+ \int_\tau^t y[\Lambda a(\xi)+Bb(\xi) + \Lambda^\beta Rr(\xi)]ud\xi = \int_\tau^t yfd\xi$$

Our hypotheses and lemmas insure that everything makes sense and using (2.25) with (2.30) plus another integration by parts there results from (2.31)

$$(2.32)\qquad u(t) = Z(t,\tau)u(\tau) + y(t,\tau)u_t(\tau) + \int_\tau^t y(t,\xi)f(\xi)d\xi$$

As in Section 1.5 for example we are making use in (2.31) and (2.32) of the hypocontinuity of separately continuous maps $E \times F \to G$ when F is barreled. Now as $\tau \to 0$, by hypocontinuity again we have (note that this is stated badly in Carroll-Wang [12])

Theorem 2.8 Assume the hypotheses of Theorem 2.6 with continuous $u/q \to 0$ and $u'/Q \to 0$ as $\tau \to 0$. Then the solution of (2.1) given by Theorem 2.6 is unique.

In the event that $q(t) > 0$ for $0 \le t \le T$ a somewhat stronger uniqueness theorem for (2.1) can be proved (cf. Carroll-Wang [12]). Assume $u(o) = u'(o) = f = 0$, with u satisfying the conclusions of Theorem 2.6, and then we define the operator in $L(H)$ by

$$(2.33)\qquad L(t,\tau) = \theta^{-1}\exp(-\lambda^\alpha z_3 \int_\tau^t s(\xi)d\xi)\theta$$

Operating on (2.1) with $L(t,\xi)$, where t has been changed to ξ in

(2.1) we obtain

$$(2.34)\qquad u'(t) = -\int_0^t L(t,\xi)[\Lambda^{\beta}Rr(\xi)+\Lambda a(\xi)+Bb(\xi)]u(\xi)d\xi$$

Since $||\Lambda u||$ and $||\Lambda^{\beta}u||$ are bounded in Theorem 2.6 we have by a well known inequality

$$(2.35)\qquad ||u'(t)|| \le \hat{c}\int_0^t (a+c_8|r|+c_9|b|)d\xi$$

$$\le \hat{c}(\int_0^t ad\xi)^{1/2}[(\int_0^t ad\xi)^{1/2} + c_8(\int_0^t \frac{|r|^2}{a}\,d\xi)^{1/2}$$

$$+ c_9(\int_0^t \frac{|b|^2}{a}\,d\xi)^{1/2}] \le c_{10}(\int_0^t a(\xi)d\xi)^{1/2}$$

(recall that $q > 0$ means $\int_0^t \frac{|r^2|}{a}\,d\xi < \infty$ and $\int_0^t \frac{|b|^2}{a}\,d\xi < \infty$). Now $Z(t,\tau)u(\tau) \to 0$ automatically in (2.32) as $\tau \to 0$ since qZ is bounded and $u \to 0$ while the term $Y(t,\tau)u_t(\tau)$ can be written

$$(2.36)\qquad Y(t,\tau)u'(\tau) = a(\tau)^{1/2}Y(t,\tau)\frac{(\int_0^\tau ad\xi)^{\delta}}{a(\tau)^{1/2}}\;\frac{u'(\tau)}{(\int_0^\tau ad\xi)^{\delta}}$$

where $\delta < 1/2$. But $Q = qa^{1/2}$ so $q > 0$ implies $a(\tau)^{1/2}Y(t,\tau) \in L_s(H,D(\Lambda^{1/2}))$ by Corollary 2.4 and Theorem 2.5 while by (2.35) $u'(\tau)/(\int_0^\tau ad\xi)^{\delta} \to 0$ as $\tau \to 0$. Hence we have

<u>Theorem 2.9</u> Let u be a solution of (2.1) satisfying the conditions of Theorem 2.6 with $q(t) > 0$ for $0 \le t \le T < \infty$ and $(\int^\tau a(\xi)d\xi)^{\delta}/a(\tau)^{1/2}$ bounded ($\delta < 1/2$). Then u is unique.

Remark 2.10 We emphasize here that $a(\cdot)$ is not required to be monotone, in contrast to Protter [2] or Krasnov [1], since by (2.11), $P(t) \geq 0$ for all λ merely implies $a' \geq -c_1|r|^2$ (examples of nonmonotone a are given in Carroll-Wang [12] and Wang [1; 2]). The condition of Protter [2] (cf. also Berezin [1], Bers [1]), phrased here in the form $tr(t)/a(t)^{1/2} \to 0$ as $t \to 0$ (with $a(\cdot)$ monotone), for solutions of a numerical version of (2.1) to be well posed in a local uniform metric, has its analogue here in the conditions on $\phi(t)$. Thus for example if $\Lambda \sim -\partial^2/\partial x^2$, $a(t) = t^m$, $\Lambda^\beta R(t)u = r(t)u_x = t^n u_x$, and $\Lambda^\alpha S(t) = s(x,t)$ with $\Lambda \sum B(t) \sim b(t)$ then it follows that $tr(t)/a(t)^{1/2} = t^{n+1-\frac{m}{2}} \to 0$ if $n > \frac{m}{2} - 1$ whereas $\int_t^T |r|^2/a \, d\xi = \int_t^T \xi^{2n-m} d\xi = (1/2n-m+1)(T^{2n-m+1} - t^{2n-m+1})$ so that $\phi(t) = k \exp(\tilde{c}_1 t^{2n-m+1}/2n-m+1) > 0$ when $n > \frac{m}{2} - \frac{1}{2}$. The L^2 condition of Krasnov [1] involves monotone $a(t) = O(t^m)$ and $r(t) = O(t^{\frac{m}{2}-1}\gamma(t))$ where $\gamma(t) \to 0$ as $t \to 0$ but Krasnov is dealing with weak solutions. Now our existence and uniqueness conditions are phrased in terms of u/q, u'/Q, and f/Q where $q^2 = \phi\psi$ and $Q^2 = \phi\psi a$ which permits a rather precise comparison between the behavior of f, u, and u' as $t \to 0$. The ψ term seems somewhat curious however since for example if $|b| \leq \hat{k} < \infty$ then $\int_t^T |b|^2/a \, d\xi \leq (\hat{k}^2/-m+1)(T^{-m+1} - t^{-m+1})$ for $a(t) = t^m$ so that $\psi(t) = O(\exp k_1 t^{-m+1}/-m+1) \to 0$ for $m > 1$ which imposes growth conditions on $b(t)$ in order that $\psi(t) > 0$. This seems to indicate that the role of the ψ term should be investigated further and some comments on this are made in Remark 2.12.

Walker, in as yet unpublished work, investigates by spectral methods (using Riesz operators R_k defined by $\partial/\partial x_k = i(-\Delta)^{1/2}R_k$) the question of directional oscillations and well posedness (cf. also Walker [1; 2; 3])

<u>Remark 2.11</u> In Wang [2] it is pointed out that if $a(\cdot)$ is monotone near $t = 0$ (locally monotone) then (2.35) can be replaced by $\|u'(t)\| \leq c_{10}t^{1/2}a(t)^{1/2}$ for small t so that in (2.36) one could write $\mathcal{Y}(t,\tau)u'(\tau) = a^{1/2}(\tau)\mathcal{Y}(t,\tau)\frac{u'(\tau)}{a^{1/2}(\tau)}$ with $u'(\tau)/a^{1/2}(\tau) \to 0$ as $\tau \to 0$. This produces a somewhat stronger version of Theorem 2.9 when $a(\cdot)$ is locally monotone. Examples are given in Wang [2] to show that $A(t) = \int_0^t a(\xi)d\xi/a(t)$ may not even be bounded for nonmonotone $a(t) = O(t^m)$.

<u>Remark 2.12</u> First we assume z_1 and $b(t)$ are real so that (2.9) becomes, after elimination of the λ^α term as before, and adding a term $c_1\lambda\int_\tau^t a(\xi)|Y|^2 d\xi$,

$$(2.37) \qquad |Y_t|^2 + \lambda a(t)|Y|^2 \leq 1 + \lambda\int_\tau^t a'(\xi)|Y|^2 d\xi$$

$$+ c_1 \int_\tau^t (|Y_\xi|^2 + \lambda a(\xi)|Y|^2)d\xi$$

$$+ c_1 \int_\tau^t \lambda^{2\beta}|r(\xi)|^2|Y|^2 d\xi - z_1 b(t)|Y|^2$$

$$+ z_1 \int_\tau^t b'(\xi)|Y|^2 d\xi$$

Now assume $|b'(t)/b(t)| \leq c_3$ on $[o,T]$ and set $\hat{c} = \max(c_1,c_3)$;

then, if $z_1 b(\xi) \geq 0$ one can write (for $2\beta \leq 1$ and $\lambda \geq 1$)

$$(2.38) \qquad \Xi(t) \leq \hat{c}\int_\tau^t \Xi(\xi)d\xi + 1 + \lambda\int_\tau^t \hat{P}(\xi)|Y|^2 d\xi$$

$$(2.39) \qquad \Xi(t) = |Y_t|^2 + \lambda a(t)|Y|^2 + z_1 b(t)|Y|^2$$

$$(2.40) \qquad \hat{P}(\xi) = a'(\xi) + c_1|r(\xi)|^2$$

Setting $\hat{E}(t,\xi) = \exp \hat{c}(t-\xi)$ we obtain from Gronwall's lemma again (cf. (2.12))

$$(2.41) \qquad \Xi(t) \leq \hat{E}(t,\tau) + \lambda\int_\tau^t \hat{P}(\xi)\hat{E}(t,\xi)|Y|^2 d\xi$$

Assuming $\hat{P} \geq 0$ Lemma 2.2 applies to (2.41) (upon omitting the term $|Y_t|^2 + z_1 b(t)|Y|^2$ in $\Xi(t)$) to yield

$$(2.42) \qquad \lambda a(t)|Y|^2 \leq \hat{E}(t,\tau)\exp\left(\int_\tau^t \frac{\hat{P}(\xi)}{a(\xi)}\,d\xi\right)$$

and as in Lemma 2.3 one obtains (cf. (2.17))

<u>Lemma 2.13</u> Assume $|b'/b| \leq c_3$ with $\hat{c} = \max(c_1, c_3)$ while $\hat{P} \geq 0$, $a \geq 0$, and $z_1 b \geq 0$. Then for $c_4 = \exp \hat{c}T$ and $\tilde{c}_1 = \max(1, c^{2\beta-1})$

$$(2.43) \qquad a(\tau)|Y|^2 \leq \frac{c_4}{\lambda}\exp\left(\tilde{c}_1\int_\tau^t \frac{|r|^2}{a}\,d\xi\right)$$

<u>Corollary 2.14</u> Defining $\phi(t,\tau)$ and $\phi(\tau)$ as in Corollary 2.4 the hypotheses of Lemma 2.13 imply

$$(2.44) \qquad \phi(\tau)a(\tau)|Y|^2 \leq c_4/\lambda$$

Thus it is possible to eliminate the ψ term in Q and q (cf. Theorems 2.5, 2.6, and 2.8, Remarks 2.10 and 2.11, and Lemma 2.7) if one assumes $z_1 b \geq 0$ with $|b'/b| \leq c_3$ in addition to $a' \geq -c_1|r|^2$ (cf. Remark 2.10). This means locally (i.e., near $t = \tau$) that if $z_1 \geq 0$ then $b \geq 0$ with either $b' \geq 0$ and $0 \leq b(t) \leq b(\tau) \exp c_3(t-\tau)$ or $b' \leq 0$ with $b(t) \geq b(\tau) \exp(-c_3(t-\tau))$; suitable oscillations in the sign of b' are of course allowed. On the other hand if $z_1 \leq 0$ then $b \leq 0$ with $|b| = -b$ so either $b' \geq 0$ with $0 \geq b(t) \geq b(\tau) \exp(-c_3(t-\tau))$ or $b' \leq 0$ with $b(t) \leq b(\tau) \exp c_3(t-\tau)$. We refer to Carroll [34] for further remarks.

Remark 2.15 In Carroll [7; 9] some weak degenerate problems are solved using Lions type energy methods. The notation is chosen here to conform to these articles rather than to earlier parts of this section and most technical details will be omitted. Thus let $\psi > 0$ be a numerical function in $C^0(o,T]$, $T < \infty$; with ψ increasing as $t \to 0$ (a priori ψ need not approach infinity but it usually will). Let $q > 0$ belong to $C^1(o,T]$ with $q(t) \to 0$ as $t \to 0$. Let $V \subset H$ be Hilbert spaces (H separable for simplicity), V dense in H with a finer topology, and $a(t,\cdot,\cdot)$ a family of continuous sesquilinear forms on $V \times V$ with $a(t,v,u) = \overline{a(t,u,v)}$ and $a(t,u,u) \geq \alpha\|u\|_V^2$. Assume $t \to a(t,u,v) \in C^1[o,T]$ for (u,v) fixed and let $t \to B(t) \in C^1(L_s(H))$ on $[o,T]$ be a family of Hermitian operators. Let $w > 0$ be a numerical function to be determined such that $w(t) \to \infty$ as $t \to 0$. Let F_s be the Hilbert space of functions u on $[o,s]$, $s \leq T$ to be determined, such that $u(o) = 0$, $\psi u' \in L^2(H)$, and $wu \in L^2(V)$ with norm $\|u\|^2_{F_s} = \int_0^s (\|wu\|_V^2 + |\psi u'|_H^2)dt$; all derivatives are taken in $D'(H)$ (cf. Carroll [14] or Schwartz [5]). Let H_s be the space of functions h satisfying $h(s) = 0$, $h/\psi \in L^2(H)$, $h'/\psi \in L^2(H)$, and $qh/w \in L^2(V)$. We define

$$\text{(2.45)} \qquad \tilde{E}_s(u,h) = \int_0^s \{qa(t,u,h) + (B(t)u',h) - (u',h')\}dt;$$

$$\tilde{L}_s(h) = \int_0^s (f,h)dt$$

where f is given with $\psi f \in L^2(H)$ (here (,) (resp. ((,))) denotes the scalar product in H (resp. V)). The first problem is to find $u \in F_s$ such that $\tilde{E}_s(u,h) = \tilde{L}_s(h)$ for all $h \in H_s$. Then since $q > 0$ on say $[s/2,T]$ one can apply standard techniques for nondegenerate problems (cf. Lions [5]) to proceed stepwise and find $u \in F_T$ satisfying $\tilde{E}_T(u,h) = \tilde{L}_T(h)$ for all $h \in H_T$. After a series of technical lemmas such a $u \in F_T$ can be found using the Lions projection theorem (see Carroll [9] for details). Another series of technical lemmas will yield uniqueness and one can state

<u>Theorem 2.16</u> Assume the conditions above with $q(t) \geq (\int_0^t d\xi/\psi^2(\xi))^{1-\varepsilon} (0 < \varepsilon < 1)$ while $Q = \lim_{t \to 0} (q'\psi^2/q)\int_0^t d\xi/\psi^2(\xi)$ exists. Then there exists a unique solution $u \in F_T$ of $\tilde{E}_T(u,h) = \tilde{L}_T(h)$ for all $h \in H_T$, based on a function $w \notin L^2 (w \in C^0(o,T])$.

We remark that w measures the rate of how rapidly $u(t) \to 0$ as $t \to 0$ but its precise description is somewhat complicated (cf. below). Another theorem based on energy methods can be obtained as follows. We define K_s to be the space of functions $k(t) = \int_0^t \phi h d\xi$ for $h \in H_s$ with suitable $\phi \in C^1[o,s]$ where $\phi > 0$ on $(o,s]$ while $\phi(t) \to 0$ as $t \to 0$. For suitable choice of the numerical function $\delta > 0$ with $\delta(t) \to \infty$ as $t \to 0$ we put a prehilbert structure on K_s with norm $||k||^2_{K_s} = \int_0^s (||\delta k||^2_V + |k'/\phi\psi|^2_H)dt$. (We note that if $v = \phi/q$ then our w above can be taken to be $w^2 = v'/v^{2-\varepsilon_1}$ $(0 < \varepsilon_1 < 1)$.) For suitable ϕ, δ as above (e.g., $\phi = \hat{c}\int_0^t d\xi/\psi^2(\xi)$ and $\delta^2 = -\tilde{c}(1/v)'$), $\hat{K}_s \subset F_s$ and one can state

<u>Theorem 2.17</u> Under the hypotheses of Theorem 2.16 there exists a unique solution $u \in \hat{K}_T$ satisfying $\tilde{E}_T(u,h) = \tilde{L}_T(h)$ for all $h \in H_T$ such that $||u||_{\hat{K}_T} \leq c(\int_0^T |\psi f|^2_H dt)^{1/2}$

3.3 Strongly regular equations. We consider first the nonlinear equation

$$(3.1) \qquad M(t)u'(t) = f(t,u(t))$$

in a separable and reflexive Banach space V. For each $t \in I_a \equiv [o,a]$ we are given a continuous linear operator $M(t) \in L(V,V')$. Denote by $B_b(u_0)$ the closed ball in V centered at u_0 with radius $b > 0$, and suppose we are given a function $f : I_a \times B_b(u_0) \to V'$. A solution of (3.1) is a function $u : I_a \to V$ which is (strongly) absolutely continuous, differentiable a.e. on I_a, has its range in $B_b(u_0)$, and satisfies (3.1) a.e. on I_a. Sufficient conditions for the Cauchy Problem for (3.1) to be well-posed are given in the following.

Theorem 3.1 Assume there is a pair of measurable functions $k(\cdot) : I_a \to (0,\infty)$ and $Q(\cdot) : I_a \to [1,\infty)$ with $Q(\cdot)/k(\cdot) \in L^1(I_a,\mathbb{R})$ such that

$$(3.2) \qquad \langle M(t)x,x\rangle \geq k(t)\|x\|_V^2 , \qquad \text{a.e. } t \in I_a,\ x \in V,$$

$$(3.3) \qquad \|f(t,x)-f(t,y)\|_{V'} \leq Q(t)\|x-y\|_V, \quad \text{a.e. } t \in I_a,\ x,y \in B_b(u_0).$$

Then any two solutions $u_1(\cdot)$, $u_2(\cdot)$ of (3.1) will satisfy the estimate

$$(3.4) \qquad \|u_1(t)-u_2(t)\|_V \leq \|u_1(o)-u_2(o)\|_V \cdot \exp\{\int_0^t (Q/k)ds\} , \qquad t \in I_a.$$

In particular, the Cauchy problem of (3.1) with $u(o) = u_0$ has at most one solution.

Proof: Since $k(t) > 0$, the coercive estimate (3.2) implies $M(t) : V \to V'$ is a bijection and $\| M(t)^{-1} \|_{L(V',V)} \leq k(t)^{-1}$. Thus, we obtain from (3.1) $\|u_1'(t)-u_2'(t)\|_V \leq k(t)^{-1} \cdot \| f(t,u_1(t))-f(t,u_2(t)) \|_{V'} \leq k(t)^{-1}Q(t) \|u_1(t)-u_2(t)\|_V$, a.e. $t \in I_a$.

Since $u_1(\cdot) - u_2(\cdot)$ is absolutely continuous with a summable derivative, we have $\|u_1(t)-u_2(t)\|_V \le \|u_1(o)-u_2(o)\|_V + \int_0^t \| u_1'(s)-u_2'(s)\|_V ds$. Therefore the bounded function $Z(t) \equiv \|u_1(t)-u_2(t)\|_V$ satisfies the inequality $Z(t) \le Z(o) + \int_0^t k(s)^{-1}Q(s)Z(s)ds$, $t \in I_a$. The estimate (3.4) follows immediately by the Gronwall inequality. QED

Theorem 3.2 In addition to the hypotheses of Theorem 3.1, assume that $t \to \langle M(t)x,y\rangle$ is measurable for each pair $x, y \in V$ and that $t \to \langle f(t,x),y\rangle$ is measurable for $x \in B_b(u_o)$ and $y \in V$. Finally, let $b_o > 0$ and $c > 0$ satisfy $\|f(t,u_o)\|_{V'} \le Q(t)b_o$, a.e. $t \in I_c$, and $\int_0^c k(t)^{-1}Q(t)dt \le b/(b_o+b)$. Then there exists a (unique) solution $u(\cdot)$ of (3.1) on I_c which satisfies $u(o) = u_o$.

Proof: We first show that for every function $u : I_c \to B_b(u_o)$ which is strongly (= weakly) measurable in V, the function $t \to M(t)^{-1} \circ f(t,u(t)) : I_c \to V$ is measurable. For each $\phi \in V'$ we have $\langle\phi,M(t)^{-1}f(t,u(t))\rangle = \langle f(t,u(t)),(M(t)^{-1})^*\phi\rangle$. The indicated adjoint is measurable when $M(t)^{-1}$ is, and so it suffices to show the measurability of each factor. To show $f(t,u(t))$ is measurable we consider first the case where $u(\cdot)$ is countably-valued; thus $u_j(t) = x_j$ for $t \in G_j$, where $\{G_j : j \ge 1\}$ is a measurable partition of I_c. Letting ϕ_j denote the characteristic function of G_j, we obtain $f(t,u(t)) = \sum\{f(t,x_j)\phi_j(t) : j \ge 1\}$ on I_c, and this is clearly measurable. The case of general $u(\cdot)$ follows from the strong continuity of $x \to f(t,x)$, since any measurable u is the strong limit of countably-valued measurable functions. To show the second factor is measurable, let $m \geq 1$ and consider the restriction of $M : I_c \to L(V,V')$ to the set J_m of those $t \in I_c$ with $k(t) \geq 1/m$. Since M is measurable on this set, there is a sequence of countably-valued functions $M_k : J_m \to M(J_m) \subset L(V,V')$ such that $\lim_{k\to\infty} M_k(t) = M(t)$ strongly for $t \in J_m$. But $M_k(t) \in M(J_m)$ implies $\|M_k(t)^{-1}\| \leq m$, so for $\phi \in V'$ we have $\|M_k(t)^{-1}\phi - M(t)^{-1}\phi\|_V = \|M_k(t)^{-1}(M(t)x - M_k(t)x)\|_V \leq m\|M(t)x - M_k(t)x\|_{V'} \to 0$ as $k \to \infty$. Hence, the restriction of M^{-1} to J_m is measurable for every $m \geq 1$, and this gives the desired result.

The proof of Theorem 3.2 now follows standard arguments. Let X be the set of $u \in C(I_c,V)$ with range in $B_b(u_0)$. For $u \in X$ the function $t \to M(t)^{-1}f(t,u(t))$ is measurable $I_c \to V$ (from above) and it satisfies the estimate $\|M(t)^{-1} \circ f(t,u(t))\|_V \leq k(t)^{-1} \cdot Q(t) \cdot (b_0+b)$ on I_c. Hence, we can define $F : X \to X$ by $[Fu](t) \equiv u_0 + \int_0^t M(s)^{-1}f(s,u(s))ds$, $t \in I_c$, and it satisfies the estimate

$$\|[Fu](t)-[Fv](t)\|_V \leq \int_0^t k(s)^{-1}Q(s)\|u(s)-v(s)\|_V \, ds, \qquad u,v \in X, \quad t \in I_c. \tag{3.5}$$

But any map F of a closed and bounded subset X of $L^\infty(I_c,V)$ into itself which satisfies (3.5) is known to have a unique fixed-point, u (cf. Carroll [14]). This fixed-point is clearly the desired solution. QED

A solution exists on the entire interval I_a in certain

situations. Two such situations are given below.

Theorem 3.3 Assume all the hypotheses of Theorem 3.1 hold with $B_b(u_0) = V$. Also, suppose $t \to \langle M(t)x,y\rangle$ and $t \to \langle f(t,x),y\rangle$ are measurable $I_a \to \mathbb{R}$ for every pair $x,y \in V$, and let $u_0 \in V$, $b_0 > 0$ be given and satisfy $\|f(t,u_0)\|_{V'} \leq Q(t)b_0$ a.e. on I_a. Then there exists a unique solution of (3.1) on I_a with $u(o) = u_0$.

Proof: Set $g(t) = b_0 \exp\{\int_0^t k(s)^{-1}Q(s)ds\}$ and let X be those functions $u \in C(I_a,V)$ for which $\|u(t)-u_0\|_V \leq g(t) - b_0$ for $t \in I_a$. If $u \in X$, the estimate $\|M(s)^{-1}f(s,u(s))\|_V \leq k(s)^{-1}Q(s)(\|u(s)-u_0\|_V + b_0)$ shows that Fu defined as above belongs to $C(I_a,V)$ and furthermore satisfies $\|[Fu](t)-u_0\|_V \leq \int_0^t k(s)^{-1}Q(s)g(s)ds = g(t)-b_0$, so $Fu \in X$. Thus F has a unique fixed-point $u \in X$ which is the solution of the Cauchy problem for (3.1). QED

Remark 3.4 The preceding results permit the leading operators to degenerate in a very weak sense at any $t_0 \in I_a$. That is, (3.2) must be maintained with $k(\cdot)^{-1} \in L^1(I_a,\mathbb{R})$. On the other hand we have placed no upper bounds on the family $\{M(t) : t \in I_a\}$; they may be singular (cf. Introduction) on a suitable set of points in I_a.

The Lipschitz condition (3.3) together with the estimate on $f(t,u_0)$ in Theorem 3.3 impose a growth rate on f(t,u) which is linear in u. Such hypotheses are frequently appropriate, especially in linear problems such as Example 1.1 which occur frequently in practice. However they are not appropriate for nonlinear wave propagation models, and these applications call for solutions global in time.

Example 3.5 The third order nonlinear equation

$$(3.6) \qquad u_t - u_{xxt} + u_x + uu_x = 0$$

provides a model for the propagation of long waves of small amplitude. The interest here is in solutions of (3.6) for all $t \geq 0$. Similarly, models with other forms of nonlinearity in addition to dispersion and dissipation occur in the form

$$(3.7) \qquad u_t - u_{xxt} + a\ \mathrm{sgn}(u) \cdot |u|^q + bu^p u_x = cu_{xx}.$$

We refer to Benjamin [1, 2], Bona [2,3, 4, 5], and Showalter [18, 19, 21] for discussion of such models and related systems.

Initial boundary value problems for (3.6) and (3.7) can be resolved in the following abstract form.

Theorem 3.6 Let $a > 0$ and assume given for each $t \in I_a \equiv [0,a]$ an operator $M(t) \in L(V,V')$, where V is a separable reflexive Banach space. Assume there is a $k > 0$ such that

$$(3.8) \qquad \langle M(t)x,x\rangle \geq k\|x\|^2, \qquad x \in V,\ t \in I_a,$$

each $M(t)$ is symmetric, and for each pair $x,y \in V$ the function $t \to \langle M(t)x,y\rangle$ is absolutely continuous with

$$(3.9) \qquad \frac{d}{dt}\langle M(t)x,x\rangle \leq m(t)\|x\|^2, \qquad \text{a.e. } t \in I_a,\quad x \in V,$$

for some $m(\cdot) \in L^1(I_a,\mathbb{R})$. Let $f: I_a \times V \to V'$ be given such that for each pair $x,y \in V$ the function $t \to \langle f(t,x),y\rangle$ is measurable, and for each bounded set B in V there is a $Q(\cdot) \in L^1(I_a,\mathbb{R})$ such that $\|f(t,x)-f(t,y)\|_{V'} \leq Q(t)\|x-y\|_V$, and $\|f(t,x)\| \leq Q(t)$, for $x,y \in B$ and a.e. $t \in I_a$. Finally, assume there exist $K(\cdot)$ and $L(\cdot) \in L^1(I_a)$ with $L(t) \geq 0$ a.e. such that $\langle f(t,x),x\rangle \leq K(t)\|x\|^2 + L(t)\|x\|$, $x \in V$. Then for each $u_0 \in V$ there exists a unique solution $u : I_a \to V$ of (3.1) with $u(0) = u_0$.

Proof: Uniqueness and local existence follow immediately from Theorem 3.1 and Theorem 3.2, respectively. The existence of a (global) solution on I_a can be obtained by standard continuation arguments if we can establish an a priori bound on a solution. But in the present situation the absolutely continuous function $\sigma(t) \equiv \langle M(t)u(t),u(t)\rangle$ satisfies (cf. Lemma 5.1) $\sigma'(t) = 2\langle f(t,u(t)),u(t)\rangle + \langle M'(t)u(t),u(t)\rangle \leq [(2K(t)+m(t))/k]\,\sigma(t) + [2L(t)/k^{1/2}]\,\sigma(t)^{1/2}$; hence we obtain

$$(3.10) \qquad \sigma(t) \leq \exp\{2\int_0^t H(\tau)d\tau\} \cdot \; [\sigma(0)^{1/2} + \int_0^t L(\tau)d\tau]^2$$

with $H(t) \equiv (K^+(t) + (1/2)m(t))/k$. The estimates (3.10) and (3.8) provide the desired bound on all solutions of (3.1) on I_a with given initial data u_0. QED

<u>Remark 3.7</u> Equations similar in form to (3.6) have been used to "regularize" higher order equations as a first step in solving them (cf. Bona [4], Showalter [17]). The advantage over

standard (e.g., parabolic) regularizations is that the order of the problem is not increased by the regularization. An important case of this is the "Yosida approximation" A_ε of a maximal accretive (linear) A in Banach space given by $A_\varepsilon = (I+\varepsilon A)^{-1}A$, $\varepsilon > 0$. Thereby, one approximates the equation $u'(t) + Au(t) = 0$ by one with A replaced by the bounded A_ε, or, equivalently, solves the equation $(I+\varepsilon A)u_\varepsilon'(t) + Au_\varepsilon(t) = 0$ and then looks for $u(t) = \lim_{\varepsilon\to 0^+} u_\varepsilon(t)$ in some sense. This technique was used for a nonlinear time-dependent equation by Kato [2] and to study the nonwell-posed backward Cauchy problem in Showalter [16, 20]; also see Brezis [3].

The preceding results immediately yield corresponding results for second order evolution equations of the form

$$(3.11)\qquad M(t)u''(t) = F(t,u(t),u'(t))$$

<u>Theorem 3.8</u> Let the family of operators $\{M(t)\}$ and the functions $k(\cdot)$ and $Q(\cdot)$ be given as in Theorem 3.2. Let $u_0, u_1 \in V$ and the function $F : I_a \times B_b(u_0) \times B_b(u_1) \to V'$ be given with $t \to \langle F(t,x_1,x_2),y\rangle$ measurable for $x_1 \in B_b(u_0)$, $x_2 \in B_b(u_1)$, $y \in V$. Suppose we have $\|F(t,x)-F(t,y)\|_{V'} \le Q(t)[\|x_1-y_1\|_V^2 + \|x_2-y_2\|_V^2]$, $\|F(t,u_0,u_1)\|_{V'} \le Q(t)$, and

$$(3.12)\qquad \|M(t)\|_{L(V,V')} \le Q(t)$$

for $x = (x_1,x_2)$ and $y = (y_1,y_2)$ in $B_b(u_0) \times B_b(u_1)$ and a.e. $t \in I_a$. Then there exists $c > 0$ and a unique $u : I_c \to V$ which is continuously differentiable with $u(t) \in B_b(u_0)$, $u'(t) \in B_b(u_1)$, u' is (strongly) absolutely continuous and differentiable a.e., (3.11) holds a.e. on I_c and $u(o) = u_0$, $u'(o) = u_1$.

The preceding follows from Theorem 3.2 applied to the equation $\mathcal{M}(t)U'(t) = f(t,U(t))$ in $\mathcal{V} \equiv V \times V$ with $\mathcal{M}(t)x \equiv [M(t)x_1, M(t)x_2]$ and $f(t,x) \equiv [M(t)x_2, F(t,x_1,x_2)]$. The estimate (3.12) limits the growth rate of the leading operator (cf. Remark 3.4). This hypothesis can be deleted when V is a Hilbert space; we need only replace M(t) by the Riesz map $V \to V'$ in the first factor of each of $\mathcal{M}(t)$ and f.

Initial boundary value problems for partial differential equations arise in the form (3.11) in various applications. Two classical cases are given below; see Lighthill [1] or Whitham [1] for additional examples of this type.

Example 3.9 The equation of S. Sobolev [1]

$$\Delta_3 u_{tt} + u_{zz} = 0 \tag{3.13}$$

describes the fluid motion in a rotating vessel. It is on account of this equation that the term "Sobolev equation" is used in the Russian literature to refer to any equation with spatial derivatives on the highest order time derivative (cf., however, Example 1.2).

Example 3.10 The equation

$$u_{tt} - a\Delta_3 u_{tt} - \Delta_3 u = f(x,t) \tag{3.14}$$

was introduced by A.E.H. Love [1] to describe transverse vibrations in a beam. The second term in (3.14) represents radial inertia in the model.

Finally, we describe how certain doubly-nonlinear evolution equations

$$(3.15) \qquad M(t,u'(t)) = f(t,u(t)), \qquad 0 \le t \le T,$$

in Hilbert space can be solved directly by standard results on monotone nonlinear operators. The technique applies as well to the corresponding variational inequality

$$(3.16) \qquad \langle M(t,u'(t)) - f(t,u(t)),\ v - u'(t)\rangle \ge 0, \qquad v \in K(t),$$
$$u'(t) \in K(t), \qquad \text{a.e. } t \in [0,T]$$

with a prescribed family of closed convex subsets $K(t) \subset V$, so we consider this more general situation.

Let V be real Hilbert space and for each $a \ge 0$ we let V_a be the Hilbert space of square summable functions from $[0,T]$ into V with the norm $\|v\|_a = (\int_0^T \|v(t)\|_V^2 e^{-2at}\,dt)^{1/2}$. For each $t \in [0,T]$ we are given a pair of (possibly nonlinear) functions $f(t,\cdot)$, $M(t,\cdot)$ from V into V'. The following elementary result is fundamental for this technique.

<u>Lemma 3.11</u> Assume $[Mv](t) \equiv M(t,v(t))$ defines a hemicontinuous $M : V_a \to V_a'$ and that there is a $k > 0$ with

(3.17) $\quad \langle M(t,x)-M(t,y),x-y\rangle \geq k\|x-y\|_V^2, \quad$ a.e. $t \in [0,T]$, $x,y \in V$.

Let $u_0 \in V$ and assume $[f(v)](t) \equiv f(t,u_0+\int_0^t v(s)ds)$ defines $f : V_a \to V_a'$ and that there is a $Q > 0$ with

(3.18) $\quad \|f(t,x)-f(t,y)\|_{V'} \leq Q\|x-y\|_V, \quad$ a.e. $t \in [0,T]$, $x,y \in V$.

Then the operator $T \equiv M - f : V_a \to V_a'$ is hemicontinuous and strongly monotone for "a" sufficiently large.

Proof: For $u,v \in V_a$ we have

$$\|f(u)-f(v)\|_{V_a'}^2 = \int_0^T e^{-2at}\| f(t,u_0+\int_0^t u) - f(t,u_0+\int_0^t v)\|^2 dt$$

$$\leq Q^2 \int_0^T(\int_0^t e^{-a(t-s)}e^{-as} \|u(s)-v(s)\|_V^2 ds)^2 dt$$

$$\leq Q^2 \int_0^T(\int_0^t e^{-2a(t-s)}ds)(\int_0^t e^{-2as} \|u(s)-v(s)\|_V^2 ds)dt$$

$$\leq Q^2 \ T[(1-e^{-2aT})/2a] \|u-v\|_a^2$$

$$\leq (Q^2T/2a) \|u-v\|_a^2$$

The condition of uniform strong monotonicity on $M(t,\cdot)$ gives $\langle Mu-Mv,u-v\rangle \geq c\|u-v\|_a^2$, $u,v \in V_a$, hence we obtain $\langle Tu-Tv,u-v\rangle \geq [c-Q(T/2a)^{1/2}]\|u-v\|_a$, $u,v \in V_a$. The desired result holds for $a > TQ^2/2c^2$. QED

Theorem 3.12 Let the hypotheses of Lemma 3.11 hold and assume in addition that we are given a family of nonempty closed convex subsets $K(t)$ of V for which the corresponding projections $P(t) : V \to K(t)$ are a measurable family in $L(V)$ (cf. Carroll [14]). Then there exists a unique absolutely continuous $u \in V_0 = L^2(0,T;V)$ with $u' \in V_0$, $u(o) = u_0$, and satisfying (3.16).

Proof: With a chosen as in Lemma 3.11, the operator $T : V_a \to V_a'$ is hemicontinuous and strongly monotone. Define $K \equiv \{v \in V_a : v(t) \in K(t), \text{ a.e. } t \in [0,T]\}$. Then K is closed, convex and nonempty in V_a, so it follows from Browder [5] (cf. Carroll [14]) that there exists a unique $w \in K$ such that

$$\text{(3.19)} \qquad \langle Tw, v-w \rangle \geq 0, \qquad\qquad v \in K.$$

Finally, the measurable family of projections $\{P(t)\}$ is used to show that (3.19) is equivalent to (3.16) with $u(t) = u_0 + \int_0^t w(s)ds$, $0 \leq t \leq T$. QED

The preceding result is far from being best possible, but serves to illustrate the applicability of the theory of monotone nonlinear operators to the situation. A similar result holds for the doubly-nonlinear equation

$$\text{(3.20)} \qquad \frac{d}{dt} M(t,u(t)) = f(t,u(t))$$

and more generally, the inequality

$$(3.21) \qquad \langle M(t,u(t)) - u_0 - \int_0^t f(s,u(s))ds,\ v-u(t)\rangle \geq 0,\ v \in K(t),$$
$$u(t) \in K(t), \qquad\qquad \text{a.e. } t \in [0,T],$$

under essentially the same hypotheses as Theorem 3.12. It would be desirable to have such results on either (3.16) or (3.21) with $M(t,\cdot)$ permitted to be degenerate, or at least not uniformly bounded (cf. Remark 3.4). We shall return to similar problems in Banach space with a single linear operator (possibly degenerate) in Section 3.6 below. Also, a variation on Theorem 3.8 (using Theorem 3.12) gives existence results for second order evolution equations and inequalities.

Remark 3.13 Theorems 3.1, 3.2 and 3.3 are from Showalter [3] and Theorem 3.4 is from Showalter [18, 19]. Theorem 3.8 was unpublished and Theorem 3.12 is due to Kluge and Bruckner [1]. For additional material on specific partial differential equations and on general evolution equations of the type considered in this section we cite the above references and Amos [1], Bardos-Brezis-Brezis [2], Bhatnager [1], Bochner-von Nuemann [1], Bona [1], Brill [1], Brown [1], Calvert [1], Coleman-Duffin-Mizel [2], Colton [1, 2, 3], Davis [1, 2], Derguzov [1], Dunninger-Levine [1], Eskin [1], Ewing [1, 2, 3], W. H. Ford [1, 2], Fox [3], Gajewski-Zacharius [1, 2, 3], Galpern [1, 2, 3, 4, 5], Grabmuller [1], Horgan-Wheeler [1], Ilin [1], Kostyučenko-Eskin [1], Lagnese [1, 2, 3, 4, 5, 6, 7, 8, 9], Lebedev [1], Levine [3, 5, 6, 7, 8, 9], Lezhnev [1], Lions [5, 9, 10], Maslennikova [1, 2, 3, 4, 5,

6], Medeiros-Menzala [1], Mikhlin [2], Miranda [1], Neves [1], Prokopenko [1], Rao [1, 2, 4, 5], Rundell [1, 2, 3], Selezneva [1], Showalter [1, 7, 8, 9, 11, 12, 13, 14], Sigillito [1, 2, 3], Stecher-Rundell [1], Ting [1, 3], Ton [1], Wahlbin [1], Zalenyak [1, 2, 3, 4]. One could also include certain references from Remarks 4.10, 5.18, 6.24 and 7.10 to which we refer for additional material.

3.4 Weakly regular equations We restrict our attention to a class of linear evolution equations of the form

$$(4.1)\qquad Mu'(t) + Lu(t) = f(t), \qquad t > 0,$$

and obtain well-posedness results when M is not (necessarily) as strong as L. Although the semigroup techniques employed here will be used for nonlinear degenerate problems in Section 3.6 below, we introduce them in the simpler situation of (4.1) where we shall be able to distinguish the analytic situation from the strongly-continuous case.

Let V_m be a complex Hilbert space with scalar-product $(\cdot,\cdot)_m$ and define the isomorphism (of Riesz) from V_m onto its antidual V_m' of conjugate-linear continuous functionals by $\langle Mx,y\rangle \equiv (x,y)_m$, $x,y \in V_m$. Suppose D is a subspace of V_m and the linear map $L : D \to V_m'$ is given. If $u_0 \in V_m$ and $f \in C((0,\infty),V_m')$ are given, we consider the problem of finding a $u(\cdot) \in C([0,\infty), V_m) \cap C^1((0,\infty), V_m)$ which satisfies (4.1) and $u(0) = u_0$.

To obtain an elementary uniqueness result, let $u(\cdot)$ be a solution of (4.1) with $f \equiv 0$ and note that $D_t(u(t),u(t))_m = -2\text{Re}\langle Lu(t),u(t)\rangle$, $t > 0$. Hence, if L is monotone then $\|u(t)\|_m = (u(t), u(t))_m^{1/2}$ is nonincreasing and a uniqueness result is obtained. Recall that L monotone means $\text{Re}\langle Lx,x\rangle \geq 0$, $x \in D$. This computation suggests that V_m is an appropriate space in which to seek well-posedness results. The equation in V_m

$$(4.2) \qquad u'(t) + M^{-1} \circ Lu(t) = M^{-1}f(t), \qquad t > 0$$

is equivalent to (4.1) and suggests consideration of the operator $A = M^{-1} \circ L$ with domain $D(A) = D$. We see from

$$(4.3) \qquad (Ax,y)_m = \langle Lx,y\rangle , \qquad x \in D,\ y \in V_m$$

that L is monotone if and only if A is accretive in the Hilbert space V_m. In this case, $-A$ generates a contraction semigroup on V_m if $I + A$ is surjective; then the Cauchy problem is well-posed for (4.2). These observations prove the following (cf. Carroll [14], Hille [2], or Kato [1]).

<u>Theorem 4.1</u> Let $M : V_m \to V_m'$ be the Riesz map of the complex Hilbert space V_m, $L : D \to V_m'$ be given monotone and linear with domain in V_m, and assume $M + L : D \to V_m'$ is surjective. Then for each $f \in C([0,\infty),V_m')$ and $u_0 \in D$ there is a unique solution $u(\cdot)$ of (4.1) with $u(o) = u_0$.

The Cauchy problem for (4.1) is solved above by a strongly continuous semigroup of contractions. This semigroup is analytic when the operator A is sectorial (Kato [1]) and we describe this situation in the following.

Theorem 4.2 Suppose M is the Riesz map of the Hilbert space V_m, V is a Hilbert space dense and continuous in V_m, and $L : V \to V'$ is continuous, linear and coercive: $\text{Re}\langle Lx,x\rangle \geq c\|x\|_V^2$, $x \in V$, for some $c > 0$. Then for every Hölder continuous $f: [0,\infty) \to V_m'$ and $u_0 \in V_m$ there is a unique, solution $u(\cdot)$ of (4.1) with $u(o) = u_0$.

Each of the two preceding results implies well-posedness of the Cauchy problem for a linear second-order evolution equation

$$(4.4) \qquad Cu''(t) + Bu'(t) + Au(t) = f(t), \qquad t > 0.$$

The idea is to write (4.4) as a first-order system in the form

$$(4.5) \qquad \begin{pmatrix} A & 0 \\ 0 & C \end{pmatrix} \begin{pmatrix} u \\ v \end{pmatrix}' + \begin{pmatrix} 0 & -A \\ A & B \end{pmatrix} \begin{pmatrix} u \\ v \end{pmatrix} = \begin{pmatrix} 0 \\ f(t) \end{pmatrix}$$

Theorem 4.3 Let V_a and V_c be complex Hilbert spaces with V_a dense and continuous in V_c, and denote by $A : V_a \to V_a'$ and $C : V_c \to V_c'$ the respective Riesz maps. Let B be linear and monotone from $D(B) \subset V_a$ into V_a' and assume $A + B + C : D(B) \to V_a'$ is surjective. Then for each $f \in C^1([0,\infty), V_c')$ and pair $u_0 \in V_a$, $u_1 \in D(B)$ with $Au_0 + Bu_1 \in V_c'$ there is a unique solution $u(\cdot) \in C([0,\infty),V_a) \cap C^1((0,\infty),V_a) \cap C^1([0,\infty),V_c) \cap C^2((0,\infty),V_c)$ of (4.4) with $u(o)) = u_0$ and $u'(o) = u_1$.

Proof: Define $V_m = V_a \times V_c$; then we have $V_m' = V_a' \times V_c'$ and the Riesz map $M : V_m \to V_m'$ is given by $M[x_1,x_2] = [Ax_1,Cx_2]$, $[x_1,x_2] \in V_m$. Let $D \equiv \{[x_1,x_2] \in V_a \times D(B) : Ax_1 + Bx_2 \in V_c'\}$, recalling $V_c' \subset V_a'$, and define $L : D \to V_m'$ by $L([x_1,x_2]) = [-Ax_2,Ax_1 + Bx_2]$, $[x_1,x_2] \in D$. With M and L so defined, the system (4.5) is clearly equivalent to $Mw'(t) + Lw(t) = [0,f(t)]$, $t > 0$. In order to apply Theorem 4.1 we need only to verify that L is monotone and M + L is surjective. But an easy computation shows $\mathrm{Re}\langle L([x_1,x_2]), [x_1,x_2]\rangle = \mathrm{Re}\langle Bx_2,x_2\rangle$ so B monotone implies L is monotone. Finally, if $f_1 \in V_a'$ and $f_2 \in V_c'$, then we can solve $(A+B+C)x_2 = f_2 - f_1$, $x_2 \in D(B)$, then set $x_1 = x_2 + A^{-1}f_1$ to obtain $[x_1,x_2] \in D$ with $(M+L)[x_1,x_2] = [f_1,f_2]$. QED

Theorem 4.4 Let the Hilbert spaces and operators $A: V_a \to V_a'$ and $C : V_c \to V_c'$ be given as in Theorem 4.3. Let $B : V_a \to V_a'$ be continuous and linear with $B + \lambda C : V_a \to V_a'$ coercive for some $\lambda > 0$. Then for every Hölder continuous $f : [0,\infty) \to V_c'$ and pair $u_0 \in V_a$, $u_1 \in V_c$, there is a unique solution $u(\cdot) \in C([0,\infty),V_a) \cap C^1((0,\infty),V_a) \cap C^1([0,\infty),V_c) \cap C^2((0,\infty),V_c)$ of (4.4) with $u(o) = u_0$ and $u'(o) = u_1$.

Proof: Following the proof of Theorem 4.3, we find that we can apply Theorem 4.2 with $V \equiv V_a \times V_a$ if we verify that $\lambda M + L$ is V-elliptic for some $\lambda > 0$. But $\lambda B + C$ being V_a-elliptic is precisely what is needed. QED

Briefly we indicate some examples of partial differential equations for which corresponding initial-boundary value problems are solved by the preceding results. The examples are far from best possible in any sense and are displayed here only to suggest the types of equations to which the results apply.

Example 4.5 Let $m(\cdot) \in L^{\infty}(0,1)$ with $m(x) > 0$ for a.e. $x \in (0,1)$ and define V_m to be the completion of $C_0^{\infty}(0,1)$ with the scalar-product $(u,v)_m \equiv \int_0^1 m(x)u(x)\overline{v(x)}dx$. Let $V = H_0^1(0,1)$ and set $\langle Lu,v\rangle \equiv \int_0^1 u'(x)\overline{v'(x)}dx$, $u,v \in V$. Then Theorem 4.2 shows that the initial-boundary value problem

$$
(4.6) \qquad \begin{cases} m(x)D_t u(x,t) - D_x^2 u(x,t) = f(x,t), & t > 0, \\ u(0,t) = u(1,t) = 0, & \\ u(x,0) = u_0(x), & 0 < x < 1, \end{cases}
$$

is well-posed for appropriate $f(\cdot,\cdot)$ and measurable u_0 with $m^{1/2}u_0 \in L^2(0,1)$. The initial condition above means $\lim_{t\to 0^+} \int_0^1 m(x)|u(x,t)-u_0(x)|^2 dx = 0$ and the equation holds at each $t > 0$ in the space V_m' of measurable functions v on $(0,1)$ with $m^{-1/2}v \in L^2(0,1)$. Note that $V_m' \subset L^2(0,1) \subset V_m$ and the equation may be elliptic on a null set (where $m(x) = 0$).

Example 4.6 Let $V_m = H_0^1(0,1)$ with the scalar-product $(u,v)_m = \int_0^1 \{u(x)\overline{v(x)} + mu'(x)\,\overline{v'(x)}\}dx$ where $m > 0$. (More general coefficients could be added as above.) Set $Lu = D_x^3 u$ on $D \equiv \{u \in H^2(0,1) \cap H_0^1(0,1) : c \cdot u'(0) = u'(1)\}$ where c is given with $|c| \leq 1$. Then $2\,\mathrm{Re}\langle Lu,u\rangle = |u'(0)|^2 - |u'(1)|^2 \geq 0, u \in D$, so L is monotone; we can show $M + L$ is surjective, so Theorem 4.1 shows the problem

$$(4.7)\qquad \begin{cases} (D_t - mD_x^2 D_t)u(x,t) + D_x^3 u(x,t) = f(x,t), & 0 < x < 1, \\ u(0,t) = u(1,t) = 0, \quad c \cdot u_x(0,t) = u_x(1,t), & t \geq 0, \\ u(x,o) = u_o(x) & \end{cases}$$

is well-posed for appropriate $f(\cdot,\cdot)$ and u_o. This equation is a linear regularized Korteweg-deVries equation (cf. Bona [4]).

<u>Example 4.7</u> Take V_m as above and let $V = H_0^2(0,1)$ with $L = D_x^4 : V \to V'$ defined by $\langle Lu,v\rangle = \int_0^1 D_x^2 u(x)\overline{D_x^2 v(x)}dx$, $u,v \in V$. Then Theorem 4.2 gives existence-uniqueness results for the meta-parabolic (Brown [1]) problem

$$(4.8)\qquad \begin{cases} (D_t - mD_x^2 D_t)u(x,t) + D_x^4 u(x,t) = f(x,t), & 0 < x < 1, \\ u(0,t) = u(1,t) = u_x(0,t) = u_x(1,t) = 0, & t > 0, \\ u(x,o) = u_o(x), & \end{cases}$$

when $u_o \in H_0^1(0,1)$ and $t \to f(\cdot,t) : [0,\infty) \to H^{-1}(0,1)$ is Hölder continuous.

<u>Example 4.8</u> Let $V_a = H_0^1(0,1)$ with $\langle Au,v\rangle = \int_0^1 u'\overline{v'}dx$, $V_c = L^2(0,1)$ with $\langle Cu,v\rangle = \int_0^1 u\bar{v}dx$; positive coefficients could be added as in Example 4.5. Let $r \in \mathbb{C}$, $b \in \mathbb{R}$ and define $B: V_a \to V_c$ by $Bu = ru + bu'$, $u \in V_a$. Since $\mathrm{Re}\langle Bu,u\rangle = \mathrm{Re}(r) \cdot \|u\|_{V_c}^2$, B is monotone whenever $\mathrm{Re}(r) \geq 0$. The operator $A + B + C : V_a \to V_a'$ is continuous, linear and coercive, hence surjective, so Theorem 4.3 shows the problem

$$(4.9)\quad \begin{cases} D_t^2u(x,t)+rD_tu(x,t)+bD_xD_tu(x,t)-D_x^2u(x,t) = f(x,t), & \\ u(0,t) = u(1,t) = 0, & t > 0, \\ u(x,0) = u_0(x),\ D_tu(x,0) = u_1(x),\ 0 < x < 1, & \end{cases}$$

is well posed with $u_0 \in H_0^1 \cap H^2(0,1)$, $u_1 \in H_0^1(0,1)$ and $t \to f(\cdot,t) \in C^1([0,\infty),L^2(0,1))$. This is a classical problem for the damped wave equation. If instead of the above choice for B we set $B = \varepsilon A$, $\varepsilon > 0$, then from Theorem 4.4 follows well-posedness for the "parabolic" problem

$$(4.10)\quad \begin{cases} D_t^2u(x,t)-\varepsilon D_x^2D_tu(x,t)-D_x^2u(x,t) = f(x,t), & 0 < x < 1, \\ u(0,t) = u(1,t) = 0, & t > 0, \\ u(x,0) = u_0(x),\quad D_tu(x,0) = u_1(x), & \end{cases}$$

of classical viscoelasticity (cf. Albertoni-Cercignani [1]). Here the data is chosen with $u_0 \in H_0^1(0,1)$, $u_1 \in L^2(0,1)$, and $t \to f(\cdot,t) : [0,\infty) \to L^2(0,1)$ Hölder continuous.

We return to consider the parabolic situation of Theorem 4.2 and describe abstract regularity results on the solution. In addition to the hypotheses of Theorem 4.2, assume given a Hilbert space H which is identified with its dual and in which V_m is dense and continuously embedded. Thus we have the inclusions $V \subset V_m \subset H = H' \subset V_m' \subset V'$. Let M^1 be the (unbounded) operator on H obtained as the restriction of $M : V_m \to V_m'$ to the domain $D(M^1) = \{u \in V_m : Mu \in H\}$. Similarly, the restriction of $L : V \to V'$ to the domain $D(L^1) = \{u \in V : Lu \in H\}$ will be denoted by L^1. Each of $D(M^1)$ and $D(L^1)$ is a Hilbert space with the induced graph-norm and they are dense and continuous in V_m and V, respectively. We shall describe a sufficient condition for the solution $u(\cdot)$ of (4.1) with $f(\cdot) = 0$ to belong to $C([0,\infty),V_m) \cap C^\infty((0,\infty),D(L^1))$. The arguments depend on the theory of interpolation spaces and fractional powers of operators (cf. Carroll [14] for references). Specifically, the operator L^1 is regularly accretive and corresponding fractional powers L^θ, $0 \le \theta \le 1$, can be defined. Their corresponding domains are related to interpolation spaces between $D(L^1)$ and H by the identities $D(L^{1-\theta}) = [D(L),H;\theta]$, $0 \le \theta \le 1$, and corresponding norms are equivalent,

<u>Theorem 4.9</u> In addition to the hypotheses of Theorem 4.2, we assume $D(L^{1-\theta}) \subset D(M^1)$ for some θ, $0 < \theta \le 1$. Then for each $u_0 \in V_m$ the solution $u(\cdot)$ of (4.1) with $f(\cdot) \equiv 0$ belongs to $C^\infty((0,\infty),D(L^1))$.

<u>Remark 4.10</u> Theorem 4.1 is new in the form given. The closely related Theorem 4.2 is from Showalter [6]; cf. Showalter [7, 11] for an earlier special case. Theorem 4.3 is the linear case of a result in Showalter [5] and Theorem 4.4 is new. Theorem 4.9 is presented in Showalter [6, pt. II] with a large class of applications to initial-boundary value problems. The additional hypothesis in Theorem 4.9 makes L strictly stronger than M; the result is false in general when $\theta = 0$. For related work we refer to Coirier [1], Krein [1], Mikhlin [2], Phillips [1] and Showalter [17].

3.5 Linear degenerate equations. We shall consider appropriate Cauchy problems for the equation

$$\frac{d}{dt}(M(t)u(t)) + L(t)u(t) = f(t) \tag{5.1}$$

whose coefficients $\{M(t)\}$ and $\{L(t)\}$ are bounded and measurable families of linear operators between Hilbert spaces. We show (essentially) that the problem is well-posed when the $\{M(t)\}$ are nonnegative, $\{L(t)+\lambda M(t)\}$ are coercive for some $\lambda > 0$, and the operators depend smoothly on t. Some elementary applications to initial-boundary value problems are presented in order to indicate the large class of problems to which the results can be applied (cf. Examples 1.1, 1.4 and 1.6). Consideration of the case of (5.1) on a product space leads to a parabolic system which contains the second order equation (cf. Example 1.2)

$$\frac{d^2}{dt^2}(C(t)u(t)) + \frac{d}{dt}(B(t)u(t)) + A(t)u(t) = f(t). \tag{5.2}$$

Certain higher order equations and systems can be handled similarly.

In order to describe the abstract Cauchy problem, let V be a separable complex Hilbert space with norm $\|v\|$; V' is the antidual, and the antiduality is denoted by $\langle f,v\rangle$. W is a complex Hilbert space containing V and the injection is assumed continuous with norm ≤ 1. $\mathcal{L}(V,W)$ denotes the space of continuous linear operators from V into W and T is the unit interval [0,1]. Assume that for each $t \in T$ we are given a continuous sesquilinear form

$\ell(t;\cdot,\cdot)$ on V and that for each pair $x,y \in V$ the map $t \to \ell(t;x,y)$ is bounded and measurable. By uniform boundedness there is a number $K_\ell > 0$ such that $|\ell(t;x,y)| \leq K_\ell \|x\| \cdot \|y\|$ for all $x,y \in V$ and $t \in T$. Standard measurability arguments then show that for any pair $u,v \in L^2(T,V)$ the function $t \to \ell(t;u(t),v(t))$ is integrable (cf. Carroll [14], p. 168). Similarly, we assume given for each $t \in T$ a continuous sesquilinear form $m(t;\cdot,\cdot)$ on W such that for each pair $x,y \in W$ the map $t \to m(t;x,y)$ is bounded and measurable. Let $\{V(t) : t \in T\}$ be a family of closed subspaces of V and denote by $L^2(T,V(t))$ the Hilbert space consisting of those $\phi \in L^2(T,V)$ for which $\phi(t) \in V(t)$ a.e. on T. Finally, let $u_0 \in W$ and $f \in L^2(T,V')$ be given. A solution of the Cauchy problem (determined by the preceding data) is a $u \in L^2(T,V(t))$ such that

$$(5.3) \qquad \int_0^1 \ell(t;u(t),v(t))dt - \int_0^1 m(t;u(t),v'(t))dt$$

$$= \int_0^1 <f(t),v(t)> dt + m(0;u_0,v(o))$$

for all $v \in L^2(T,V(t)) \cap H^1(T,W)$ with $v(1) = 0$.

A family $\{a(t;\cdot,\cdot) : t \in T\}$ of continuous sesquilinear forms on V is called regular if for each pair $x,y \in V$ the function $t \to a(t;x,y)$ is absolutely continuous and there is an $M(\cdot) \in L^1(T)$ such that for $x,y \in V$ we have

$$(5.4) \qquad |a'(t;x,y)| \leq M(t)\|x\| \|y\|, \qquad \text{a.e.} \quad t \in T.$$

The following "chain rule" will be used below.

Lemma 5.1 Let $\{a(t;\cdot,\cdot) : t \in T\}$ be a regular family on V. Then for each pair $u,v \in H^1(T,V)$ the function $t \to a(t;u(t),v(t))$ is absolutely continuous and its derivative is given by $D_t a(t;u(t),v(t)) = a'(t;u(t),v(t)) + a(t;u'(t),v(t)) + a(t;u(t),v'(t))$, a.e. $t \in T$.

Proof: Define $\alpha(t) \in L(V,V)$ by $(\alpha(t)x,y)_V = a(t;x,y)$, $x,y \in V$. Fix $x \in V$ and let $\{ y_n : n \geq 1\}$ be dense in V. For each $n \geq 1$ define $(\dot{\alpha}(t)x,y_n)_V = D_t(\alpha(t)x,y_n)_V$, a.e. $t \in T$. (The estimate (5.4) shows $(\dot{\alpha}(t)x,y)_V$ is defined and continuous at every $y \in V$ and a.e. $t \in T$.) The map $\dot{\alpha}(\cdot)x$ is weakly, hence strongly, measurable and the estimate (5.4) shows it is in $L^1(T,V)$. The weak absolute continuity of $t \to \alpha(t)x$ then shows $\alpha(t)x = \alpha(o)x + \int_0^t \dot{\alpha}(s)x\, ds$, $t \in T$. Thus $\alpha(\cdot)x$ is strongly absolutely continuous and strongly differentiable a.e. on T.

Let $u \in H^1(T,V)$. For each $v \in V$ we have $(\alpha(t)u(t),v)_V = (u(t),\alpha^*(t)v)_V$ absolutely continuous, since the above discussion applies as well to the adjoint $\alpha^*(t)$. Hence, $t \to \alpha(t)u(t)$ is weakly absolutely continuous. The strong differentiability of $\alpha(\cdot)$ from above implies $D_t[\alpha(t)u(t)] = \dot{\alpha}(t)u(t) + \alpha(t)u'(t)$, a.e. $t \in T$, hence the indicated strong derivative is in $L^1(T,V)$. From this it follows that $\alpha(\cdot)u(\cdot)$ is strongly absolutely continuous and the desired result now follows easily. QED

Our first result gives existence of a solution with an a priori estimate.

Theorem 5.2 Let the Hilbert spaces $V(t) \subset V \subset W$, sesquilinear forms $m(t;\cdot,\cdot)$ and $\ell(t;\cdot,\cdot)$ on W and V, respectively, $u_0 \in W$ and $f \in L^2(T,V')$ be given as above. Assume that $\{m(t;\cdot,\cdot) : t \in T\}$ is a regular family of Hermitian forms on W: $m(t;x,y) = \overline{m(t;y,x)}$, $x,y \in W$, a.e. $t \in T$, and $m(o;x,x) \geq 0$ for $x \in W$. Assume that for some real λ and $c > 0$

$$(5.5) \qquad \mathrm{Re}\,\ell(t;x,x) + \lambda m(t;x,x) + m'(t;x,x) \geq c\|x\|_V^2, \qquad x \in V(t), \quad \text{a.e. } t \in T.$$

Then there exists a solution u of the Cauchy problem, and it satisfies $\|u\|_{L^2(T,V)} \leq \text{const.}(\|f\|^2_{L^2(T,V')} + m(o;u_0,u_0))^{1/2}$, where the constant depends on λ and c.

Proof: Note first that by a standard change-of-variable argument we may replace $\ell(t;\cdot,\cdot)$ by $\ell(t;\cdot,\cdot) + \lambda m(t;\cdot,\cdot)$. Therefore we may assume without loss of generality that $\lambda = 0$ in (5.5): $2\mathrm{Re}\,\ell(t;x,x) + m'(t;x,x) \geq c\|x\|^2$, $x \in V(t)$, a.e. $t \in T$. Now define $H = L^2(T,V(t))$ with the norm $\|u\|_H^2 = \int_0^1 \|u(t)\|^2 dt$ and let $F \equiv \{\phi \in H : \phi' \in L^2(T,W), \phi(1) = 0\}$ with the norm $\|\phi\|_F^2 = \|\phi\|_H^2 + m(o;\phi(o),\phi(o))$. For $u \in H$ and $\phi \in F$ we define $E(u,\phi) = \int_0^1 \ell(t;u(t),\phi(t))dt - \int_0^1 m(t;u(t),\phi'(t))dt$ and $L(\phi) = \int_0^1 \langle f(t),\phi(t)\rangle dt + m(o;u_0,\phi(o))$. Then $E : H \times F \to \mathbb{C}$ is sesquilinear, $L : F \to ₵$ is conjugate-linear, and we have the estimates $|E(u,\phi)| \le K_\ell \|u\|_H \|\phi\|_H + K_m \|u\|_H \|\phi'\|_{L^2(T,W)}$ and $|L(\phi)| \le \|f\|_{L^2(T,V')} \cdot \|\phi\|_H + m(o;u_o,u_o)^{1/2} m(o;\phi(o),\phi(o))^{1/2} \le (\|f\|^2_{L^2(T,V')} + m(o;u_o,u_o))^{1/2} \|\phi\|_F$. These imply that $u \to E(u,\phi)$ is continuous $H \to ₵$ for each ϕ in F and that $L : F \to ₵$ is continuous.

Finally, for ϕ in F we have from Lemma 5.1 and 5.5 with $\lambda = 0$, $2\mathrm{Re}E(\phi,\phi) = \int_0^1 2\mathrm{Re}\ell(t;\phi(t),\phi(t))dt + \int_0^1 \{m'(t;\phi(t),\phi(t)) - D_t m(t;\phi(t),\phi(t))\}dt \ge c\|\phi\|_H^2 + m(o;\phi(o),\phi(o)) \ge \min(c,1)\|\phi\|_F^2$. These estimates show that the projection theorem of Lions [5, p. 37] applies here. Thus, there is a $u \in H$ such that $E(u,\phi) = L(\phi)$ for all $\phi \in F$ and it satisfies $\|u\|_H \le (2/\min(c,1))\|L\|_{F'}$. But this is precisely the desired result. QED

Remark 5.3 The hypotheses of Theorem 5.2 not only permit $m(t;x,x)$ to vanish but also to actually take on negative values. This will be illustrated in the examples.

Remark 5.4 An easy estimate shows that (5.5) holds (with λ replaced by $2\lambda + \alpha$) if we assume that for some $c > 0$ and $\alpha \ge 0$

(5.6) $$m'(t;x,x) + \alpha m(t;x,x) \ge 0, \qquad x \in V(t), \qquad \text{a.e. } t \in T,$$

(5.7) $$\mathrm{Re}\ell(t;x,x) + \lambda m(t;x,x) \ge c\|x\|_V^2.$$

This pair of conditions is thus stronger than (5.5) but occurs frequently in applications. Note that (5.6) and $m(o;x,x) \ge 0$ imply $m(t;x,x) \ge 0$ for all $t \in T$.

We present an elementary uniqueness result with a rather severe restriction of monotonicity on the subspaces $\{V(t)\}$. This hypothesis is clearly "ad hoc" and is a limitation of the technique which is not indicative of the best possible results. In particular, much better uniqueness and regularity results for the very special case of $m(t;x,x) = (x,x)_W$ have been obtained (cf. Carroll [14], Carroll-Cooper [35], Carroll-State [36], Lions [5]); the novelty here is in the forms $m(t;\cdot,\cdot)$ determining the leading (possibly degenerate) operators $\{M(t)\}$.

Theorem 5.5 Let the Hilbert spaces $V(t) \subset V \subset W$, sesquilinear forms $\ell(t;\cdot,\cdot)$ and $m(t;\cdot,\cdot)$ on V and W, respectively, $u_0 \in W$ and $f \in L^2(T,V')$ be given as above. Assume that $\text{Re}\, m(t;x,x) \geq 0$, $x \in V(t)$, a.e. $t \in T$, that $\{\ell(t;\cdot,\cdot) : t \in T\}$ is a regular family of Hermitian forms on $V(t)$:

$$(5.8) \qquad \ell(t;x,y) = \overline{\ell(t;y,x)}, \qquad x,y \in V(t), \quad \text{a.e. } t \in T$$

and for some real λ and $c > 0$, $\ell(t;x,x) + \lambda \text{Re}\, m(t;x,x) \geq c\|x\|^2$, $x \in V(t)$, a.e. $t \in T$. Finally, assume that the family of subspaces $\{V(t) : t \in T\}$ is decreasing:

$$(5.9) \qquad t > \tau, \quad t,\tau \in T \text{ imply } V(t) \subset V(\tau).$$

Then there is at most one solution of the Cauchy problem.

Proof: Let $u(\cdot)$ be a solution of the Cauchy problem with $u_0 = 0$ and $f(\cdot) = 0$. By linearity it suffices to show that

$u(\cdot) = 0$. Let $s \in (0,1)$ and define $v(t) = -\int_t^s u(\tau)d\tau$ for $t \in [o,s]$ and $v(t) = 0$ for $t \in [s,1]$. Then $v \in L^2(T,V)$ and (5.9) shows $v(t) \in V(t)$ for each $t \in T$. Also, $v'(t) = -u(t)$ for $t \in (o,s)$ and $v'(t) = 0$ for $s \in (0,1)$, so we have $v' \in L^2(T,V(t)) \subset L^2(T,W)$ and $v(1) = 0$. Since $u(\cdot)$ is a solution we have by (5.3) $\int_0^s \ell(t;v'(t),v(t))dt - \int_0^s m(t;u(t),u(t))dt = 0$. Lemma 5.1 and (5.8) give us the identities

$$\int_0^s 2\mathrm{Re}\, m(t;u(t),u(t))dt = \int_0^s \{D_t\ell(t;v(t),v(t)) - \ell'(t;v(t),v(t))\}dt,$$

$$\int_0^s \{2\mathrm{Re}\, m(t;u(t),u(t)) + \ell'(t;v(t),v(t))\}dt + \ell(o;v(o),v(o)) = 0.$$

As before, we may assume $\ell(t;x,x) \geq c\|x\|^2$, $x \in V(t)$, $t \in T$. Define $W(t) = \int_0^t u(\tau)d\tau$; then $W(s) = -v(o)$ and $W(t) - W(s) = v(t)$ for $t \in (o,s)$ and we obtain the estimate

$$c\|W(s)\|^2 \leq \ell(o;W(s),W(s)) + \int_0^s 2\mathrm{Re}\, m(t;u(t),u(t))dt$$

$$= -\int_0^s \ell'(t;v(t),v(t))dt \leq \int_0^s M(t)\|v(t)\|^2 dt$$

$$\leq 2\int_0^s M(t)\{\|W(t)\|^2 + \|W(s)\|^2\}dt.$$

Choose $s_0 > 0$ so that $2\int_0^{s_0} M(t)dt < c$. Then for $s \in [o,s_0]$ we have $\|W(s)\|^2 \leq \{2/(c - 2\int_0^{s_0} M(t)dt)\}\int_0^s M(t)\|W(t)\|^2 dt$. From the Gronwall inequality we conclude that $W(s) = 0$ for $s \in [o,s_0]$, hence $u(s) = 0$ for $s \in [o,s_0]$. Since $M(\cdot)$ is integrable on T we could use the absolute continuity of the integral $\int M(t)dt$ to choose $s_0 > 0$ in the above so that $\int_\tau^{\tau+s_0} M(t)dt < c/2$ for every $\tau \geq 0$ with $\tau + s_0 \leq 1$. Then we apply the above argument a finite number of times to obtain $u(\cdot) = 0$ on T. QED

A fundamental problem with degenerate evolution equations is to determine the precise sense in which the initial condition is attained. A consideration of the two cases of $m(t,x,y) \equiv (x,y)_W$ and $m(t;x,y) \equiv 0$ shows (see below) that we may have the case of $\lim_{t\to 0^+} u(t) = u_0$ in an appropriate space or, respectively, that $\lim_{t\to 0^+} u(t)$ does not necessarily exist in any sense. These remarks illustrate the importance of the following.

<u>Theorem 5.6</u> Let $u(\cdot)$ be a solution of the Cauchy problem and assume the subspaces $\{V(t) : t \in T\}$ are initially decreasing: there is a $t_0 \in (0,1]$ such that (5.9) holds for $t,\tau \in [0,t_0]$. Then for every $\tau \in (0,t_0)$ and $v \in V(\tau)$ we have $t \to m(t;u(t),v)$ is in $H^1(0,\tau)$ (hence is absolutely continuous with distribution derivative in $L^2(0,\tau)$) and $m(0;u(0) - u_0,v) = 0$. Thus, if $\bigcup\{V(\tau) : \tau > 0\}$ is dense in W, then $m(o;u(o) - u_0,u(o) - u_0) = 0$.

Proof: Define $v(t) = \overline{\phi(t)v}$ for $t \in (0,\tau)$ and $v(t) = 0$ for $t \in [\tau,1]$. Then $v(\cdot) \in L^2(T,V(t)) \cap H^1(T,W)$ and $v(1) = 0$, so
$$\int_0^\tau \ell(t;u(t),v)\phi(t)dt - \int_0^\tau m(t;u(t),v)\phi'(t)dt = \int_0^\tau \langle f(t),v\rangle\phi(t)dt.$$
That is, we have the identity

$$(5.10) \qquad \ell(\cdot;u(\cdot),v) + D_t m(\cdot;u(\cdot),v) = \langle f(\cdot),v\rangle$$

in the space of distributions on (o,τ) for every $v \in V(\tau)$. The first and last terms of (5.10) are in $L^2(o,\tau)$ and so, then, is the second and we thus have pointwise values a.e. in (5.10) and hence the equation

$$(5.11) \qquad \int_0^\tau \ell(t;u(t),v)g(t)dt + \int_0^\tau D_t m(t;u(t),v)g(t)dt$$

$$= \int_0^\tau \langle f(t),v\rangle g(t)dt$$

for $g \in L^2(o,\tau)$ and $v \in V(\tau)$.

Suppose now that $\phi \in H^1(o,\tau)$ is given with $\phi(\tau) = 0$. Define $v(\cdot)$ as above so as to obtain from (5.3) $\int_0^\tau \ell(t;u(t),v)\phi(t)dt - \int_0^\tau m(t;u(t),v)\phi'(t)dt = \int_0^\tau \langle f(t),v\rangle\phi(t)dt + m(o;u_0,v)\phi(o)$ where $v \in V(\tau)$. Since (5.11) holds with $g(t) = \phi(t)$, we obtain from these $\int_0^\tau \{D_t m(t;u(t),v)\phi(t) + m(t;u(t),v)\phi'(t)\}dt = -m(o;u_0,v)\phi(o)$. The integrand is the derivative of the function $t \to m(t;u(t),v)\phi(t)$ in $H^1(o,\tau)$ and $\phi(\tau) = 0$, so the desired result follows. QED

Another situation which arises frequently in applications and to which the above technique is applicable is the following.

Theorem 5.7 Let $u(\cdot)$ be a solution of the Cauchy problem and assume there is a closed subspace V_0 in V with $V_0 \subset \bigcap\{V(t) : t \in T\}$. Define the two families of linear operators $\{L(t)\} \subset \mathcal{L}(V,V_0')$ and $\{M(t)\} \subset \mathcal{L}(W,V_0')$ by $\langle L(t)x,y\rangle = \ell(t;x,y)$, $x \in V$, $y \in V_0$, $t \in T$; $\langle M(t)x,y\rangle = m(t;x,y)$, $x \in W$, $y \in V_0$, $t \in T$. Then in the space of V_0'-valued distributions on T we have (5.1) and the function $t \to M(t)u(t) : T \to V_0'$ is continuous.

The proof of Theorem 5.7 follows from (5.10) with $v \in V_0$ and the remark that each term in (5.1), as well as $M(\cdot)u(\cdot)$, is in $L^2(T,V_0')$.

Our last result concerns variational boundary conditions. Suppose we have the situation of Theorem 5.7 and also that (5.9) holds. Let H be a Hilbert space in which V is continuously imbedded and V_0 is dense. We identify H' with H by the Riesz theorem and hence obtain $V_0 \hookrightarrow H \hookrightarrow V_0'$ and the identity $\langle h,v\rangle = (h,v)_H$ for $h \in H$, $v \in V_0$. Assume $f \in L^2(T,H)$, $\tau > 0$ and $v \in V(\tau)$. Then from (5.11) and (5.1) we obtain $\int_0^\tau \ell(t;u(t),v)\phi(t)dt + \int_0^\tau D_t m(t;u(t),v)\phi(t)dt = \int_0^\tau (L(t)u(t) + D_t(M(t)u(t)),v)_H\phi(t)dt$ for each $\phi \in C_0^\infty(o,\tau)$. This gives us the following.

<u>Theorem 5.8</u> Assume the situation of Theorem 5.7 and that (5.9) holds. Let H be given as above and $f \in L^2(T,H)$. Then for each $\tau > 0$ and $v \in V(\tau)$ we have in $L^2(o,\tau)$

$$\ell(t;u(t),v) + D_t m(t;u(t),v) = (L(t)u(t) + D_t M(t)u(t),v)_H. \tag{5.12}$$

We briefly indicate how the preceding theorems can give corresponding results for second order evolution equations which are "parabolic". These results are time dependent analogues of Theorem 4.4.

Theorem 5.9 Let V_a and V_c be separable complex Hilbert spaces with V_a dense and continuously imbedded in V_c. Let $\{a(t;\cdot,\cdot) : t \in T\}$, $\{b(t;\cdot,\cdot) : t \in T\}$ and $\{c(t;\cdot,\cdot) : t \in T\}$ be families of continuous sesquilinear forms on V_a, V_a, and V_c, respectively. Define corresponding families of linear operators $\{A(t)\}$ and $\{B(t)\} \subset \mathcal{L}(V_a,V_a')$ and $\{C(t)\} \subset \mathcal{L}(V_c,V_c')$ as in Theorem 5.7. Assume that $\{a(t;\cdot,\cdot) : t \in T\}$ and $\{c(t;\cdot,\cdot) : t \in T\}$ are regular families of Hermitian forms with $a(o;x,x) \geq 0$ and $c(o;x,x) \geq 0$ for $x \in V_a$. Assume that for some λ and $c > 0$

$\lambda a(t;x,x) + a'(t;x,x) \geq c\|x\|^2_{V_a}$,

$$2\mathrm{Re}\, b(t;x,x) + \lambda c(t;x,x) + c'(t;x,x) \geq c\|x\|^2_{V_a}, \tag{5.14}$$

for each $x \in V_a$ and a.e. $t \in T$. Then for $u_o \in V_a$, $v_o \in V_c$, and $f(\cdot),g(\cdot) \in L^2(T,V_a')$ given, there exists a pair $u(\cdot),v(\cdot) \in L^2(T,V_a)$ which satisfies the system

$$\begin{cases} D_t(A(t)u(t)) - A(t)v(t) = -f(t) \\ D_t(C(t)v(t)) + A(t)u(t) + B(t)v(t) = g(t) \end{cases} \tag{5.15}$$

in the sense of V_a'-valued distributions on T and satisfies the initial conditions $A(o)u(o) = A(o)u_o$, $C(o)v(o) = C(o)v_o$.

Proof: This follows directly from Theorem 5.2 and Theorem 5.7 with $m(t;\bar{x},\bar{y}) \equiv a(t;x_1,y_1) + c(t;x_2,y_2)$ for $\bar{x},\bar{y} \in W \equiv V_a \times V_c$ and $\ell(t;\bar{x},\bar{y}) = a(t;x_1,y_2) - a(t;x_2,y_1) + b(t;x_2,y_2)$ for $\bar{x},\bar{y} \in V \equiv V(t) \equiv V_o \equiv V_a \times V_a$. QED

There are two natural reductions of the first order system (5.15) to a second order equation, and we shall describe these. First, we set $g(\cdot) \equiv 0$ in Theorem 5.9 and note than that both $C(\cdot)v(\cdot)$ and $D_t(C(\cdot)v(\cdot)) + B(\cdot)v(\cdot)$ belong to $H^1(T,V_a')$. Thus $v(\cdot)$ is a solution of the second order equation

$$(5.16) \qquad D_t(D_t(C(t)v(t)) + B(t)v(t)) + A(t)v(t) = f(t)$$

and the initial conditions $(Cv)(o) = C(o)v_o$, $(D_t(Cv) + Bv)(o) = -A(o)u_o$. Second, we set $f(\cdot) \equiv 0$ in Theorem 5.9 and assume $A(t)$ is independent of t, i.e., $A(t) \equiv A$ for some necessarily coercive and Hermitian $A \in L(V_a,V_a')$. From (5.15) it follows that $u(\cdot) \in H^1(T,V_a)$ and is a solution of the second order equation

$$(5.17) \qquad D_t(C(t)u'(t)) + B(t)u'(t) + Au(t) = g(t)$$

and the initial conditions $u(o) = u_o$, $(C(\cdot)u')(o) = C(o)v_o$.

We shall present some elementary applications of the preceding results to boundary value problems for partial differential equations in the form (5.2). These examples will illustrate some limitations on the types of problems to which the theorems apply.

Example 5.10 For our first example we choose $V_o = V(t) = V = H_0^1(T)$, $W = \{\phi \in H^1(T) : \phi(1) = 0\}$ and $H = L^2(T)$. Recall the estimate

$$(5.18) \qquad \sup\{|\phi(x)| : x \in T\} \leq \|\phi'\|_{L^2(T)}, \qquad \phi \in W.$$

Let $\alpha : T \to T$ be absolutely continuous and define the Hermitian forms $m(t;\phi,\psi) = \int_0^{\alpha(t)} \phi(x)\overline{\psi(x)}dx$, $\phi,\psi \in W$, $t \in T$. Then $m(\cdot;\phi,\psi)$ is absolutely continuous and satisfies $m'(t;\phi,\phi) = |\phi(\alpha(t))|^2\alpha'(t)$, $\phi \in W$, a.e. $t \in T$. Since $\alpha' \in L^1(T)$, the estimate (5.18) shows that this is a regular family of forms. For $\phi,\psi \in V$, we set $\ell(t;\phi,\psi) = \int_0^1 \phi'(x)\overline{\psi'(x)}dx$, $t \in T$. From the well-known inequality

$$(5.19) \qquad \pi\|\phi\|_{L^2(T)} \leq \|\phi'\|_{L^2(T)}, \qquad \phi \in H_0^1(T),$$

it follows that the preceding family satisfies (5.7) with $\lambda = 0$. Finally, let $u_0 \in W$ and $F \in L^2(T \times T)$ be given and set $f(t) = F(\cdot,t)$, $t \in T$. Then the hypotheses of Theorem 5.5 (on uniqueness) are satisfied. Suppose additionally there is a number σ, $0 < \sigma < 2$, such that

$$(5.20) \qquad \alpha'(t) \geq \sigma - 2, \qquad \text{a.e. } t \in T.$$

Then (5.18) shows that $2\ell(t;\phi,\phi) + m'(t;\phi,\phi) \geq (2-\sigma)\|\phi'\|^2_{L^2(T)}$, $\phi \in V$, so by (5.19) it follows that (5.5) holds with $\lambda = 0$, hence, the hypotheses of Theorem 5.2 (on existence) are satisfied.

Let $u(\cdot,t) = u(t)$ be the solution of the Cauchy problem. Then $u(x,t)$ is the unique generalized solution of the elliptic-parabolic boundary value problem, $u_t - u_{xx} = F(x,t)$, $0 < x < \alpha(t)$; $-u_{xx} = F(x,t)$, $\alpha(t) < x < 1$; $u(0,t) = u(1,t)$, $t \in T$; $u(x,o) = u_0(x)$, $0 < x < \alpha(o)$. The solution $u(t)$ belongs to $H^2(T)$ for each $t \in T$, so $u(\cdot,t)$ and $u_x(\cdot,t)$ are continuous across the curve $x = \alpha(t)$, $t \in T$.

Remark 5.11 The estimate (5.20) permits $\alpha(\cdot)$ to decrease, but not very rapidly, whereas the estimate (5.6) holds if and only if $\alpha'(t) \geq 0$ a.e. on T. The curve $x = \alpha(t)$ is noncharacteristic a.e. on T; the example $\alpha(t) = 1/2 + [(t - 1/2)/4]^{1/3}$ shows it may actually be characteristic at certain points.

Remark 5.12 The conclusions of Lemma 5.1 hold if we assume only that $a(\cdot;x,y)$ is absolutely continuous for each pair $x,y \in V$. Hence we need not assume an estimate like (5.4) on the family $\{m(t;\cdot,\cdot)\}$ in order to obtain Theorem 5.2. One can prove this assertion as follows: (1) use the closed graph and uniform boundedness theorems to obtain an estimate

$$\int_0^1 |a'(t;x,y)|\,dt \leq K\|x\|\,\|y\|, \qquad x,y \in V; \tag{5.21}$$

(2) approximate u' in $L^2(T,V)$ by simple functions and use the Lebesgue theorem with (5.21) to obtain the result for the special case of constant $v \in V$; approximate v' by step functions and use the results of (1) and (2) to obtain the general result. The details are standard but involve some lengthy computations.

Example 5.13 Our second example is similar but allows the equation to be of Sobolev type in portions of $T \times T$. Choose the spaces and the forms $\ell(t;\cdot,\cdot)$ as before. For $\phi,\psi \in W$, define $m(t;\phi,\psi,) = \int_0^{\alpha(t)} \phi(x)\overline{\psi(x)}\,dx + \int_0^{\beta(t)} \phi'(x)\overline{\psi'(x)}\,dx$, $t \in T$, where α

and β map T into itself, both are absolutely continuous, and β is nondecreasing. For $\phi,\psi \in W$, $m(\cdot;\phi,\psi)$ is absolutely continuous so Remark 5.12 applies here. Using the estimate $\sup\{|\phi(s)|^2 : 0 \leq s \leq \beta(t)\} \leq \int_0^{\beta(t)} |\phi'(x)|^2 dx$, $\phi \in V$, one can show that the estimate (5.5) holds if $\alpha'(\cdot)$ has an essential lower bound (possibly negative) and if there is a number σ, $0 < \sigma < 2$, such that $\alpha'(t) \geq \sigma - 2$ where $\alpha(t) > \beta(t)$. Thus, for each $u_0 \in W$ and $F \in L^2(T \times T)$ it follows from Theorems 5.2 and 5.5 that there exists a unique generalized solution of the elliptic-parabolic-Sobolev boundary value problem, $u_t - u_{xxt} - u_{xx} = F$, $0 < x < \alpha(t)$, $x < \beta(t)$; $-u_{xxt} - u_{xx} = F$, $\alpha(t) < x < \beta(t)$; $u_t - u_{xx} = F$, $\beta(t) < x < \alpha(t)$; $-u_{xx} = F$, $\alpha(t) < x$, $\beta(t) < x < 1$; $u(0,t) = u(1,t) = 0$, $t \in T$; $u(x,0) = u_0(x)$, $0 < x < \max\{\alpha(0), \beta(0)\}$. Examples (e.g., $\alpha(t) \equiv 0$) show that if β is permitted to decrease, then a solution exists only if the initial data u_0 satisfies a compatibility condition. This is to be expected since the lines "x=constant" are characteristic for the third order Sobolev equation.

Example 5.14 We shall seek conditions on the function $m(x,t)$ which are sufficient to apply our abstract results above to the boundary value problem $\partial/\partial t(m(x,t)u(x,t)) - u_{xx} = F(x,t)$, $(x,t) \in T \times T$; $u(0,t) = u(1,t) = 0$; $m(x,0)(u(x,0)-u_0(x)) = 0$, a.e. $x \in T$. We choose the spaces $V_0 = V(t) = V = H_0^1(T)$ and the forms $\ell(t;\cdot,\cdot)$ as above and let $H = W = L^2(T)$. Assume the real valued $m(\cdot,\cdot) \in L^\infty(T \times T)$ is such that $m(x,\cdot)$ is absolutely continuous for a.e. $x \in T$ and satisfies $|m_t(x,t)| \leq M(t)$ for a.e. $t \in T$, where $M(\cdot) \in L^1(T)$. Then it follows that the family of Hermitian forms $m(t;\phi,\psi) \equiv \int_0^1 m(x,t)\phi(x)\overline{\psi(x)}dx$, $\phi,\psi \in W$, is regular. Let $u_0 \in H$, $F \in L^2(T \times T)$, and set $f(t) = F(\cdot,t)$. Theorems 5.6 and 5.7 show that the boundary value problem above is the realization of our abstract Cauchy problem.

In order to obtain uniqueness of a generalized solution from Theorem 5.5, note that the forms $\ell(t;\cdot,\cdot)$ are coercive over V, so Theorem 5.5 is applicable if and only if additionally we assume

$$(5.22) \qquad m(x,t) \geq 0 \qquad\qquad x,t \in T,$$

since this is equivalent to $m(t;\cdot,\cdot) \geq 0$ for all $\phi \in W$, $t \in T$. To obtain existence from Theorem 5.2 it suffices to assume

$$(5.23) \qquad \begin{aligned} &m(x,o) \geq 0, \qquad\qquad \text{a.e. } x \in T, \text{ and} \\ &\text{ess inf}\{m_t(x,t) : (x,t) \in T \times T\} > -2\pi^2, \end{aligned}$$

for then we obtain the estimate $m'(t;\phi,\phi) \geq (\sigma - 2\pi^2)\int_0^1 |\phi|^2$, $\phi \in V$, for some number σ, $0 < \sigma < 2\pi^2$. Then from (5.19) we obtain (5.5) with $\lambda = 0$ and $c = \sigma/2\pi^2$. Thus, (5.22) gives uniqueness and (5.23) implies existence of a generalized solution of the problem.

Clearly neither of the conditions (5.22), (5.23) implies the other. In particular (5.23) permits $m(x,t)$ to be negative;

then the equation is backward parabolic. This is the case for the equation

$$\frac{\partial}{\partial t}((\pi^2/4)(1-2t)^3 u(x,t)) = u_{xx} \tag{5.24}$$

for which $m_t(x,t) = -(3\pi^2/2)(1-2t)^2$; hence, (5.23) is satisfied and existence follows. But $m(\cdot,\cdot)$ is nonnegative only when $0 \leq t \leq 1/2$, so uniqueness is claimed only on this smaller interval. To see that uniqueness does not hold beyond $t = 1/2$, note that $u(x,t) = (1-2t)^{-3} \exp\{1-(1-2t)^{-2}\} \sin(\pi x)$ is a solution of (5.24); a second solution is obtained by changing the values of $u(x,t)$ to zero for all $t > 1/2$.

The leading coefficient in the equation

$$\frac{\partial}{\partial t}((\pi^2/N)(1-t^2)^{1/2} u(x,t)) = u_{xx} \qquad (N > 0) \tag{5.25}$$

satisfies (5.22), hence uniqueness follows, but $m_t(x,t) > -2\pi^2$ only on the interval $[0, 1-(4N)^{-2})$. Hence, the existence of a solution follows on the interval $[0, 1-(4N)^{-2}-\sigma]$ for any $\sigma > 0$; choosing N large makes the interval of existence as close as desired to $[0,1]$. However the function $u(x,t) = (1-t)^{-1/2} \exp(2N(1-t)^{1/2}) \sin(\pi x)$ is the unique solution of the problem with $u_0(x) = \sin(\pi x)$ on any interval $[0, 1-\sigma]$, $\sigma > 0$, and this function does not belong to $L^2(T \times T)$. Hence, there is no solution on the entire interval.

<u>Example 5.15</u> Let G be a smooth bounded domain in $\mathbb{R}^n$ and $r(x)$ be a smooth positive function on G such that for those x

near the boundary ∂G, $r(x)$ equals the distance from x to ∂G. Parabolic equations of the form

$$(5.26) \qquad tu_t + r(x)L(t)u = F(x,t), \qquad x \in G, \quad t \in T$$

arise in mathematical genetics (cf. Kimura and Ohta [1], Levikson and Schuss [1]). They are typically degenerate in t (at $t = 0$) and degenerate in x (at $x \in \partial G$); $L(t)$ is a strongly uniformly elliptic operator. We indicate how such problems can be included in the preceding theorems. Suppose we are given real-valued absolutely continuous $\ell_{ij}(t)(1 \le i,j \le n)$ and $\ell_j(t)(0 \le j \le n)$ with $\ell_0(t) \geq 0$ and $\sum_{i,j=1}^{n} \ell_{ij}(t) z_i \bar{z}_j \geq c(|z_1|^2 + \ldots + |z_n|^2)$, $c > 0$, $z \in \mathbb{C}^n$. Take $V_0 = V = H_0^1(G)$ and define a regular family of coercive forms by $\ell(t;\phi,\psi) = \int_G \{\sum_{i,j=1}^{n} \ell_{ij}(t) D_i\phi D_j\psi + \sum_{j=0}^{n} \ell_j(t) D_j\phi \cdot \psi\} dx$, $\phi,\psi \in V$. The corresponding elliptic operators are given by

$$(5.27) \qquad L(t) = -\sum_{i,j=1}^{n} \ell_{ij}(t) D_j D_i + \sum_{j=0}^{n} \ell_j(t) D_j, \qquad t \in T.$$

Let $m(x,t)$ be given as in Example 5.14, set $W = H_0^1(G)$, and define $m(t;\phi,\psi) = \int_G (m(x,t)/r(x)^2)\phi(x)\overline{\psi(x)}dx$, $\phi,\psi \in W$. From the inequality

$$(5.28) \qquad \|\phi/r\|_{L^2} \leq C\|\phi\|_W, \qquad \phi \in W,$$

it follows that $m(t;\cdot,\cdot)$ is a regular family of Hermitian forms on W. Let $F \in L^2(G \times T)$ and define $f(t) = F(\cdot,t)/r(\cdot)$ for $t \in T$. From (5.28) it follows that $f \in L^2(T,V')$. Finally, assume $u_0 \in$

$H_0^1(G)$. From Theorems 5.6 and 5.7 we see that our abstract Cauchy problem is equivalent to the initial-boundary value problem

$$(5.29)\quad \begin{cases} \frac{\partial}{\partial t}\left(\frac{m(x,t)}{r(x)}u(x,t)\right) + r(x)L(t)u(x,t) = F(x,t), & x \in G, t \in T \\ u(s,t) = 0, & s \in \partial G, \\ m(x,o)(u(x,o)-u_0(x)) = 0, & x \in G. \end{cases}$$

(The first equation has been multiplied by $r(x)$ and the last by $r(x)^2 > 0$.) Existence and uniqueness follow from the additional assumptions $m(x,t) \geq 0$, $m_t(x,t) \geq 0$, a.e. $x \in G$, $t \in T$. These are not best possible for existence; they can be weakened as in Example 5.14.

Remark 5.16 The equation (5.29) permits the leading operator to be singular and the second to be degenerate in the spatial variable, while the leading operator may be degenerate in time.

Remark 5.17 A minor technical difficulty arises from the form in which (5.26) is given with the time derivative on u, not on (tu). We first note that $tu_t = (tu)_t - u$ but see that the resulting second operator $L(t) - u/r$ might not be coercive. However, in the change of variable $\tau = t^\alpha$ the leading term is given by $tu_t = \alpha\tau u_\tau = \alpha(\tau u)_\tau - \alpha u$ and the second operator resulting as above is given by $L(\tau^{1/\alpha}) - \alpha u/r$. From (5.28) it follows that this last operator is coercive for $\alpha > 0$ sufficiently small.

Remark 5.18 Theorems 5.2, 5.5, 5.6, 5.7 and 5.8 are from Showalter [2]. Theorem 5.9 which contains (5.16) and (5.17) is new. See Showalter [2] for examples in higher dimension with elliptic-parabolic interface or boundary conditions and other constraints on the time-differential along a submanifold or portion of the boundary. For additional material on specific partial differential equations or evolution equations of the type considered in this section we refer to Baiocchi [1], Browder [4], Cannon-Hill [2], Fichera [2], Ford [1], Ford-Waid [2], Friedman-Schuss [1], Gagneux [1], Glusko-Krein [1], Kohn-Nirenberg [1], Lions [5, 9], Oleinik [2], Schuss [1], Višik [1]. Also see references given in preceding sections for related linear problems and in the following sections for nonlinear problems.

3.6 Semilinear degenerate equations. We shall present some methods for obtaining existence-uniqueness results for the abstract Cauchy problem for the equation

$$\frac{d}{dt}(Mu(t)) + N(u(t)) = f(t), \qquad 0 \le t \le T, \tag{6.1}$$

where M is continuous, self-adjoint and monotone. The most we shall assume on the nonlinear N is that it is monotone, hemicontinuous, bounded and coercive. Specifically, let V be a reflexive real Banach space with dual V' and let N be a function from V to V'. Then N is said to be monotone if

$$\langle N(u) - N(v), u-v\rangle \ge 0, \qquad u,v \in V, \tag{6.2}$$

hemicontinuous if for each triple $u,v,w \in V$ the function $s \to \langle N(u+sv),w\rangle$ is continuous from $\mathbb{R}$ to $\mathbb{R}$, bounded if the image of each bounded set in V is a bounded set in V', and coercive if for each $v \in V$

$$\lim_{\|u\|\to\infty} (\langle N(u+v),u\rangle/\|u\|) = +\infty . \tag{6.3}$$

Let H be a Hilbert space in which V is dense and continuously imbedded. Identify H and its dual and thereby obtain the inclusions $V \subset H \subset V'$. We suppose M is a continuous, self-adjoint and monotone operator of H into H; thus the square root $M^{1/2}$ of M is defined. Let $p \geq 2$, $1/p + 1/q = 1$, $0 < T < \infty$, and define $\mathcal{V} = L^p(o,T;V)$, $\mathcal{H} = L^2(o,T;H)$, $\mathcal{V}' = L^2(o,T;V')$. Let $L_0 u = dMu/dt$ for $u \in D(L_0) = \{u \in \mathcal{V} : du/dt \in \mathcal{H}\}$ and denote by $L : D(L) \in \mathcal{V}'$ the closure of L_0 in $\mathcal{V} \times \mathcal{V}'$. A standard computation shows that if $u \in D(L)$ then $M^{1/2}u \in C(o,T;H)$, the space of continuous H-valued functions on $[o,T]$.

<u>Lemma 6.1</u> For $u \in \mathcal{V}$ and $f \in \mathcal{V}'$, the following are equivalent:

(a) $u \in D(L)$, $M^{1/2}u(o) = 0$, and $Lu = f$,

(b) $-\int_0^T \langle dMv/dt,u\rangle\, dt = \int_0^T \langle f,v\rangle\, dt$

for all $v \in \mathcal{V}$ with $dv/dt \in \mathcal{H}$ and $v(T) = 0$.

Proof: Clearly (a) implies (b): we need only observe that the identity in (b) holds for $u \in D(L_0)$, than pass to the general case of $u \in D(L)$ by continuity. Conversely, if (b) is true then we define $u_n(t) = n\int_0^t e^{n(s-t)}u(s)ds$, $f_n(t) = n\int_0^t e^{n(s-t)}f(s)ds$. Then $u_n \in D(L_0)$, $L_0 u_n = f_n$, and we have $(u_n, f_n) \to (u,f)$ in $V \times V'$. QED

The basic idea here is to consider the linear operator $\Lambda : D(\Lambda) \to V'$ defined by $\Lambda u = Lu$ and $D(\Lambda) = \{u \in D(L) : M^{1/2}u(0) = 0\}$. That is, $\Lambda = (d/dt)M$ with zero initial condition. Clearly Λ is closed, densely-defined and linear. Lemma 6.1 shows Λ is the adjoint of the operator $-Md/dt$ with domain $\{v \in V : dv/dt \in H, v(T) = 0\}$. That is, Λ is the adjoint of a monotone operator, so Λ^*, being the closure of a monotone operator, is necessarily monotone. Since the above implies that Λ is maximal monotone by a theorem of Brezis [1] we obtain the following from Browder [1]: for $N : V \to V'$ monotone, hemicontinuous, bounded and coercive, the operator $\Lambda + N$ is surjective. The preceding remarks give the following result.

Theorem 6.2 Let the spaces $V \subset H \subset V'$, the self-adjoint and monotone $M \in L(H)$ and the monotone, hemicontinuous, bounded and coercive $N : V \to V'$ be given. For each $f \in L^q(0,T;V')$ there exists a $u \in L^p(0,T;V)$ such that

$$(6.4) \qquad -\int_0^T \langle Mdv/dt, u\rangle + \int_0^T \langle Nu, v\rangle = \int_0^T \langle f, v\rangle$$

for all $v \in L^p(0,T;V)$ with $dv/dt \in L^2(0,T;H)$ and $v(T) = 0$. Also $M^{1/2}u \in C(0,T;H)$ and $M^{1/2}u(0) = 0$. The solution is unique if N is strictly monotone, i.e., if $\langle Nx-Ny, x-y\rangle > 0$, $x, y \in V$, $x \neq y$.

Remark 6.3 The condition expressed by (6.4) is equivalent to the Cauchy problem

$$\text{(6.5)} \qquad \begin{aligned} &\frac{d}{dt}Mu(t) + N(u(t)) = f(t) \\ &(M^{1/2}u)(o) = 0 \end{aligned}$$

in which the equations hold in $L^q(o,T;V')$ and in H, respectively.

Remark 6.4 The extension of Theorem 6.2 to the case of initial data $u_0 \in H$ (not necessarily homogeneous) is given in Brezis [1]. Then a term $(M^{1/2}u_0, M^{1/2}v(o))_H$ is added to (6.4) and the initial condition in (6.5) becomes $M^{1/2}u(o) = M^{1/2}u_0$. Also, the initial condition in Theorem 6.2 can be replaced by the periodic condition $M^{1/2}u(o) = M^{1/2}u(T)$ by an obvious modification of the operator Λ.

We consider briefly the Cauchy problem that arises when the solution is constrained to a specified convex set. This leads to a variational inequality. For such problems, the "natural" hypotheses on the nonlinear operator $N : V \to V'$ is that it be pseudomonotone: if $\lim u_j = u$ in V weakly and $\limsup \langle N(u_j), u_j - u\rangle \leq 0$, then for every $v \in V$

$$\text{(6.6)} \qquad \langle N(u), u-v\rangle \leq \liminf \langle N(u_j), u_j - v\rangle.$$

Remark 6.5 The pseudomonotone operators include the semi-monotone operators of Browder [6], the operators of calculus-of-variations introduced by Leray and Lions [1], and the demi-monotone operators of Brezis [4]. The pseudomonotone operators are precisely those which are most natural for elliptic variational inequalities.

To indicate the relationship between Theorem 6.2 and results to follow, we shall prove that every hemicontinuous and monotone operator $N : V \to V'$ is pseudomonotone. So, suppose the net $\{u_j\}$ in V satisfies $\lim u_j = u$ (weakly) and $\limsup \langle N(u_j), u_j - u\rangle \leq 0$. Since N is monotone it follows that $\langle N(u_j), u_j - u\rangle \geq \langle N(u), u_j - u\rangle$. The right side converges to zero so we have $\lim \langle N(u_j), u_j - u\rangle = 0$. Let $v \in V$. For $0 < t < 1$ and $w = (1-t)u + tv$ we have by monotonicity $\langle N(u_j) - N(w), u_j - w\rangle \geq 0$ and this shows $t\langle N(u_j), u-v\rangle + \langle N(u_j), u_j - u\rangle + t\langle N(w), v-u\rangle \geq \langle N(w), u_j - u\rangle$. Taking the lim inf and dividing by t gives $\liminf \langle N(u_j), u_j - v\rangle \geq \langle N(w), u-v\rangle$. The result follows by using hemicontinuity to let $t \to 0$ above.

A variational inequality problem corresponding to (6.5) and a given convex subset K of V is the following: find $u \in K$ with $M^{1/2}u(o) = 0$ for which

$$(6.7) \qquad \int_0^T \langle \frac{d}{dt}(Mu) + N(u) - f, v-u\rangle dt \geq 0$$

for all $v \in K$. A technical difficulty with this formulation is that the functional equation (6.7) might not imply that $(d/dt)Mu \in V'$, so we shall weaken the problem appropriately. If u is a solution of (6.7) and if $v \in K$ with $(d/dt)Mv \in V'$ then we obtain

$\int_0^T \langle \frac{d}{dt}(Mv)+Nu-f,v-u\rangle dt = \int_0^T \langle \frac{d}{dt}(Mu)+Nu-f,v-u\rangle dt + \int_0^T \langle \frac{d}{dt}M(v-u),v-u\rangle dt$ $\geq (1/2)\langle M(v-u),v-u\rangle\big|_0^T$. The last term is nonnegative if we require $Mv(o) = Mu(o) = 0$. Under hypotheses to be given below, we shall consider the (weak) variational inequality

$$(6.8) \qquad u \in \mathcal{K}, \int_0^T \langle \frac{d}{dt}(Mv)+Nu-f,v-u\rangle dt \geq 0$$

for all $v \in \mathcal{K}$ such that $(d/dt)Mv \in \mathcal{V}'$ and $Mv(o) = 0$.

Let the spaces $V \subset H \subset V'$ and $\mathcal{V} \subset \mathcal{H} \subset \mathcal{V}'$ be given as above. Let $K(t)$ be a closed convex subset of V for each $t \in [o,T]$; assume $0 \in K(t)$ and $K(s) \subset K(t)$ for $0 \leq s \leq t \leq T$. Define $\mathcal{K} = \{v \in \mathcal{V} : v(t) \in K(t) \text{ a.e.}\}$.

The semigroup of right shifts given by

$$(6.9) \qquad [G(s)u](t) = \begin{cases} 0, & 0 < t < s \\ u(t-s), & s < t \leq T \end{cases} \qquad u \in \mathcal{V}'$$

is strongly continuous on each of $\mathcal{V}$, $\mathcal{H}$ and $\mathcal{V}'$, it is a contraction semigroup on $\mathcal{H}$, and it satisfies

$$(6.10) \qquad G(s)\mathcal{K} \subset \mathcal{K}$$

by the assumption that the convex sets are increasing. Denote by $-L$ the generator of $\{G(s) : s \geq 0\}$ on $\mathcal{V}(\mathcal{H},\mathcal{V}')$ and its domain by $D(L,\mathcal{V})$ (respectively, $D(L,\mathcal{H})$, $D(L,\mathcal{V}')$). Thus, we have $L = d/dt$ with null initial condition in each of $\mathcal{V}$, $\mathcal{H}$, and $\mathcal{V}'$. From (6.10) and the representation for the resolvent of $-L$ by the formula $(I+\varepsilon L)^{-1} = \int_0^\infty \varepsilon^{-1}\exp(-t/\varepsilon)\, G(t)dt$ we obtain

(6.11) $\quad (I + \varepsilon L)^{-1}K \subset K, \qquad\qquad \varepsilon > 0.$

Suppose $M \in L(V,V')$ is nonnegative and self-adjoint and define $M \in L(\mathcal{V},\mathcal{V}')$ by $[Mv](t) = Mv(t)$. Then we consider the operator defined by $\Lambda = L \circ M$, $D(\Lambda) = \{v \in \mathcal{V} : Mv \in D(L,\mathcal{V}')\}$. The plan is to apply Corollary 39 of Brezis [4] to solve the abstract variational inequality

(6.12) $\quad u \in K, \quad \langle \Lambda v + Nu - f, v - u\rangle \geq 0$

for all $v \in D(\Lambda) \cap K$. The problems (6.8) and (6.12) are certainly equivalent with our choice of data and we are led to the following result.

Theorem 6.6 Let the spaces $V \subset H \subset V'$ and the self-adjoint and monotone $M \in L(V,V')$ be given as above. Let K be the closed convex set in $\mathcal{V}$ obtained as above from an increasing family $\{K(t) : 0 \leq t \leq T\}$ of closed convex subsets of V, and assume $0 \in K$. Let $N : \mathcal{V} \to \mathcal{V}'$ be pseudomonotone, bounded and coercive. Then for each $f \in L^q(0,T;V')$ there exists a solution of (6.8). The solution is unique if N is strictly monotone.

Proof: The result follows immediately from Brezis [4] after we verify that the linear Λ is nonnegative (= monotone) and compatible with K. Note first that for those $v \in \mathcal{V}$ with $Mv \in \mathcal{V}'$ we have $\int_0^T \langle \frac{d}{dt}Mv, v\rangle dt = (1/2)\langle Mv(s), v(s)\rangle|_0^T$. Thus for $v \in D(\Lambda)$ it follows

(6.13) $$\int_0^T \langle\Lambda v,v\rangle dt = (1/2)\langle Mv(T),v(T)\rangle \geq 0.$$

That Λ is compatible with K means that for each $u \in K$ there exists $\{u_n\} \subset K \cap D(\Lambda)$ such that $\lim(u_n) = u$ in V and $\limsup \langle\Lambda u_n,u_n-u\rangle \leq 0$. Thus, let $u \in K$ and define for $n \geq 1$ $u_n = (I+n^{-1}L)^{-1}u$. We have $\lim (u_n) = u$, $u_n(t) = n\int_0^t e^{n(s-t)}u(s)ds$, and (6.11) shows $u_n \in K$. The representation above gives $Mu_n(t) = n\int_0^t e^{n(s-t)}Mu(s)ds$, hence, $LMu_n = MLu_n$. Finally, since M is monotone we have $\langle\Lambda u_n,u_n-u\rangle = -\langle LMu_n,n^{-1}Lu_n\rangle = -n^{-1}\langle MLu_n,Lu_n\rangle \leq 0$, so Λ is compatible with K. QED

Remark 6.7 If we change the domain of L appropriately, we can replace the initial condition implicit in (6.8) by a corresponding periodic condition.

We return now to the Cauchy problem for the evolution equation (6.1) where M is given as in Theorem 6.6 but with weaker restrictions on the nonlinear operator N. Specifically, we shall assume $N : V \to V'$ is of type M: if $\lim (u_j) = u$ in V weakly, $\lim (Nu_j) = f$ in V' weakly, and $\limsup \langle Nu_j,u_j\rangle \leq \langle f,u\rangle$, then $Nu = f$.

Remark 6.8 The operators of type M include weakly continuous operators (i.e., continuous from V weakly into V' weakly) and the pseudomonotone operators, hence, the operators of Remark 6.5.

We show that pseudomonotone operators are of type M. Let

$\{u_j\}$ be a net (or sequence if V is separable) as in the definition above of "type M" and assume N is pseudomonotone. Then $\limsup \langle Nu_j, u_j - u\rangle \le 0$, so for every $v \in V$ we obtain from (6.6) $\langle Nu, u-v\rangle \le \liminf \langle Nu_j, u_j - v\rangle \le \langle f, u-v\rangle$. But $v \in V$ is arbitrary so $Nu = f$ follows from this inequality.

Much of the notation of Theorem 6.6 will be used to prove the following result.

<u>Theorem 6.9</u> Let the spaces $V \subset H \subset V'$ and the monotone self-adjoint $M \in L(V,V')$ be given as in Theorem 6.6. Let $N : V \to V'$ be type M, bounded and coercive. Then for each $f \in L^q(o,T;V')$ there exists a solution $u \in L^p(o,T;V)$ of the Cauchy problem

$$(6.14)\qquad \begin{aligned} &\frac{d}{dt}(Mu) + N(u) = f \\ &(Mu)(o) = 0. \end{aligned}$$

If N is strictly monotone (cf. Theorem 6.2) the solution of (6.14) is unique.

Proof: We give a finite-difference approximation to (6.14) of implicit type. Thus, (6.14) is approximated by a corresponding family of "stationary" problems (cf. Chapter II.7 of Lions [10]). Note that the Cauchy problem (6.14) is equivalent to the equation in V'

$$(6.15)\qquad \Lambda(u) + N(u) = f, \qquad\qquad \Lambda = L \circ M$$

where -L is the generator of the strongly-continuous semigroup (6.9) in V' and $M \in L(V,V')$ is obtained (pointwise) from M as before.

To approximate (6.15) we let $s > 0$ and consider the equation in V'

$$(6.16) \qquad [(I-G(s))/s]Mu_s + Nu_s = f.$$

Since $m(x,y) = \langle Mx,y\rangle$ is a semi-scalar-product on V we obtain the estimate $\int_0^T \langle G(s)Mv,v\rangle dt = \int_s^T m(v(t-s),v(t)dt \le \int_0^T \langle Mv,v\rangle dt$, $v \in V$, which shows that the operator $[I-G(s)]M$ is monotone (cf. 6.13). Thus, the operator $[(I-G(s))/s] \circ M + N$ is of type M, bounded and coercive. This implies by Brezis [4] or Lions [10] that it is surjective, hence, (6.16) has a solution u_s for each $s > 0$. From (6.16) we obtain $\langle Nu_s,u_s\rangle \le \langle f,u_s\rangle \le \|f\|\cdot\|u_s\|$, so the coercivity of N shows that $\{u_s : s > 0\}$ is bounded in V. Since N and M are bounded and V is reflexive there is a subset (which we denote also by $\{u_s : s > 0\}$) for which $\lim (u_s) = u$, $\lim N(u_s) = g$ and $\lim M(u_s) = Mu$ weakly in the appropriate spaces.

Let L^* denote the adjoint of the operator L in V'. We apply (6.16) to any $v \in D(L^*,V)$ and take the limit to obtain $\langle Mu,L^*v\rangle + \langle g,v\rangle = \langle f,v\rangle$, $v \in D(L^*,V)$. This shows $Mu \in D(L,V')$ and that

$$(6.17) \qquad LMu + g = f$$

Since N is of type M, it suffices for existence to show

lim sup $\langle Nu_s, u_s - u\rangle \leq 0$. But from (6.16) and the monotonicity of $[(I-G(s))/s]M$ we obtain $\langle Nu_s, u_s - u\rangle \leq \langle f, u_s - u\rangle - \langle [(I-G(s))/s]Mu, u_s - u\rangle$. Since $Mu \in D(L, V')$ the left factor of the last term converges strongly and this gives the desired result. Uniqueness follows easily from the equivalence of (6.14) and (6.15); we need only note that Λ is monotone. QED

Remark 6.10 The extension to nonhomogeneous initial data is given in Bardos-Brezis [1] with relevant regularity results on the solution. The problem with periodic conditions Mu(o) = Mu(T) is solved similarly.

Our next class of results will be obtained from the generation theory of nonlinear semigroups and its extensions to evolution equations with multivalued operators. The linear semigroup theory is inadequate for degenerate problems, but its presentation above in Section 4 provided an elementary motivation for the techniques and constructions to follow. Specifically, it indicates the "optimal" hypotheses from which one might expect the Cauchy problem to be well-posed and it suggests the correct space in which to look for such results. Also, we shall not present here a nonlinear version of the analytic situation of Theorem 4.2 and Theorem 4.4.

Let E be a vector space with algebraic dual E^* and let $M : E \to E^*$ be nonnegative (monotone) and self-adjoint. Let K be the kernel of M and $q : E \to E/K$ the corresponding quotient map. The symmetric bilinear function $m : E \times E \to \mathbb{R}$ given by $m(x,y) =$

<Mx,y> determines a scalar product m_0 on E/K by

$$m_0(q(x),q(y)) = m(x,y), \qquad x,y \in E. \tag{6.18}$$

Let W be the Hilbert space completion of E/K with m_0, and denote by m_0 the (extended) scalar product on W.

Let E' be the Hilbert space obtained as the strong dual of the space E with the seminorm induced by m. Note that since E/K is dense in W we may identify the dual spaces $(E/K)' = W'$. Let $q^* : W' \to E'$ be the continuous dual of $q : E \to W$. Since q has dense range, q^* is necessarily injective. Also, each $g \in E'$ vanishes on K, hence, $g = f \circ q$ for some $f \in (E/K)'$. This shows q^* is surjective, and it follows from (6.18) that q^* is norm-preserving. Finally, if $M_0 : W \to W'$ is the Riesz isomorphism determined by $\langle M_0 x,y\rangle = m_0(x,y)$, then we obtain the identity

$$M = q^* M_0 q \tag{6.19}$$

relating the functions considered above.

Assume we are given for each $t \in [o,T]$ a nonempty set $D(t) \subset E$ and a (not necessarily linear) function $N(t) : D(t) \to E'$. Then for each such t we define a multivalued function or relation $N_0(t)$ on $q[D(t)] \times W'$ as the composition

$$N_0(t) = (q^*)^{-1} \circ N(t) \circ q^{-1}. \tag{6.20}$$

Finally, we define the composite relation $A(t) = M_0^{-1} \circ N_0(t)$ on $W \times W$ with domain $q[D(t)]$ for each $t \in [o,T]$. Note that $N_0(t)$

and A(t) are functions if and only if $x,y \in D(t)$ and $Mx = My$ imply $N(t)x = N(t)y$. The following diagram illustrates the various relationships.

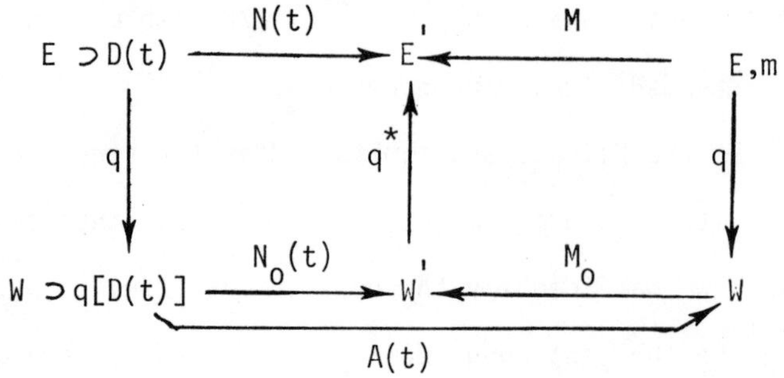

We shall consider the semilinear evolution equation

$$\frac{d}{dt}(Mu(t)) + N(t,u(t)) = 0, \qquad 0 \leq t \leq T. \tag{6.21}$$

A solution of (6.21) is a function $u : [o,T] \to E$ such that $Mu : [o,T] \to E'$ is absolutely continuous, hence, differentiable a.e., $u(t) \in D(t)$ for all t, and (6.21) is satisfied a.e. on $[o,T]$. The Cauchy problem is to find a solution of (6.21) for which $(Mu)(o)$ is specified in E'.

We follow the idea from Section 4 of reducing (6.21) to a "standard" evolution equation with M = identity. If u is a solution of (6.21) and we define $v = q \circ u$, then (6.19) and (6.20) imply

$$(q^* M_0 v)' = -N(t,u(t)) \in - q^* N_0(t,v(t)). \tag{6.22}$$

Since q^* and M_0 are linear isometries, we see that $v(t) \in q(D(t))$ for $t \in [o,T]$, $v : [o,T] \to W$ is absolutely continuous, and

$$(6.23) \qquad v'(t) \in -A(t,v(t))$$

is satisfied a.e. on $[o,T]$. We call such a $v(\cdot)$ a solution of (6.23). Conversely, suppose v is a solution of (6.23). Then for each t at which (6.23) holds, (6.20) shows that $q(u(t)) = v(t)$ for some $u(t) \in D(t)$ and (6.22) holds. By choosing $u(t) \in D(t)$ with $q(u(t)) = v(t)$, hence, $Mu(t) = q^* M_o v(t)$, for all remaining $t \in [o,T]$, we obtain a solution u of (6.21). This proves the following.

Lemma 6.11. If v is a solution of (6.23), then for each $t \in [o,T]$ there is a $u(t) \in D(t)$ such that u is a solution of (6.21). Conversely, for each solution u of (6.21), the function $v = q \circ u$ is a solution of (6.23).

Corollary 6.12 Let $u_o \in D(o)$. There exists a solution v of (6.23) with $v(o) = q(u_o)$ if and only if there exists a solution u of (6.21) with $(Mu)(o) = Mu_o$. There is at most one solution v of (6.23) with $v(o) = q(u_o)$ if and only if for every pair of solutions u_1, u_2 of (6.21) with $(Mu_1)(o) = (Mu_2)(o) = Mu_o$ it follows that $Mu_1(t) = Mu_2(t)$ for all $t \in [o,T]$, hence, $N(t,u_1(t)) = N(t,u_2(t))$ a.e. on $[o,T]$.

Results on the Cauchy problem for (6.21) can now be obtained from corresponding results for (6.23). Turning first to the question of uniqueness, we find that a sufficient condition for the m_o-seminorm on the difference of two solutions to be nonincreasing is that each $A(t)$ be accretive: if $[x_1,w_1]$ and $[x_2,w_2]$

belong to A(t), then $m_0(w_1-w_2,x_1-x_2) \geq 0$. The success of the above construction for which (6.21) and (6.23) correspond to one another is reflected in the following.

Lemma 6.13 The relation A(t) is accretive if and only if the function N(t) is monotone.

Proof: Let $w_1 \in A(x_1)$ and $w_2 \in A(x_2)$. Choose $u_1,u_2 \in D(t)$ with $x_j = q(u_j)$ and $N(t,u_j) = q^*M_0w_j$, j = 1,2. Then we have

$$(6.24) \qquad m_0(w_1-w_2,x_1-x_2) = \langle N(t,u_1)-N(t,u_2),u_1-u_2\rangle$$

so A(t) is accretive if N(t) is monotone. Conversely, if $u_1,u_2 \in D(t)$ there is a unique pair $w_1,w_2 \in W$ with $N(t,u_j) = q^*M_0w_j$, j = 1,2. Then $[q(u_j),w_j] \in A(t)$ and (6.24) holds. Hence N(t) is monotone if A(t) is accretive. QED

Remark 6.14 N(t) is strictly monotone if and only if it is injective and A(t) is a strictly accretive function.

Theorem 6.15 Let N(t) be monotone and M + N(t) strictly monotone for each $t \in [o,T]$. Then for each $u_0 \in D(o)$ there is at most one solution $u(\cdot)$ of (6.21) with $(Mu)(o) = Mu_0$.

Proof: Since A(t) is accretive for each t by Lemma 6.13, uniqueness holds for (6.23). The result then follows from Corollary 6.12. QED

We consider now the existence of solutions. Specifically,

we shall recall the existence results of Crandall and Pazy [1] as they apply to the rather special Hilbert space situation of (6.23). (The monograph of H. Brezis [3] provides an excellent introduction to this topic with an extensive bibliography; cf. also the recent monograph of Browder [7].) The equation (6.23) has a (unique) solution v with v(o) specified in q(D(o)) under the following hypotheses:

Each A(t) is accretive and $I + \lambda A(t)$ is surjective for all $\lambda > 0$.

It follows that $J_\lambda(t) \equiv (I+\lambda A(t))^{-1}$ is a function defined on all of W.

The domain of A(t), q[D(t)], is independent of t; we denote it by q[D].

There is a monotone $g : [o,\infty) \to [o,\infty)$ such that

$$\|J_\lambda(t,x)-J_\lambda(s,x)\|_W \le \lambda|t-s|g(\|x\|_W)$$
$$\cdot(1+ \inf\{\|y\|_W : y \in A(s,x)\}), \quad t,s \ge 0,\ x \in W,$$
$$0 < \lambda \le 1.$$

The preceding results with Corollary 6.12 lead to the following.

Theorem 6.16 Let M be nonnegative and symmetric from the linear space E to its dual E^*. Denote by E' the dual of the linear topological space E with the seminorm $\langle Mx,x\rangle^{1/2}$; E' is a Hilbert space with norm given by $\|f\|_{E'} = \sup\{|\langle f,x\rangle| : x \in E, \langle Mx,x\rangle \le 1\}$. For each $t \in [o,T]$ assume we are given a (possibly nonlinear) $N(t) : D(t) \to E'$ with domain $D(t) \subset E$. Assume further that for each t, N(t) is monotone and $M + \lambda N(t) : D(t) \to E'$ is surjective; M[D(t)] is independent of t; and for some monotone increasing function $g : [o,\infty) \to [o,\infty)$ we have

$$(6.25) \qquad \|N(t,w)-N(s,w)\|_{E'} \leq |t-s|g(\langle Mw,w\rangle)(1+\|N(t,w)\|_{E'}),$$

$$0 \leq t,\ s \leq T, \quad w \in D(t).$$

Then for each $u_0 \in D(o)$ there exists a solution u of (6.21) with $(Mu)(o) = Mu_0$.

Proof: From Corollary 6.12 we see that it suffices to show (6.23) has a solution v with $v(o) = q(u_0)$, and this will follow if we verify the hypotheses on $\{A(t)\}$ above. First note that each A(t) is accretive by Lemma 6.13. Also, M[D(t)] independent of t implies the same for q[D(t)]. To determine the range of $I + \lambda A(t)$ we have $(q^* M_0)(I+\lambda A(t)) = q^* M_0 + \lambda N_0(t) = q^* M_0 + \lambda N(t)q^{-1} = (M+\lambda N(t))q^{-1}$ on q[D(t)]. The range of the above is E' and $q^* M_0$ is a bijection, so $I + \lambda A(t)$ is necessarily onto W. The estimate on $J_\lambda(t,x)$ follows from (6.25) and we refer to Showalter [5] for the rather lengthy computation. QED

Remark 6.17 In contrast to preceding results of this section, N need not be coercive or even defined everywhere on V. Thus it may be applied to examples where N corresponds to a not necessarily elliptic differential operator (cf. Theorem 4.1 and Example 4.6).

Second order evolution equations with (possibly degenerate) operator coefficients on time derivatives can be resolved by the preceding results. These contain as a special case a first order semilinear equation which is nonlinear in the time derivative.

Theorem 6.18 Let A and C be symmetric continuous linear operators from a reflexive Banach space V into its dual V'; assume C is monotone and A is coercive. Denote by V_c' the (Hilbert space) dual of V with the seminorm $\langle Cx,x\rangle^{1/2}$. Let $B : V \to V'$ be a (possibly nonlinear) monotone and hemicontinuous function. Then for each pair $u_1,u_2 \in V$ with $Au_1 + B(u_2) \in V_c'$ there exists a unique $v \in L^1(o,T;V)$ such that $Cv : [o,T] \to V_c'$ is absolutely continuous, $\frac{d}{dt}(Cv) + B(v) : [o,T] \to V'$ is (a.e. equal to) an absolutely continuous function, $Cv(o) = Cu_2$, $[\frac{d}{dt}(Cv)+B(v)](o) = Au_1$, and

$$(6.26) \qquad \frac{d}{dt}\{\frac{d}{dt}(Cv(t)) + B(v(t))\} + Av(t) = 0, \quad \text{a.e. } t \in [o,T].$$

Proof: On the product space $E \equiv V \times V$ consider $M[x,y] = [Ax,Cy]$, hence, $E' = V' \times V_c'$. Define $D = \{[x,y] \in V \times D(B) : Ax + B(y) \in V_c'\}$ and $N : D \to E'$ by $N[x,y] = [-Ay,Ax+B(y)]$. Then apply Theorem 6.16 to obtain existence of a solution $u = [w,v]$ of (6.21). Note that $C + \lambda B + \lambda^2 A$ is monotone, hemicontinuous and coercive, hence onto V', and this shows $M + \lambda N$ is surjective (cf. proof of Theorem 4.3). The second component v of this solution u with $Mu = M[u_1,u_2]$ is the solution of (6.26). Uniqueness follows from Corollary 6.12 since $M[x_1,y_1] = M[x_2,y_2]$ and $N[x_1,y_1] = N[x_2,y_2]$ imply $[x_1,y_1] = [x_2,y_2]$. QED

Remark 6.19 The first component of the solution u of (6.21) in the preceding proof is the unique absolutely continuous $w : [0,T] \to V$ with $B(\frac{dw}{dt}) : [0,T] \to V_c'$ absolutely continuous and satisfying $w(0) = u_1$, $Bw'(0) = Bu_2$,

$$(6.27) \qquad B(\tfrac{dw}{dt}) + Aw(t) \in V_c', \qquad t \in [0,T]$$

$$(6.28) \qquad \frac{d}{dt}C(\tfrac{dw}{dt}) + B(\tfrac{dw}{dt}) + Aw(t) = 0, \qquad \text{a.e. } t \in [0,T].$$

Remark 6.20 The special case that results from choosing C = 0 is of interest and should be compared with (6.21). Specifically, (6.26) and (6.27) become first order equations in which the time derivative acts on the nonlinear term. Also, the invertibility of A was never used in solving (6.26). It is sufficient for existence of a solution of (6.26) that $B + \lambda A$ be coercive for each $\lambda > 0$; however, uniqueness may then be lost (e.g., take A = 0).

We briefly sketch some applications of the results of this section to initial-boundary value problems. Let G be a bounded domain in $\mathbb{R}^n$ with smooth boundary ∂G, $W^p(G)$ be the Sobolev space of $u \in L^p(G)$ with each $D_j u = \partial u/\partial x_j \in L^p(G)$, $1 \le j \le n$, and $D_0 u = u$. Let functions $N_j(x,y)$ be given, measurable in $x \in G$ and continuous in $y \in \mathbb{R}^{n+1}$. Suppose for some $C > 0$, $c > 0$, and $g \in L^q$, $q = p/(p-1)$, $p \ge 2$, we have

$$(6.29) \qquad |N_k(x,y)| \le C \sum_{j=0}^{n} |y_j|^{p-1} + g(x),$$

$$(6.30)\qquad \sum_{j=0}^{n} (N_j(x,y)-N_j(x,z))(y_j-z_j) \geq 0$$

$$(6.31)\qquad \sum_{j=0}^{n} N_j(x,y)y_j + g(x) \geq c|y|^p,$$

for $x \in G$, $y,z \in \mathbb{R}^{n+1}$, and $0 \leq k \leq n$. Let $Du = \{D_j u : 0 \leq j \leq n\}$, let V be a subspace of $W^p(G)$ containing $C_0^\infty(G)$, and define $N : V \to V'$ by

$$(6.32)\qquad \langle Nu,v\rangle = \sum_{j=0}^{n} \int_G N_j(x,Du(x))D_j v(x)dx, \qquad u,v \in V.$$

The restriction of Nu to $C_0^\infty(G)$ is the distribution on G given by

$$(6.33)\qquad \tilde{N}u = -\sum_{j=1}^{n} D_j N_j(\cdot,Du) + N_0(\cdot,D\phi).$$

The divergence theorem gives the (formal) Green's formula $\langle Nu-\tilde{N}u,v\rangle = \int_{\partial} \frac{\partial u}{\partial N} \cdot v(s)ds$, $v \in V$, where the conormal derivative on ∂G is given by $\frac{\partial u}{\partial N} = \sum_{j=1}^{n} N_j(\cdot,Du)n_j$ and $(n_1, \ldots, n_n)$ is the unit outward normal. The operator $N : V \to V'$ is bounded, hemicontinuous, monotone and coercive; cf. Browder [3, 6], Carroll [14] or Lions [10] for details and construction of more general operators.

<u>Example 6.21</u> Let $m(\cdot) \in L^\infty(G)$, $m_0(x) \geq 0$, a.e. $x \in G$, and define $M : V \to V'$ by

$$(6.34)\qquad \langle Mu,v\rangle = \int_G m(x)u(x)v(x)ds, \qquad u,v \in V.$$

Let $u_0 \in V$ with $N(u_0) = m^{1/2}h$ for some $h \in L^2(G)$. Then each of

Theorems 6.2, 6.6, and 6.16 gives existence of the solution to the elliptic-parabolic equation

$$(6.35)\qquad D_t(m(x)u(x,t)) + \tilde{N}(u(x,t)) = 0$$

with initial condition $m(x)(u(x,0)-u_o(x)) = 0$, $x \in G$, and a boundary condition depending on V. If $V = W_o^p(G)$, the Dirichlet boundary condition is obtained, while the (nonlinear) Neumann condition $\partial u/\partial N = 0$ results when $V = W^p(G)$. The third boundary condition is obtained by adding to (6.32) a boundary integral. The assumption that $m(\cdot)$ is bounded may be relaxed; cf. Bardos-Brezis [1] or Showalter [5]. If we add the boundary integral $\int_{\partial G} u(s)v(s)ds$ to (6.34) and if V is the space of functions in $W^p(G)$ which are constant on ∂G, we obtain the boundary condition of fourth type (cf. Adler [1])

$$\begin{cases} u(s,t) = f(t), & s \in \partial G \\ f'(t) + \int_{\partial G} \dfrac{\partial u(s,t)}{\partial N}\, ds = 0, & t \geq 0. \end{cases}$$

This problem is solved by Theorems 6.9 and 6.16. By adding appropriate terms to (6.32) and (6.34) involving integrals over portions of the boundary (or over a manifold S of dimension n-1 in $\bar{G}$) one obtains solutions of equation (6.35) subject to degenerate parabolic constraints

$$\begin{aligned} &D_t(m(s)u(s,t)) - \text{div}(a(s)\text{grad } u(s,t)) \\ &\quad = -\frac{\partial u(s,t)}{\partial N}, \qquad s \in S, \end{aligned}$$

where the indicated divergence and gradient are in local

coordinates on S and the coefficients are nonnegative (cf. Showalter [2, 5]).

Example 6.22 Choose $V = W_0^2(G)$ and define

$$\langle Au,v\rangle = \sum_{j=1}^{n} \int_G D_j u(x) D_j v(x)\,dx, \qquad u,v \in V,$$

$$\langle Cu,v\rangle = \sum_{j=0}^{n} \int_G c_j(x) D_j u(x) D_j v(x)\,dx,$$

where each $c_j \in L^\infty(G)$ is nonnegative, $0 \le j \le n$. Let $B = N$ be given by (6.32) and assume (6.29), (6.30) with $1 < p \le 2$. Then Theorem 6.18 gives a unique solution of the equation

$$(6.36) \qquad D_t\{D_t[c_0 v - \sum_{j=1}^{n} D_j(c_j D_j v)] + \tilde{N}(v)\} - \sum_{j=1}^{n} D_j^2 v = 0,$$

with Dirichlet boundary condition and appropriate initial conditions. Note that $\tilde{N}$ is given by (6.33). Special cases of (6.35) include the wave equation (4.9), the viscoelasticity equation (4.10), parabolic equations (4.6), and Sobolev equations with first or second order time derivatives.

Example 6.23 A special case of (6.36) has been of considerable interest. Specifically, we choose $C \equiv 0$, $N_j = 0$ for $1 \le j \le n$, and $N_0(x,s) = m(x)|s|^{p-1}\operatorname{sgn}(s)$ where $m \in L^\infty(G)$ is nonnegative and $1 < p \le 2$. This gives the equation

$$(6.37) \qquad D_t(m(x)|v(x,t)|^{p-1}\operatorname{sgn} v(x,t)) - \Delta v(x,t) = 0.$$

The change of variable $u = |v|^{p-1}\operatorname{sgn}(v)$ puts this in the form

$$(6.38) \qquad D_t(m(x)u(x,t)) - (q-1)\sum_{j=1}^{n} D_j(|u|^{q-2}D_ju(x,t)) = 0$$

with $q - 2 = (2-p)/(p-1) \geq 0$. This equation arises in certain diffusion processes; it is "doubly-degenerate" since the leading coefficient may vanish and the power of u may vanish independently. Note that the second operator is not monotone so the equation is not in the form of (6.21).

Remark 6.24 Theorem 6.2 is from Brezis [1]. See Lagnese [10] for a variation on Theorem 6.6 and corresponding perturbation results on such problems. Theorem 6.9 is contained in the results of Bardos-Brezis [1] along with other types of degenerate evolution equations not covered here (cf. Lagnese [5] for corresponding perturbation results). Theorems 6.15, 6.16 and 6.18 are from Showalter [5]. We refer to Aronson [1], Dubinsky [1], Lions [10] and Raviart [1] for additional results on (6.38) and, specifically, to Strauss [1] for a direct integration of semilinear equations of the form (cf. (6.37))

$$(6.39) \qquad D_t(N(u)) + A(t)u(t) = f(t)$$

with nonlinear N and time-dependent linear $\{A(t)\}$ in Banach spaces. Cf. Brezis [1], Kamenomotskaya [1], and Ladyženskaya-Solonnikov-Uralceva [2] for a reduction of certain (Stefan) free-boundary value problems to the form (6.39), hence, (6.36), and Lions-Strauss [8] and Strauss [2, 3] for additional results on (6.28). Crandall [3] and Konishi [1] have studied special cases of (6.39) by different methods.

3.7 Doubly nonlinear equations. We briefly consider some first order evolution equations of the form

$$\frac{d}{dt}M(u(t)) + N(u(t)) = 0 \tag{7.1}$$

where (possibly) both operators are nonlinear. Some sort of degeneracy is desirable (and necessary for applications) so we specifically do not make assumptions on M of strong monotonicity as was done in Theorem 3.12. Moreover, we can include certain related variational inequalities (e.g., by letting one of the operators be a subdifferential) so we permit the operators to be multivalued functions or relations as considered in Section 3.6.

Let B be a real reflexive Banach space, let D(M) and D(N) be subsets of B and suppose $M : D(M) \to B$ and $N : D(N) \to B$ are multivalued functions. That is, $M \subset D(M) \times B$ and $N \subset D(N) \times B$. (When M and N are functions, we identify them with their respective graphs.) We call the pair of functions u,v a solution to the differential "inclusion"

$$v(t) \in M(u(t)), \qquad -v'(t) \in N(u(t)), \qquad \text{a.e. } t \geq 0, \tag{7.2}$$

if $u(t) \in D(M) \cap D(N)$, $v : [0,\infty) \to B$ is Lipschitz continuous, hence, strongly-differentiable a.e., and (7.2) holds. If M and N are functions, then (7.1) is clearly satisfied a.e. on $[0,\infty)$, whereas if only M is a function we replace the equality symbol in

(7.1) by an inclusion. Let $D(A) \equiv M[D(M) \cap D(N)]$ denote the indicated image in B and define a relation $A : D(A) \to B$ by the composition $A = N \circ M^{-1}$. That is, $(x,y) \in A$ if and only if for some $z \in D(M) \cap D(N)$ we have $(z,x) \in M$ and $(z,y) \in N$. Thus, if the pair (u,v) is a solution of (7.2), then it follows that

$$(7.3) \qquad -v'(t) \in A(v(t)), \qquad\qquad \text{a.e. } t \geq 0,$$

so we call a Lipschitz function $v : [o,\infty) \to B$ a solution of (7.3) if $v(t) \in D(A)$ and (7.3) holds. Conversely, if v is a solution of (7.3), the definition of A shows there exists for each $t \geq 0$ a $u(t) \in D(M) \cap D(N)$ for which the pair u,v is a solution of (7.2).

<u>Theorem 7.1</u> Let M and N be multivalued operators on the real reflexive Banach space B; assume M + N is onto B and that

$$(7.4) \qquad \begin{cases} \|(x_1-x_2)+s(y_1-y_2)\| \geq \|x_1-x_2\|, & \text{for } s > 0, \\ \text{and } (z_j,y_j) \in N,\ (z_j,x_j) \in M, & j = 1,2. \end{cases}$$

Then for each $u_0 \in D(M) \cap D(N)$ and $v_0 \in M(u_0)$, there exists a solution pair u,v of (7.2) with $v(o) = v_0$.

Proof: By the preceding remarks it suffices to show (7.3) has a solution. From (7.4) it follows that the same estimate holds for $s > 0$ whenever $(x_j,y_j) \in A$; this is precisely the statement that A is accretive (cf. Crandall-Liggett [2]).

Furthermore, it is easy to check that $(x,y) \in I + A$ if and only if for some $z \in D(M) \cap D(N)$ we have $(z,y-x) \in N$ and $(z,x) \in M$, so it follows that the range of $I + A$ equals the range of $M + N$. Thus $I + A$ is hyperaccretive and it follows from results of Crandall-Liggett [2] that (7.3) has a unique solution for each $v_0 \in D(A)$. QED

Remark 7.2 If u_1,v_1 and u_2,v_2 are solutions of (7.2), then it follows $\| v_1(t)-v_2(t) \| \leq \|v_1(0)-v_2(0)\|$. If $v_1(0) = v_2(0)$ and either M^{-1} or N^{-1} is a function, then the solution pair u,v is unique.

Remark 7.3 In a Hilbert space H with inner product $(\cdot,\cdot)_H$, the condition (7.4) is equivalent to

$$(7.5)\quad \begin{cases} (x_1-x_2, y_1-y_2)_H \geq 0 & \text{whenever } (z_j,x_j) \in M \\ \text{and } (z_j,y_j) \in N & \text{for } j = 1,2. \end{cases}$$

For linear functions this is the right angle condition (cf. Grabmuller [1], Lagnese [2, 4, 8], and Showalter [6, 14, 15, 16, 20]).

The difficulty in applying Theorem 7.1 to initial-boundary value problems arises from the assumption (7.4) which relates the operators M and N to each other. A desirable alternative which we shall describe is to place hypotheses on each of the operators independently. The results of Section 3.6 were obtained when the leading (linear) operator was symmetric. The nonlinear analogue of this is that this operator be a differential or, more generally, a subdifferential. Recall that the subdifferential at $x \in E$ of an extended-real-valued lower semicontinuous convex function ϕ on the locally convex space E is the set $\partial\phi(x)$ of those $u \in E'$ verifying

$$\phi(y) - \phi(x) \geq \langle u, y-x \rangle, \qquad y \in E.$$

(Cf. Brezis [3], Ekeland-Teman [1], Lions [10].) Some examples will be given below. In general, the subgradient $\partial\phi$ is a multivalued operator from (a subset of) E into E'.

<u>Theorem 7.4</u> Let V_1 and V_2 be real separable reflexive Banach spaces with $V_1 \subset V_2$, V_1 is dense in V_2, and assume the inclusion is compact. Let ϕ_1 and ϕ_2 be convex continuous (extended) real-valued functions on V_1 and V_2, respectively. Assume their subdifferentials $N \equiv \partial\phi_1$ and $M \equiv \partial\phi_2$ are bounded (cf. Section 3.6) and that N is "coercive" in the sense that $\liminf \{\phi_1(u)/|u|_{V_1}^p : |u|_{V_1} \to \infty\} > 0$ for some p, $1 < p < \infty$. Suppose we are given $u_0 \in V_1$, $v_0 \in M(u_0)$, and $f \in L^\infty(0,T;V_1')$ with $df/dt \in L^q(0,T;V_1')$ where $1/p + 1/q = 1$. Then there exists a pair of functions $u \in L^\infty(0,T;V_1)$, $v \in L^\infty(0,T;V_2')$ with $dv/dt \in L^\infty(0,T;V_1')$ satisfying

$$(7.6) \qquad v(t) \in M(u(t)), \qquad f(t) - v'(t) \in N(u(t))$$

a.e. on $[0,T]$ and $v(0) = v_0$.

Remark 7.5 The compactness assumption above limits application of Theorem 7.4 to parabolic problems; cf. examples below.

We illustrate some applications of the preceding results through some examples of initial-boundary value problems.

Example 7.6 Let $V = W_0^{1,p}(G)$, the closure of $C_0^\infty(G)$ in $W^{1,p}(G)$, $p \geq 2$, and define the nonlinear elliptic operator $T : V \to V'$ by (6.32), where the functions $N_j(x,y)$ satisfy (6.29), (6.30) and (6.31). For $j = 1,2$, let $m_j(\cdot) \in L^\infty(G)$ be given with $m_j(x) \geq 0$, a.e. $x \in G$, and let $k > 0$. Set $H = L^2(G)$ so $V \subset H \subset V'$, and define $D(M) \equiv \{u \in V : m_1 u + m_2 T(u) \in H\}$, $M(u) \equiv m_1 u + m_2 T(u)$; $D(N) \equiv \{u \in V : (k/m_1(\cdot))Tu \in H\}$, $N(u) \equiv (k/m_1(\cdot))T(u)$. Then M and N are (single-valued) nonlinear operators in the Hilbert space H. To apply Theorem 7.1, we first verify (7.4) in its equivalent form (7.5). For $u,v \in D(M) \cap D(N)$ we have $(Mu-Mv,Nu-Nv)_H = k\langle T(u)-T(v),u-v\rangle + \int_G (km_2(x)/m_1(x))(T(u)-T(v))^2 \cdot dx \geq 0$, so (7.4) holds. To show $M + N$ is onto H, let $w \in H$ and consider the equation

$$(7.7) \qquad [(m_1)^2/(m_1 m_2+k)]u + T(u) = [m_1/(m_1 m_2+k)]w.$$

The coefficients all belong to $L^\infty(G)$, so the operator on the left side of (7.7) is monotone, hemicontinuous and coercive from V to V', hence, surjective, so there is a solution $u \in V$ of (7.7). Since $m_1(\cdot)$ and $m_2(\cdot)$ are bounded, it follows from (7.7) that $u \in D(M) \cap D(N)$ and that $M(u) + N(u) = w$. Thus, Theorem 7.1 gives existence of a weak solution of an appropriate initial-boundary value problem containing the equation

$$(7.8) \qquad D_t\{m_1(x,t)u(x,t)+m_2(x)T[u]\} + \frac{k}{m_1(x)}T[u] = 0$$

where T[u] is given by (6.33). Note that the coefficients m_1 and m_2 may vanish over certain subsets of G.

Example 7.7 We apply Theorem 7.4 with $V_1 = W_0^{1,p}(G)$ and $V_2 = L^\alpha(G)$, $\alpha \leq p$, $\phi_1(u) = 1/p \int_G \sum_{j=1}^n |D_j u(x)|^p dx$, $\phi_2(u) = 1/\alpha \int_G |u(x)|^\alpha dx$, and appropriate $f : [o,T] \to W^{-1,q}(G)$. The corresponding subdifferentials are given by $\partial\phi_1(u) = \{-\sum_{j=1}^n D_j(|D_j u|^{p-2}D_j u)\}$, $u \in V_1$; $\partial\phi_2(u) = \{|u|^{\alpha-2}u\}$, $u \in V_2$, so we obtain existence of a solution of an initial-boundary value problem for the equation

$$(7.9) \qquad D_t(|u|^{\alpha-2}u) - \sum_{j=1}^n D_j(|D_j u|^{p-2}D_j u) = f(x,t).$$

Example 7.8 Let V_1, V_2, ϕ_1 and f be given as above but define $\phi_2(u) = 1/\alpha \int_G |u^+(x)|^\alpha dx$, $u \in V_2$, where $u^+(x) = 0$ when $u(x) < 0$ and $u^+(x) = u(x)$ when $u(x) \geq 0$. Theorem 7.4 gives existence for an initial-boundary value problem for

$$(7.10) \qquad D_t(u^+)^{\alpha-1} - \sum_{j=1}^n D_j(|D_j u|^{p-2}D_j u) = f(x,t).$$

Example 7.9 Let V_2 be a Hilbert space and set $\phi_2(u) = \max\{1/2, |u|_H^2/2\}$. The subdifferential is given by

$$M(u) = \begin{cases} 0, & |u|_H < 1, \\ \{\lambda u : 0 \leq \lambda \leq 1\}, & |u| = 1, \\ u, & |u|_H > 1. \end{cases}$$

If V_1 and N are given as in Theorem 7.4, there is then a solution pair u,v which satisfies the abstract "parabolic" equation $-u'(t) + f(t) \in N(u(t))$ where $|u(t)| > 1$ and the "elliptic" equation $f(t) \in N(u(t))$ where $|u(t)| < 1$.

Remark 7.10 Theorem 7.1 is an extension of the corresponding result of Showalter [4] when M and N are functions. Theorem 7.4 and Examples 7.7, 7.8 and 7.9 are from Grange-Mignot [1]. See Barbu [1] for related abstract results and examples where M and N are subdifferentials which are assumed related by an assumption similar to (7.4) but distinct from it. The equation (7.9) was studied directly in Raviart [2]; cf. Chapter IV.1.3 of Lions [10]. A doubly-nonlinear parabolic-hyperbolic system is discussed by Volpert-Hudjaev [1], and parabolic systems are considered by Cannon-Ford-Lair [3]. These last two references indicate how certain doubly-nonlinear equations arise in applications; for applications of equations of the type (7.8), cf. Gajewski-Zacharius [2].

Chapter 4

Selected Topics

4.1 Introduction. In this chapter we will discuss briefly some further questions arising in the study of singular or degenerate equations. Some of the presentation will be in a more "classical" spirit and we will follow the original papers in describing problems and results. Proofs will often only be sketched or even omitted entirely, mainly for reasons of space and time, but suitable references will be provided (without attempting to be exhaustive in this respect); further relevant information can often be obtained from the bibliographies in these references. The material considered by no means covers all possible questions and has been selected basically to be representative of the historical development and to reflect some current research trends in the area. Thus we will deal for example with Huygens' principle, radiation problems, initial-boundary value problems, nonwellposed problems, etc., among other questions; the section title will indicate the problem under consideration.

4.2 Huygens' principle. Referring back to (1.4.8) and (1.4.10) with $A_x = -\Delta$ let us suppose $m + p = \frac{n}{2} - 1$ with T a function so that $w^{m+p}(t)$ is given in terms of a surface mean value $\mu_x(t) * T$ of $T(\xi)$ over the sphere $|x-\xi| = t$ (here we also want $m \geq -1/2$ to insure the uniqueness of w^m). Thus if $n - (2m+1) = 2p+1$ is an odd positive integer the value $w^m(t) = w^m(x,t)$ depends only on the values of $T(\xi)$ on the surface

$|x-\xi| = t$, as was noticed by Diaz-Weinberger [2] and Weinstein [4]. To picture this one can draw the retrograde "light" cone from $(x,t) \in \mathbb{R}^n \times \mathbb{R}^+$ to the singular hyperplane $t = 0$ and look at the intersection. We note here that for the wave equation where $m = -1/2, n = 2p + 1$ is required above which reproduces a well known fact. This phenomenon furnishes an example of what is known as the minor premise in Huygens' principle. In general Huygens' principle can be enunciated in various versions involving other features as well (cf. Baker-Copson [1], Courant-Hilbert [1], Hadamard [1], Lax-Phillips [1]) but we shall simply say here that a linear second order partial differential operator $L(D_{x_k}, D_t)$ $(k = 1,\ldots,n)$ is of Huygens' type in a region $\Omega \subset \mathbb{R}^n \times \mathbb{R}$ if the value $u(x,t)$ of the solution of $Lu = 0$ at a point $(x,t) \in \Omega$ depends only upon the Cauchy data on the intersection of the space like initial manifold with the surface of the characteristic conoid with vertex (x,t). This formulation and others play an important role in wave propagation for example and the matter has been studied extensively. (In addition to the above references we cite here only some more or less recent work by Bureau [1], Douglis [1], Fox [1], Günther [1; 2; 3; 4; 5], Helgason [4], Lagnese [11; 12; 13], Lagnese-Stellmacher [14], Solomon [1; 2], and Stellmacher [1].) In dealing with Huygens' principle for singular problems we will follow Solomon [1; 2] and only make a few remarks about other work on singular problems later.

Thus let us consider the operator

(2.1) $E^m = \partial^2/\partial t^2 + \frac{2m+1}{t}\,\partial/\partial t + c(t)$

where Rem > - 1/2 or m = - 1/2 and $c(\cdot)$ is a complex valued function belonging to $L^1(o,b)$ $(0\le t\le b<\infty)$. We deal here with the singular Cauchy problem in D'

(2.2) $E^m u^m = \Delta u^m$; $u^m(0) = T \in D'$; $u_t^m(0) = 0$

where Δ is the Laplace operator in $\mathbb{R}^n$ and since $c(\cdot) \in L^1$, u^m is required only to satisfy the differential equation almost everywhere (a.e.). Solomon [1; 2] employs the Fourier technique developed in Chapter 1 and looks for resolvants $Z_x^m(t)$ satisfying (2.2) with $T = \delta$. Setting $\lambda^2 = \sum_1^n y_k^2 = |y|^2$ one then looks for a function $\hat{Z}^m(y,t) = F_x Z_x^m(t)$ satisfying

(2.3) $E^m\hat{Z}^m + \lambda^2\hat{Z}^m = 0$; $\hat{Z}^m(y,o) = 1$; $\hat{Z}_t^m(y,o) = 0$

A solution $\hat{Z}^m(y,t)$ of (2.3) is obtained by converting (2.3) into an integral equation as in Chapter 1 (cf. (1.3.24)) and using a Neumann series such as (1.3.25) (cf. also Section 1.5). Using techniques similar to those of Chapter 1 the solution is obtained in the following form (see Solomon [1], pp. 226-231).

Theorem 2.1 The unique solution of (2.3) for $t \in [o,b]$ and $y \in \mathbb{R}^n$ can be written as

(2.4) $\hat{Z}^m(y,t) = \hat{R}^m(y,t) + \hat{W}^m(y,t)$

where $\hat{R}^m$ is given by (1.3.6) with $z = t|y|$ and $\hat{W}^m$ is a "smooth" perturbation satisfying the estimate

$$(2.5)\qquad |\hat{W}^m(y,t)| \leq \frac{k_0}{|y|}\,(1+z^{\text{Rem}+\frac{1}{2}})^{-1}\int_0^t |c(\xi)|\,d\xi$$

for $|y| > \lambda_0$ and $o \leq t \leq b$ (k_0 and λ_0 are independent of (y,t)). The corresponding unique resolvant $Z_x^m(t) \in E'$ has support contained in the ball $|x| \leq t$ and $u^m(t) = Z_x^m(t) * T$ is the unique solution of (2.2) in D'.

Remark 2.2 The estimate (2.5) is useful in determining the order of the distribution $Z_x^m(t)$; in particular for Rem + 1/2 > n - 1 it follows that $Z_x^m(t)$ will be continuous in x for $0 < t \leq b$.

We will say now that (2.2) is of Huygens' type if and only if the support of $Z_x^m(t)$ is concentrated on the sphere $|x| = t$. Note here that this definition depends on the choice of initial data when $m = -1/2$ and $n = 1$ for example since the one dimensional wave operator is not a Huygens' operator in the sense previously delimited for arbitrary Cauchy data u(x,o) and $u_t(x,o)$. Solomon uses the distributions $\delta^{(\nu)}(|x|^2-t^2)$, ν = 0, 1, 2, ..., defined for t > 0 and n > 1 by (cf. Gelfand-Šilov [5])

$$(2.6)\qquad \langle\delta^{(\nu)}(|x|^2-t^2),\,\phi(x)\rangle = \frac{(-1)^\nu}{2}\int\{\left(\frac{1}{2\rho}\frac{\partial}{\partial\rho}\right)^\nu[\rho^{n-2}\tilde{\phi}]\}_{\rho=t}\,d\Omega_n$$

Here $|x| = \rho$, θ denotes the n - 1 angular variables $(\theta_1, \ldots, \theta_{n-1})$ in polar spherical coordinates with $\tilde{\phi}(\rho,\theta) = \phi(x)$, and Ω_n is the surface of the unit sphere in $\mathbb{R}^n$. For n = 1 one writes

(2.7) $$\delta^{(\nu)}(|x|^2-t) = \frac{1}{2|t|}\left(\frac{1}{2x}\,\partial/\partial x\right)^{\nu}[\delta(x+t)+\delta(x-t)]$$

clearly the distributions $\delta^{\nu}(|x|^2-t^2)$ are concentrated on the sphere $|x| = t$. Then the main result in Solomon [1] is given by

Theorem 2.3 A necessary and sufficient condition for (2.2) to be of Huygens' type is that $n - (2m+1) = 2N + 1$ with $N \geq 0$ an integer while there exists a resolvant $Z_x^m(t)$ of the form

(2.8) $$Z_x^m(t) = c_m\ t^{-2m}\sum_{\nu=0}^{N} a_{\nu}(t,m)\delta^{(N-\nu)}(|x|^2-t^2)$$

where $c_m = (-1)^N\pi^{-n/2}\Gamma(m+1)$, $a_0 = 1$, and for $\nu = 1, \ldots, N$ (if $N > 0$) one has $E^{-m}(a_N) = 0$ with

(2.9) $$4(ta_{\nu}' + \nu a_{\nu}) = E^{-m}(a_{\nu-1})$$

Remark 2.4 We recall from Gelfand-Šilov [5] that for ν = 0, 1, 2, . . .

(2.10) $$F\delta^{(\nu)}(|x|^2-t^2) = \hat{c}(\nu,n)|y|^{2\nu-n+2}z^{1/2(n-2)-\nu}J_{\frac{1}{2}(n-2)-\nu}(z)$$

where $c(\nu,n) = \frac{1}{2}(-1/2)^{\nu}(2\pi)^{n/2}$ and $z = t|y|$. Then examples to illustrate Theorem 2.3 can be given as follows. Suppose first that $Z_x^m(t)$, expressed by (2.8), has precisely one term (i.e., $a_{\nu} = 0$ for $\nu \geq 1$). Then by (2.9) $E^{-m}(a_0) = c(t) = 0$ since a_0 = 1 and we are in the EPD situation. Let N be the integer involved in (2.8) with of necessity $n - (2m+1) = 2N + 1$ an odd positive integer as indicated. To check the form of $Z_x^m(t)$ given by (2.8) in this case we use (2.10) to obtain $Z^m(y,t) = c_m t^{-2m} a_0 F\delta^{(N)}$.

$(|x|^2-t^2)$ and since $\frac{1}{2}(n-2) - N = m$ this yields $\hat{Z}^m(y,t) = 2^m\Gamma(m+1)z^{-m}J_m(z) = \hat{R}^m(y,t)$ as required by (1.3.6). Next suppose that (2.8) has no more than two terms (i.e., $a_\nu = 0$ for $\nu \geq 2$). From (2.9) and the requirement $E^{-m}(a_1) = 0$ we have $ta_1' + a_1 = E^{-m}(a_0) = c(t)$ or $c(t) = (ta_1)'$ and $E^{-m}(a_1) = a_1'' + \frac{(1-2m)}{t}a_1' + c(t)\, a_1 = 0$. Combining these two equations one has $a_1'' + \frac{(1-2m)}{t} a_1' + (ta_1' + a_1)a_1 = t^{-1}[ta_1'-2ma_1 + \frac{t^2}{2}a_1^2]' = 0$ or $ta_1' - 2ma_1 + \frac{t^2}{2} a_1^2 = \alpha_0$ for α_0 any constant. Making a change of variables $a_1(t) = \frac{2}{t} w'/w$ this becomes $w'' - \frac{2m+1}{t}w' - \frac{\alpha_0}{2} w = 0$. Now in terms of w we have $a_1(t) = \frac{2}{t}(\log w)'$ with $c(t) = 2(\log w)''$ and suitable w can be found as follows. If $\alpha_0 = 0$ we get $w(t) = \alpha + \beta t^{2m+2}$ where α and β are arbitrary constants but may depend on m. In order that the resulting a_1 be such that (2.8) satisfies the resolvant initial conditions we must have $\alpha \neq 0$ while for $m = -1/2$ it is necessary that $\beta = 0$. If $\alpha_0 \neq 0$ the solutions w of the differential equation for w will be of the form $w = \exp(-\tau/2)\, F(-\frac{k}{2}; -k; \tau)$ where $\tau = t\alpha^{1/2}$, $k = 2m+1$, and $F(\gamma; \delta; t)$ is any solution of Kummer's confluent hypergeometric equation $tF'' + (\delta - t)F' + \gamma F = 0$. Upon examination of the behavior of such $F(\gamma;\delta;t)$ as $t \to 0$ it follows that for the resulting a_1 to satisfy the appropriate initial conditions F must be taken in the form $F = \alpha\, {}_1F_1(-\frac{k}{2}; -k; \tau) + \beta F_2(-\frac{k}{2}; -k; \tau)$ where F_2 is any solution of the Kummer equation independent of ${}_1F_1$, $\alpha \neq 0$, and when $m = -1/2$ (i.e. $k = 0$) $\beta = 0$. Further observations can be found in Solomon [1;2].

Remark 2.5 We will deal here explicitly only with some work of Fox [1] on the solution and Huygens' principle for singular Cauchy problems of the type

$$(2.11)\qquad u^m_{tt} + \frac{2m+1}{t} u^m_t - \sum_{k=1}^{n} (u^m_{x_k x_k} + \frac{\lambda_k}{x_k^2} u^m) = L^m_\lambda u^m = 0$$

with data $u^m(x,o) = T(x)$ and $u^m_t(x,o) = 0$ given on the singular hyperplane $t = o$. Except for some results of Günther (loc. cit.) most of the other work on Huygens' principle mentioned earlier deals with nonsingular Cauchy problems so we will not discuss it here; Günther's work on the other hand is expressed in a more geometric language which requires some background not assumed or developed in this book and hence the details will be omitted. Thus let $\Gamma = \Gamma(x,t;\xi) = t^2 - \sum_{k=1}^{n} (x_k - \xi_k)^2$ be the square of the Lorentzian distance between a point $(x,t) \in \mathbb{R}^n \times \mathbb{R}$ and a point (ξ,o) in the singular hyperplane $t = o$. We denote by $(2.11)_{-m}$ the equation (2.11) with index -m and observe that if u^{-m} satisfies $(2.11)_{-m}$ then $u^m = t^{-2m} u^{-m}$ satisfies (2.11) (cf. Remark 1.4.7). Fox composes the transformations $z_k = \Gamma/4x_k\xi_k$ and $y_k = (\xi_k/x_k)\,\Gamma$ $(k = 1, \ldots, n)$ to produce new variables (z,Γ,y) in place of (x,t,ξ) (suitable regions are of course delineated). A solution of $(2.11)_{-m}$ is sought in the form $u^{-m} = \Gamma^{m-\frac{n}{2}}\, V(z)$ where $V(z)$ is to be determined. After some computation one shows that such u^{-m} are solutions if $V(z)$ satisfies a certain system of n ordinary differential equations and such a $V(z)$ can be found in the form $V = F_B$ where F_B is the Lauricella function given for

$|z_k| < 1$ by (cf. Appell - Kampé de Fériet [1])

$$(2.12) \qquad F_B = F_B\left(\alpha, 1-\alpha, m - \frac{n}{2} + 1, z\right)$$
$$= \sum_{\rho_1,\ldots,\rho_s=0}^{\infty} \frac{1}{(m-\frac{n}{2}+1, \sum_1^s \rho_k)} \prod_{k=1}^{s} \frac{(\alpha_k,\rho_k)(1-\alpha_k,\rho_k)z_k^{\rho_k}}{\rho_k!}$$

where $\alpha_k = \frac{1}{2} + \frac{1}{2}(1-4\lambda_k)^{1/2}$ and $(\ell,q) = \Gamma(\ell+q)/\Gamma(\ell)$. The series defining F_B converges uniformly in (α,m,z) on closed bounded regions where $|z_k| < 1$ and $m - \frac{n}{2} + 1 \neq 0, -1, -2, \ldots$. One notes that since α_k and $1 - \alpha_k$ are complex conjugates for λ_k real, the products $(\alpha_k,\rho_k)(1-\alpha_k,\rho_k)$ are real, and thus F_B is real when m and the λ_k are real. If $\mathrm{Re}\, m > \frac{n}{2} - 1$ and $|x_k| > |t| > 0$ it makes sense then to consider as a possible solution of the singular Cauchy problem for (2.11) the function.

$$(2.13) \qquad u_\lambda^m(x,t,T) = k_m \int_{S(x,t)} T(\xi) v_\lambda^m(x,t,\xi)\, d\xi$$

where $k_m = \Gamma(m+1)/\pi^{n/2}\Gamma(m-\frac{n}{2}+1)$ and $v_\lambda^m(x,t,\xi) = |t|^{-2m}\Gamma^{m-\frac{n}{2}} F_B$ is a solution of (2.11). We assume here that $T(\cdot)$ is suitably differentiable and denote by $S(x,t)$ the region of the singular hyperplane $t = 0$ cut out by the characteristic conoid with vertex (x,t) (thus $S(x,t)$ is an n-dimensional sphere of radius $|t|$ and center x in this hyperplane determined by $\Gamma(x,t,\xi) \geq 0$). As described, with $|x_k| > |t| \geq 0$, $S(x,t)$ does not intersect any of the hyperplanes $x_k = 0$ and lies in an octant of the hyperplane $t = 0$. Note here also that the factor $\Gamma(m-\frac{n}{2}+1)^{-1}$ in k_m can be played off against $(m-\frac{n}{2}+1, \sum_1^s \rho_k)^{-1}$ in F_B to remove any difficulties when $m-\frac{n}{2}+1 = -\rho$. Now for $\mathrm{Re}\, m > \frac{n}{2} + 1$ one checks easily that $L^m_\lambda u^m_\lambda = k_m \int_{S(x,t)} T(\xi) L^m_\lambda v^m_\lambda(x,t,\xi)d\xi = 0$ when $T(\cdot)$ is continuous, since all boundary integrals which arise will vanish. Using analytic continuation and after several pages of interesting analysis Fox shows that (for $m \neq -1, -2, \ldots$) if $T(\cdot) \in C^{\{2+p\}}$, where p is the smallest integer such that $\text{Re}\, m \geq \frac{n}{2} - p - 1$, then u^m_λ, defined by (2.13), satisfies the singular Cauchy problem for (2.11) ($\{2+p\} = 0$ if $2 + p \leq 0$). Uniqueness of the solution $u^m_\lambda(x,t,T)$ is proved for $2m + 1 \geq 0$ and it is shown that Huygens' principle holds when, for some nonnegative integer p, the parameters m and λ_k satisfy $m = \frac{n}{2} - p - 1$, $\lambda_k = \alpha_k(1-\alpha_k)$ where $\alpha_k \in \{1,\ldots,p+1\}$, and $\sum_1^n \alpha_k \leq n + p$. For $m \neq -1, -2, \ldots$ the solution u^m_λ, although not unique for $\text{Re}\, m < -1/2$, is the unique analytic continuation of the unique solutions u^m_λ with $2m + 1 \geq 0$ and the criteria above for Huygens' principle remain valid (in fact necessary and sufficient).

4.3 The generalized radiation problem of Weinstein. One can read extensively on the subject of radiation for the wave equation and we mention for example Courant-Hilbert [1], Lax-Phillips [1], Sommerfeld [1], and references there. The usual formulation (cf. Weinstein [6]) prescribes a function $f(\cdot)$ with $f(t) = 0$ for $t < 0$ and asks for a function u satisfying $u_{tt} = \Delta u$ in $\mathbb{R}^n \times \mathbb{R}$ with $u(x,o) = u_t(x,o) = 0$ and such that

$$(3.1) \qquad \lim_{r\to 0} \int \frac{\partial u}{\partial r} d\sigma_n = -\omega_n f(t)$$

where $r^2 = |x|^2 = \sum_1^n x_i^2$, $\omega_n = 2\pi^{n/2}/\Gamma(n/2)$, and the integral in

(3.1) is over a sphere of radius r in $\mathbb{R}^n$. When $n = 3$ for example one has the well-known solution $u(x,t) = f(t-r)/r$ and one is led to look for solutions which depend only on t and r. This entails a radial form of Δ and yields a singular problem in two variables

$$(3.2) \qquad u_{tt} = u_{rr} + \frac{n-1}{r} u_r$$

with $u = 0$ for $r \geq t$ and for $r = 0$, u should have a singularity of the form $r^{-(n-2)}$. Now replace $n - 1$ by $2m + 1$ and consider a generalized radiation problem

$$(3.3) \qquad u_{tt}^m = u_{rr}^m + \frac{2m+1}{r} u_r^m$$

where we recall that if u^m satisfies (3.3) with index m then $u^{-m} = r^{2m}u^m$ satisfies (3.3) with index $-m$ (cf. Remark 1.4.7). If $2m + 1 = n - 1$ then $2m = n - 2$ so that a solution u^{-m} finite near $r = 0$ corresponds to a solution u^m with a singularity of the desired type as $r \to 0$. Thus, thinking of suitably negative values of Rem (see Theorem 3.2 for the precise range Rem < 0), one looks for a solution $u^m(t,r)$ of (3.3) satisfying

$$(3.4) \qquad u^m(t,o) = f(t); \; u^m(t,t) = 0$$

where $f(t) = 0$ for $t < 0$. This type of problem has been studied in particular by Diaz and Young [1; 8], Lieberstein [1; 2], Lions [4], Suschowk [1], Weinstein [6], and Young [6; 12].

We will follow Lions [4] here in treating the generalized

radiation problem because of the generality of his approach and the use of transmutation operators, which serves to augment Section 1.7. First, as a sketch of the procedure, we consider for Re$m > -1/2$

$$(3.5) \qquad v^m(t,r) = \frac{2}{\Gamma(m+1/2)} \int_r^\infty z(z^2-r^2)^{m-1/2}\, u^m(t,z)dz$$

Then, if u^m satisfies (3.3) with $u^m(t,r) = 0$ for $t \le r$, it follows that $v^m_{tt} = v^m_{rr}$ with $v^m(t,r) = 0$ for $t \le r$. Consequently $v^m(t,r) = F^m(t-r)$ where $F^m(s) = 0$ for $s \le 0$. Now (cf. below) (3.5) can be inverted in the form

$$(3.6) \qquad u^m(t,r) = \frac{2}{\Gamma(-m-1/2)} \int_r^\infty z(z^2-r^2)^{-m-3/2}\, v^m(t,z)dz$$

For Re$m < -1/2$ (3.6) is valid in the usual sense while both (3.5) and (3.6) can be extended by analytic continuation to every $m \in \mathbb{C}$. Putting $F^m(t-z)$ in (3.6) and integrating by parts one obtains

$$(3.7) \qquad u^m(t,r) = \frac{1}{\Gamma(-m+1/2)} \int_r^\infty (z^2-r^2)^{-m-1/2}(F^m)'(t-z)dz$$

For Re$m \ge 0$ such $u^m(t,r)$ do not converge when $r \to 0$ unless F^m, and hence u^m, is identically zero. For Re$m < 0$, on the other hand, as $r \to 0$

$$(3.8) \qquad u^m(t,r) \to \frac{1}{\Gamma(-m+1/2)} \int_0^\infty z^{-2m-1}(F^m)'(t-z)dz$$

so that, recalling the definition of the R - L integral in Section 1.6 (cf. (1.6.20) for example),

$$(3.9)\qquad u^m(t,o) = \frac{\Gamma(-2m)}{\Gamma(-m+1/2)} I^{-2m}(F^m)' = f(t)$$

is required. Since $I^{-2m}(F^m)' = I^{-2m-1}(F^m)$ we have F^m determined uniquely by $F^m = (\Gamma(-m+1/2)/\Gamma(-2m))\, I^{2m+1}(f)$ and putting (3.9) in (3.7) one obtains

$$(3.10)\qquad u^m(t,r) = \frac{1}{\Gamma(-2m)} \int_r^\infty (z^2-r^2)^{-m-1/2}(I^{2m}f)(t-z)dz$$

which is equivalent to Weinstein's solution (cf. Weinstein [6]) and demonstrates uniqueness also.

In Lions [4] the formal calculations above are rephrased and justified in terms of transmutation operators as follows. (We recall from Section 1.7 that B transmutes P into Q if QB = BP.) Now Let $\Omega = (o, \infty)$ and $D'_-(\Omega) = D_+(\Omega)'$ be the space of distributions on Ω with support limited to the right (cf. Schwartz [1] and note that $D_+(\Omega)$ signifies C^∞ functions with not necessarily compact support but equal to zero near the origin). For simplicity we will occasionally use a function notation for distributions in what follows in writing for example $\langle T,\phi\rangle = \langle T(x),\phi(x)\rangle$. Then for $T \in D'(\Omega)$ set $MT(\xi) = T(\sqrt{\xi})$ which means that $\langle MT,\phi\rangle = \langle T(x), 2x\phi(x^2)\rangle$ for $\phi \in D(\Omega_\xi)$; it is easy to see that M is an isomorphism $D'(\Omega_x) \to D'(\Omega_\xi)$ (or $D'_-(\Omega_x) \to D'_-(\Omega_\xi)$) with inverse $M^{-1}S(x) = S(x^2)$ (i.e., $\langle M^{-1}S,\psi\rangle = \langle S(\xi), \frac{1}{2}\xi^{-1/2}\psi(\sqrt{\xi})\rangle$ for $\psi \in D(\Omega_x)$). A simple calculation yields $MD_x^2T = (4\xi D_\xi^2 + 2D_\xi)\, MT$ for $T \in D'(\Omega_x)$ (here $D_x = d/dx$, etc.). Now referring to the standard spaces D',

D'_+, and D'_- over $\mathbb{R}$ as in Schwartz [1], we set for $p \in \mathbb{C}$, $Y_p = (1/\Gamma(p))\ Pf(x^{p-1})|_{x>0} \in D'_+$ (the notation Pf is explained in Schwartz [1]). Then $Y_p = 0$ for $x < 0$, $Y_{-\ell} = D^\ell \delta$ for ℓ a nonnegative integer, $p \to Y_p : \mathbb{C} \to D'_+$ is analytic and $Y_p * Y_q = Y_{p+q}$. Set now $Z_p(x) = Y_p(-x)$ so that $Z_p \in D'_-$, $Z_p = 0$ for $x > 0$, $Z_{-\ell} = (-1)^\ell D^\ell \delta$ for ℓ a nonnegative integer, etc.; in particular $DZ_p = -Z_{p-1}$ and $xZ_p = -pZ_{p+1}$. Finally one notes that $T \to S*T$ is a continuous linear map $D'_-(\Omega) \to D'_-(\Omega)$ when $S \in D'_-$ is zero for $x > 0$ (formally $(S*T)(x) = \int_x^\infty S(x-y)T(y)dy$) and we denote by $Z_p^* \in L(D'_-(\Omega), D'_-(\Omega))$ the map $T \to Z_p * T$.

Now let $L_m = D_x^2 + \frac{2m+1}{x} D_x$ ($L_m = L_m^x$ in Chapter 1) and note that $L_m : D'_-(\Omega) \to D'_-(\Omega)$ is continuous. For $m \in \mathbb{C}$ Lions constructs a transmutation operator $\mathcal{H}_m = M^{-1} \circ Z^*_{-m-1/2} \circ M$ of D^2 into L_m on $D'_-(\Omega)$ (i.e., $\mathcal{H}_m D^2 = L_m \mathcal{H}_m$) such that $\mathcal{H}_m$: $D'_-(\Omega) \to D'_-(\Omega)$ is an isomorphism and $m \to \mathcal{H}_m$ is an entire analytic function with values in $L(D'_-(\Omega), D'_-(\Omega))$. The inverse $H_m = \mathcal{H}_m^{-1} = M^{-1} \circ Z^*_{m+1/2} \circ M$, which necessarily satisfies $D^2 H_m = H_m L_m$ on $D'_-(\Omega)$, enjoys similar properties and in function notation one can write

$$(3.11)\qquad \mathcal{H}_m T(x) = \frac{2}{\Gamma(-m-1/2)} \int_x^\infty y(y^2-x^2)^{-m-3/2} T(y)dy;\ \text{Re}\, m < -1/2$$

$$(3.12)\qquad H_m T(x) = \frac{2}{\Gamma(m+1/2)} \int_x^\infty y(y^2-x^2)^{m-1/2} T(y)dy;\ \text{Re}\, m > -1/2$$

For other values of $m \in \mathbb{C}$ one extends these formulas by analytic continuation. It is important to note that if $T \in D'_-(\Omega)$ is zero for $x > a_T$ then $\mathcal{H}_m T = H_m T = 0$ for $x > a_T$. Lions phrases the generalized radiation problem as follows.

Problem 3.1 Let $P \subset \mathbb{R} \times \mathbb{R}$ be the open half plane (t,r) where $r > 0$ and $Q \subset P$ be the set $Q = \{(t,r) \in P;\ t \geq r\}$. For $\text{Re}\, m < 0$ find $u^m \in D'(P)$ satisfying (3.3) with supp $u^m \subset Q$ and $u^m(\cdot,r) \to f$ in D'_+ as $r \to 0$ where $f \in D'_+$ is given with $f = 0$ for $t < 0$.

The problem is restricted to $\text{Re}\, m < 0$ since it is shown not to be meaningful for other $m \in \mathbb{C}$ (cf. Lions [4]). There results

Theorem 3.2 Problem 3.1 has a unique solution u^m depending continuously in $D'(P)$ on $f \in D'_+$ (f as described).

Formally the proof goes as follows. One writes $v^m = (1 \otimes H_m)u^m$ so that v^m is given by (3.5) for $\text{Re}\, m > -1/2$ say (cf. (3.12)). Applying $1 \otimes H_m$ to (3.3) one has $v^m_{tt} = v^m_{rr}$ and $v^m \in D'(P)$ has its support in Q. Consequently $v^m(t,r) = F^m(t-r)$ as before with $F^m \in D'$ and $F^m(s) = 0$ for $s < 0$. Since H_m is an isomorphism with $H_m^{-1} = \mathcal{H}_m$ we have $u^m = (1 \otimes \mathcal{H}_m)v^m$ and for $\text{Re}\, m < -1/2$ say we obtain formally by (3.11) (cf. (3.6))

$$(3.13)\qquad u^m(t,r) = \frac{2}{\Gamma(-m-1/2)} \int_r^\infty z(z^2-r^2)^{-m-3/2} F^m(t-z)\,dz$$

$$= (S^r_m * F^m)(t)$$

where $\langle S^r_m,\phi\rangle = (2/\Gamma(-m-1/2))\int_r^\infty z(z^2-r^2)^{-m-3/2}\phi(z)\,dz$ for $r > 0$, $\text{Re}\, m < -1/2$, and $\phi \in D_-$. The function $m \to S^r_m$ can in fact be extended by analytic continuation to be entire with values in

D_+' and, for $\mathrm{Re}\, m < 0$, as $r \to 0$, $S_m^r \to (2\Gamma(-2m-1)/\Gamma(-m-1/2))\, Y_{-2m-1}$ in D_+'. Hence from (3.13) as $r \to 0$

$$(3.14) \qquad u^m(\cdot,r) \quad \frac{2\Gamma(-2m-1)}{\Gamma(-m-1/2)}\, Y_{-2m-1} * F^m$$

and this limit must equal f (cf. (3.9)). Consequently the explicit solution of Problem 3.1 for $\mathrm{Re}\, m < 0$ is

$$(3.15) \qquad u^m(t,r) = \frac{\Gamma(-m-1/2)}{2\Gamma(-2m-1)}\, S_m^r * Y_{2m+1} * f$$

which can be rewritten in a form equivalent to (3.10) (cf. Lions [4]).

4.4 Improperly posed problems. Improperly posed problems such as the Dirichlet problem for the wave equation or the Cauchy problem for the Laplace equation have been studied for some time and have realistic applications in physics. For a nonexhaustive list of references see e. g., Abdul-Latif-Diaz [1], Agmon-Nirenberg [3], Bourgin [1], Bourgin-Duffin [2], Brezis-Goldstein [5], Dunninger-Zachmanoglou [2], Fox-Pucci [2], John [2], Lavrentiev [2], Levine [1; 5; 10; 11], Levine-Murray [12], Ogawa [1;2], Pucci [2], Sigillito [4], and a recent survey by Payne [3] (cf. also - Symposium on nonwellposed problems and logarithmic convexity, Springer, Lecture notes in mathematics, Vol. 316, 1973). In particular there has been considerable investigation of improperly posed singular problems by Diaz-Young [10], Dunninger-Levine [1], Dunninger-Weinacht [3], Travis [1], Young [4; 5; 7; 8; 11], etc. and we will discuss some of this latter work.

The most general abstract results seem to appear in Dunninger-Levine [1] and we will sketch their technique first. Let E be a real Banach space and A a closed densely defined linear operator in E (as is well-known the use of spaces E over $\mathbb{R}$ instead of $\mathbb{C}$ involves no loss of generality). We shall denote by $\sigma_p(A)$ the point spectrum of A.

Definition 4.1 The E valued function $u^m(\cdot)$ is said here to be a strong solution of

$$(4.1) \qquad u^m_{tt} + \frac{2m+1}{t} u^m_t + Au^m = 0$$

on (o,b) if $u^m(\cdot)$ is norm continuous on (o,b), $u^m(\cdot) \in C^2(E)$ weakly, and for any $e' \in E'$

$$(4.2) \qquad \langle u^m_{tt}, e' \rangle + \frac{2m+1}{t} \langle u^m_t, e' \rangle + \langle Au^m, e' \rangle = 0$$

on (o,b), where $\langle u^m_{tt}, e' \rangle = \frac{d^2}{dt^2} \langle u^m, e' \rangle$, etc., and writing $\rho(t) = t^{2m+1} \, \| Au^m(t) \|$ for $m > -1/2$ with $\rho(t) = t \, \| Au^m(t) \|$ for $m < -1/2$, one has $\rho(\cdot) \in L^1(o,b)$.

The nonsingular case m = 1/2 is encompassed in forthcoming work of Dunninger-Levine and will not be discussed here. Set now $p = |m|$ and let $\{\lambda_n\}$ denote the positive roots of $J_p(\sqrt{\lambda_n}\, b) = 0$ for $m \geq 0$ or $m < -1/2$; if $-1/2 < m < 0$ let $\{\lambda_n\}$ denote the positive roots of $J_{-p}(\sqrt{\lambda_n}\, b) = 0$. Then there results

<u>Theorem 4.2</u> Let u^m be a strong solution of (4.1) as in Definition 4.1 with $u^m(b) = 0$. If $m > -1/2$ assume $t^{2m+1}(\| u_t^m\| + t\|u^m\|) \to 0$ as $t \to 0$ while for $m < -1/2$ assume that $t\| u_t^m\| + \|u^m\| \to 0$ as $t \to 0$. Then $u^m \equiv 0$ if and only if the sequence $\{\lambda_n\}$ satisfies $\lambda_n \notin \sigma_p(A)$ for all n.

The proof can be sketched as follows. Consider the eigenvalue problem

$$(4.3) \qquad \psi'' + \frac{1}{t}\psi' + (\lambda - \frac{p^2}{t^2})\,\psi = 0$$

$$(4.3) \qquad \psi(b) = 0;\ t^{1/2}\psi(t) \text{ bounded}$$

and take for $m \geq 0$ or $m < -1/2$ the solutions

$$(4.5) \qquad \psi_{n,p}(t) = J_p(\sqrt{\lambda_n}t);\ J_p(\sqrt{\lambda_n}b) = 0$$

while for $-1/2 < m < 0$ choose the solutions

$$(4.6) \qquad \psi_{n,-p}(t) = J_{-p}(\sqrt{\lambda_n}t);\ J_{-p}(\sqrt{\lambda_n}b) = 0$$

Assume first that, for some n, $\lambda_n \in \sigma_p(A)$ with $Av_n = \lambda_n v_n$ and set

$$(4.7) \qquad u^m(t) = \begin{cases} t^{-m}J_p(\sqrt{\lambda_n}t)v_n; & m \geq 0 \text{ or } m < -1/2 \\ t^{-m}J_{-p}(\sqrt{\lambda_n}t)v_n; & -1/2 < m < 0 \end{cases}$$

Such u^m are strong solutions of (4.1) in the sense of Definition 4.1 and their behavior as $t \to 0$ is correct, as specified in the hypotheses of Theorem 4.2. Hence $\lambda_n \in \sigma_p(A)$ for some n implies the existence of nontrivial u^m satisfying the hypotheses of

Theorem 4.2 which means that $u^m \equiv 0$ in Theorem 4.2 implies $\lambda_n \notin \sigma_p(A)$ for all n.

On the other hand suppose $\lambda_n \notin \sigma_p(A)$ for all n and denote by ψ_n the appropriate functions $\psi_{n,p}$ or $\psi_{n,-p}$ from (4.5) - (4.6). Suppose u^m is a strong solution of (4.1) satisfying the conditiosn of Theorem 4.2 and for some $\delta \in (o,b)$ set

$$(4.8) \qquad v_n^m(\delta) = \int_\delta^b t^{m+1}\psi_n(t)u^m(t)dt$$

The $u_n^m(\delta)$ are well defined by virtue of the hypotheses on u^m and one notes easily that, as $t \to 0$, $t^{m+1}\psi_n(t) = O(t^{2m+1})$ for $m > -\frac{1}{2}$ while $t^{m+1}\psi_n(t) = O(t)$ for $m < -\frac{1}{2}$. Hence $t \to t^{m+1}\psi_n(t)\|u^m(t)\| \in L^1(o,b)$ and $v_n^m = \lim v_n^m(\delta)$ exists in the norm topology of E as $\delta \to 0$. Since A is closed it can be passed under the (Riemann type) integral sign in (4.8) so that $v_n^m(\delta) \in D(A)$ and furthermore $y_n^m = \lim A\, v_n^m(\delta)$ exists as $\delta \to 0$ under our hypotheses on $\rho(t)$ in Definition 4.1 with $Av_n^m = y_n^m$ by the closedness of A. For $e' \in E'$ fixed a routine calculation yields now

$$(4.9) \qquad \langle Av_n^m(\delta), e'\rangle = \int_\delta^b t^{m+1}\psi_n(t)\langle Au^m(t), e'\rangle dt$$

$$= -\int_\delta^b t^{m+1}\psi_n(t)\{\frac{d^2}{dt^2}\langle u^m(t), e'\rangle + \frac{2m+1}{t}\frac{d}{dt}\langle u^m(t), e'\rangle\}dt$$

$$= [t^{2m+1}(t^{-m}\psi_n(t))'\langle u^m(t), e'\rangle - t^{m+1}\psi_n(t)\langle u^m(t), e'\rangle']\Big|_\delta^b$$

$$+ \lambda_n \int_\delta^b t^{m+1}\psi_n(t)\langle u^m(t), e'\rangle dt$$

and in view of the hypotheses one obtains as $\delta \to 0$ $\langle Av_n^m, e' \rangle = \lim \lambda_n \langle v_n^m(\delta), e' \rangle = \lambda_n \langle v_n^m, e' \rangle$. Therefore $\langle (A-\lambda_n)v_n^m, e' \rangle = 0$ for any $e' \in E'$ which implies $Av_n^m = \lambda_n v_n^m$ which by our assumptions means $v_n^m = 0$. But one knows that the Bessel functions ψ_n form a complete orthogonal set on (o,b) relative to the weight function t(cf. Titchmarsh [3]) and hence from $\langle v_n^m, e' \rangle = \int_0^b t^{m+1}\psi_n(t)\langle u^m(t), e' \rangle dt = 0$ for all n it follows that $\langle u^m(t), e' \rangle = 0$ for any $e' \in E'$ from which results $u^m(t) \equiv 0$.

Other boundary conditions of the type $u_t^m(b) + \alpha u^m(b) = 0$ are also considered in Dunninger-Levine [1] and uniqueness for weaker type solutions is also treated. One notes that very little is required of the operator A in the above result. The technique does not immediately extend to more general locally convex spaces E because of the norm conditions.

We go now to some more concrete problems which are related to Theorem 4.2 in an obvious way. First we sketch some results of Young [4] on uniqueness of solutions for the Dirichlet problem relative to the singular hyperbolic operator defined by $L_m u = u_{tt} + \frac{2m+1}{t}u_t - \sum_1^n \sum_1^n (a^{ij}u_{x_i})_{x_j} + cu$ where c and the a^{ij} are suitably smooth functions of $x = (x_1, \ldots, x_n)$ and m is real. Let $\Omega \subset \mathbb{R}^n$ be bounded and open, $Q = \Omega \times (o,b)$, $c(x) \geq 0$ for $x \in \Omega$, and assume the matrix (a^{ij}) is symmetric and positive definite; sufficient smoothness of the boundary Γ of Ω will be taken for granted.

Theorem 4.3 Let $L_m u^m = 0$ in Q with $u^m = 0$ on $\Gamma x[o,b)$ and $u^m(x,o) = 0$. Suppose $m > -\frac{1}{2}$ and assume $u^m \in C^2(Q) \cap C^1(\overline{Q})$. Then $u^m \equiv 0$ in $\overline{Q}$.

The proof uses the fact that any solution u^m of $L_m u^m = 0$ belonging to C^2 for $t > 0$ and to C^1 for $t \geq 0$ necessarily satisfies $u^m_t(x,o) = 0$ for any $m \neq -\frac{1}{2}$ (cf. Fox [1], Walter [1]). Now integrate the identity

$$(4.10) \qquad 2u_t L_m u = (u_t^2 + \sum_{i,j} a^{ij} u_{x_i} u_{x_j} + cu^2)_t + \frac{2(2m+1)}{t} u_t^2 - 2\sum_{i,j} (a^{ij} u_{x_i} u_t)_{x_j}$$

over the region $Q_s = \Omega x(o,s)$, $s \leq b$, and use the divergence theorem to obtain, when $L_m u = 0$,

$$(4.11) \qquad \int_{\partial Q_s} [(u_t^2 + \sum_{i,j} a^{ij} u_{x_i} u_{x_j} + cu^2)\nu_t - 2\sum_{i,j} a^{ij} u_{x_i} u_t \nu_j] d\sigma + 2(2m+1)\int_{Q_s} \frac{1}{t} u_t^2 dxdt = 0$$

where $\nu = (\nu_1, \ldots, \nu_n, \nu_t)$ is the exterior normal on ∂Q_s = boundary Q_s. Putting our u^m in (4.11) and using the boundary conditions plus $u^m_t(x,o) = 0$ we have the equation $\int_\Omega [(u^m_t)^2 + \sum a^{ij} u^m_{x_i} u^m_{x_j} + c(u^m)^2]|_{t=s} dx + 2(2m+1) \int_{Q_s} \frac{1}{t}(u^m_t)^2 dxdt = 0$. Since (a^{ij}) is positive definite, $c \geq 0$, and $m > -\frac{1}{2}$ both of these integrals are zero and by continuity the first term then vanishes for $0 \leq s \leq b$. Hence $u^m \equiv 0$ in $\overline{Q}$ since $u^m(x,o) = 0$.

Theorem 4.4 If $m \leq -\frac{1}{2}$ then any solution $u^m \in C^2(Q) \cap C^1(\overline{Q})$ of $L_m u^m = 0$, with $u^m = 0$ on the boundary of Q, vanishes identically if and only if $J_{-m}(\sqrt{\lambda_n}b) \neq 0$, where the λ_n are the nonzero eigenvalues of the problem $\lambda v = -\sum_{i,j}(a^{ij}v_{x_i})_{x_j} + cv$ in Ω with $v = 0$ on Γ.

There is an obvious relation between Theorems 4.4 and 4.2 where A corresponds to the operator determined by $Aw = cw - \sum_{i,j}(a^{ij}w_{x_i})_{x_j}$ in $E = L^2(\Omega)$ with $w = 0$ on Γ. The proof of Theorem 4.4 goes as follows. First suppose there is a nonzero eigenvalue λ_n such that $J_{-m}(\sqrt{\lambda_n}b) = 0$ with corresponding eigenfunction $v_n(x)$. Then $u^m(x,t) = t^{-m}J_{-m}(\sqrt{\lambda_n}t)v_n(x)$ satisfies $L^m u^m = 0$ and the specified boundary conditions. Conversely if $J_{-m}(\sqrt{\lambda_n}b) \neq 0$ for any λ_n one integrates the identity $wL_m u - uM_m w = (u_t w - uw_t + \frac{2m+1}{t}uw)_t - \sum_j [\sum_i a^{ij}(u_{x_i}w - uw_{x_i})]_{x_j}$ over the cylinder Q_s^b enclosed by Q between the planes $t = s$ and $t = b$ ($0 < s < b$). Here M_m is the formal adjoint of L_m given by $M_m w = w_{tt} - (2m+1)(w/t)_t - \sum(a^{ij}w_{x_i})_{x_j} + cw$. By the divergence theorem one obtains

$$(4.12) \qquad \int_{Q_s^b}(wL_m u - uM_m w)\,dxdt$$
$$= \int_{\partial Q_s^b}[(u_t w - uw_t + \frac{2m+1}{t}uw)\nu_t - \sum a^{ij}(u_{x_i}w - uw_{x_i})\nu_j]d\sigma$$

Now let $L_m u^m = 0$ with $u^m = 0$ on the boundary of Q and set $w^m(x,t) = t^{m+1}J_{-m}(\sqrt{\lambda_n}t)v_n(x)$ where λ_n is a nonzero eigenvalue of the problem indicated in the statement of Theorem 4.4. It is easily checked that $M_m w_m = 0$ so that the left hand side in (4.12) vanishes and since $w^m = 0$ on Γ we have

(4.13) $$\int_\Omega \left(u_t^m w^m - u^m w_t^m + \frac{2m+1}{t} u^m w^m\right)\Big|_{t=s}^{t=b} dx = 0$$

Now as $s \to 0$ w_t^m and $\frac{1}{t}w^m$ are bounded with $u^m(x,o) = 0$ (and $u_t^m(x,o) = 0$ for $m \neq -\frac{1}{2}$); the evaluation in (4.13) at $t = s$ tends to zero and we obtain

(4.14) $$b^{m+1} J_{-m}(\sqrt{\lambda_n} b) \int_\Omega u_t^m(x,b) v_n(x) dx = 0$$

since $u^m(x,b) = 0$. Consequently for $n = 1, 2, \ldots$ $\int_\Omega u_t^m(x,b) v_n(x) dx = 0$ and by the completeness of the v_n (cf. Hellwig [3]) it follows that $u_t^m(x,b) = 0$. Now one may integrate (4.10) over Q_s^b and since $u^m(x,b) = u_t^m(x,b) = 0$ there results

(4.15) $$\int_\Omega [(u_t^m)^2 + \sum a^{ij} u_{x_i}^m u_{x_j}^m + c(u^m)^2]\Big|_{t=s} dx$$

$$= 2(2m+1) \int_{Q_s^b} \frac{1}{t}(u_t^m)^2 dx dt$$

Since $m \leq -\frac{1}{2}$ it follows that for any s, $0 \leq s \leq b$, $\int_\Omega [(u_t^m)^2 + \sum a^{ij} u_{x_i}^m u_{x_j}^m + c(u^m)^2]\Big|_{t=s} dx = 0$ and the conclusion of the theorem is immediate.

Young also treats the Neumann problem by similar techniques. Travis [1] uses a different method with more general boundary conditions and breaks up the range of m into parts corresponding the the Weyl limit circle and limit point situations (cf. Coddington-Levinson [1]). Thus one considers the singular eigenvalue problem

$$(4.16) \qquad (t^{2m+1}\phi')' + t^{2m+1}\lambda\phi = 0; \qquad 0 < t < b;$$

$$\int_0^b t^{2m+1}|\phi(t)|^2 dt < \infty; \qquad \phi(b) = 0$$

It can be shown that for $m \leq -1$ or $m \geq 1$ the problem (4.16) has a pure point spectrum and the eigenfunctions form a complete orthogonal set on (o,b) relative to the weight function t^{2m+1}. Furthermore if f has finite t^{2m+1} norm on (o,b) (i.e., $\int_0^b t^{2m+1}|f(t)|^2 dt < \infty$) then the Fredholm alternative (cf. Hellwig [3]) holds for the nonhomogeneous equation $(t^{2m+1}\phi')' + \lambda t^{2m+1}\phi = t^{2m+1}f$. In the limit circle case $-1 < m < 1$ there is always a pure point spectrum with a complete orthogonal set of eigenfunctions ϕ_n (relative to the weight function t^{2m+1}) for the problem

$$(4.17) \qquad (t^{2m+1}\phi')' + t^{2m+1}\lambda\phi = 0; \qquad \phi(b) = 0;$$

$$\lim_{t\to 0} t^{2m+1} W(\phi,\overline{\psi},t) = 0$$

where $W(\phi,\psi,t) = \phi\psi' - \phi'\psi$ and $\psi = \psi(t,\lambda_0)$ is a possibly complex valued solution of the differential equation in (4.17) for $\lambda = \lambda_0$ satisfying $t^{2m+1}W(\psi,\overline{\psi},t) \to 0$ as $t \to 0$ (such ψ exist). The Fredholm alternative also holds again for the corresponding nonhomogeneous problem.

We take now $a^{ij}(\cdot) \in C^1$ in a bounded open set $\Omega \subset \mathbf{R}^n$ with sufficiently smooth boundary Γ where $a^{ij} = a^{ji}$ and $\sum a^{ij}\xi_i\xi_j \geq \mu|\vec{\xi}|^2$ while $c(\cdot)$ is continuous in Ω. Let $\partial/\partial\nu$ denote the transverse derivative $\partial/\partial\nu = \sum a^{ij}(x)\cos(\nu,x_j)\,\partial/\partial x_i$ where ν is the exterior normal to Γ and let $L_m u = u_{tt} + \frac{2m+1}{t}u_t - \sum(a^{ij}u_{x_i})_{x_j} + cu$ as before.

Theorem 4.5 Let $m \notin (-1,1)$ and $L_m u^m = 0$ in $\bar{\Omega} \times (o,b)$ with $u^m \in C^2(\bar{\Omega} \times (o,b))$, $\partial u^m/\partial\nu + \sigma(x)u^m = 0$ on $\Gamma \times (o,b)$, and $u^m(x,b) = 0$ for $x \in \Omega$ while $\int_0^b t^{2m+1}|u^m(x,t)|^2 dt < \infty$. Then $u^m \equiv 0$ if and only if $J_p(\sqrt{\lambda_n}b) \neq 0$ where $p = |m|$ and $\{\lambda_n\}$ is the set of positive eigenvalues of $-\sum(a^{ij}v_{x_i})_{x_j} + cv = \lambda v$ in Ω with $\partial v/\partial\nu + \sigma v = 0$ on Γ.

The proof is similar in part to that of Theorem 4.4. Thus, if there exists a positive eigenvalue λ_n of the elliptic problem indicated in the statement of Theorem 4.5 such that $J_p(\sqrt{\lambda_n}b) = 0$ then set $u^m(x,t) = t^{-m}J_p(\sqrt{\lambda_n}t)v_n(x)$ where v_n is the eigenfunction corresponding to λ_n. Evidently u^m satisfies $L_m u^m = 0$ and the required boundary conditions. On the other hand suppose $J_p(\sqrt{\lambda_n}b) \neq 0$ for all λ_n as described. Let u^m be a solution of $L_m u^m = 0$ satisfying the conditions stipulated in Theorem 4.5. Let $\phi_n(t)$ and $v_n(x)$ be the normalized eigenfunctions for the (4.16) and the elliptic eigenvalue problem in the statement of Theorem 4.5 respectively. Then the set $\{\phi_n(t)v_k(x)\}$ is a complete orthonormal set for the real Hilbert space $H = L_m^2(\Omega \times (o,b)) = \{h; \int_0^b \int_\Omega t^{2m+1}h^2(x,t)dxdt < \infty\}$ and since $u^m \in H$ we can write $u^m(x,t) = \sum\alpha_{ij}\phi_j(t)v_i(x)$ with the standard formula for α_{ij}. Now define $h_i(t) = \int_\Omega u^m(x,t)v_i(x)dx$ so that

$(t^{2m+1}h_i'(t))' + \lambda_i t^{2m+1}h_i(t) = 0$, $h_i(b) = 0$, where λ_i is the eigenvalue corresponding to v_i. By the Parseval-Plancherel formula $\int_0^b t^{2m+1}h_i^2(t)dt = \sum|\alpha_{ij}|^2 < \infty$ and therefore $h_i(t) \equiv 0$ if λ_i is not an eigenvalue for the problem (4.16), i.e., if $J_p(\sqrt{\lambda_i}b) \not\equiv 0$. Consequently $u^m(x,t) \equiv 0$ since the $v_i(x)$ are complete in $L^2(\Omega)$ and u^m is continuous in $\Omega \times (o,b)$.

Remark 4.6 We remark first (cf. Travis [1]) that for $m \leq -1$ the problem treated in Theorem 4.5 is equivalent to $L_m u^m = 0$, $\partial u^m/\partial\nu + \sigma u^m = 0$ on $\Gamma \times (o,b)$ and $u^m(x,b) = u^m(x,o) = 0$.

Remark 4.7 Let us note in passing that, in general, if L is a singular second order ordinary differential operator in t such that the numerical initial value problem $Lu + \lambda_n u = 0$, $u(o) = 1$, $u_t(o) = 0$, has a solution $u_n(t)$, when λ_n is an eigenvalue of the (closed densely defined) operator A in a Hilbert space E, while the corresponding set of eigenvectors $\{e_n\}$ of A forms a complete orthonormal set in E, then formally the E valued function $w(t) = \sum a_n u_n(t)e_n$ satisfies $Lw + Aw = \sum a_n(Lu_n+\lambda_n u_n)e_n = 0$ with $w(o) = \sum a_n e_n$ and $w_t(o) = 0$. Choosing the a_n to be the Fourier coefficients of $w(o) = e \in D(A)$ one has a formal solution of $Lw + Aw = 0$ with $w(o) = e$ and $w_t(o) = 0$. Uniqueness questions could then be handled in terms of the L^2 formal adjoint L' of L and the operator A^*. Indeed given that $Ae = A(\sum a_n e_n) = \sum a_n Ae_n = \sum a_n\lambda_n e_n$ for $e \in D(A)$ with $a_n = (e,e_n)$ we have $(Ae,e_k) = \sum a_n\lambda_n(e_n,e_k) = a_k\lambda_k = \lambda_k(e,e_k)$ so

$e_k \in D(A^*)$ with $A^* e_k = \bar{\lambda}_k e_k$. Hence A^* has a complete orthonormal set of eigenvectors e_k with eigenvalues $\bar{\lambda}_k$. Assume then that the problem $L' v + \bar{\lambda}_n v = 0$, $v(b) = 0$, $v_t(b) = 1$ has a solution $v_n(t)$ on $[o,b]$ for any b. Here, given $Lu = u'' + \alpha u' + \beta u$ with $u(o) = u'(o) = 0$ and v a C^2 function satisfying $v(b) = 0$ and $v'(b) = 0$ we define $L' v = v'' - (\alpha v)' + \beta v$ and observe that $\int_0^b u'' v\, dt = -u(b) + \int_0^b uv''\, dt$ while $\int_0^b \alpha u' v dt = -\int_0^b u(\alpha v)' dt$. Now write $\hat{w} = \sum b_n v_n(t) e_n$ where $\hat{e} = \hat{w}'(b) = \sum b_n e_n$ is an arbitrary element of some dense set (e.g., $D(A^*)$). Then if $(L+A)w = 0$ with $w(o) = w'(o) = 0$ we consider formally the equation

$$(4.18) \qquad 0 = \int_0^b ((L+A)w,\hat{w})dt = \sum b_n \int_0^b (Lw, v_n e_n)dt$$

$$+ \sum b_n \int_0^b (w, v_n A^* e_n)dt = \sum b_n \{-(w(b),e_n)$$

$$+ \int_0^b (w, L' v_n e_n)dt\} + \sum b_n \int_0^b (w, \bar{\lambda}_n v_n e_n)dt$$

$$= \sum b_n \int_0^b (w, (L' + \bar{\lambda}_n) v_n e_n)dt - \sum b_n (w(b), e_n)$$

$$= -(w(b), \sum b_n e_n) = -(w(b), \hat{e})$$

It follows that $w(b) = 0$ for any b and hence $w \equiv 0$. The calculations have been formal but they apply to realistic situations.

<u>Theorem 4.8</u> Assume $m \in (-1,1)$, $L_m u^m = 0$, $u^m \in C^2(\bar{\Omega} \times (o,b))$, $\partial u^m/\partial \nu + \sigma(x) u^m = 0$ on $\Gamma \times (o,b)$, $u^m(x,b) = 0$, and $\lim t^{2m+1} W(u^m, \bar{\psi}, t) = 0$ as $t \to 0$ where ψ is a possibly complex

valued solution of the differential equation in (4.17) for some $\lambda_0 = \lambda$ satisfying $\lim t^{2m+1} W(\psi,\overline{\psi},t) = 0$ as $t \to 0$. Then $u^m \equiv 0$ if and only if $\ell_n \neq \lambda_k$ where ℓ_n and λ_k are the eigenvalues of (4.17) and the elliptic problem in the statement of Theorem 4.5 respectively.

The proof here is similar to that of Theorem 4.5. Finally Travis [1] states some results related to Theorems 4.3 and 4.4 and some comparison is instructive (see Remark 4.10).

<u>Theorem 4.9</u> Let $u \in C^2(\overline{\Omega}\times(o,b))$ satisfy $L_m u^m = 0$ and $\partial u^m/\partial\nu + \sigma(x)u^m = 0$ on $\Gamma \times (o,b)$, while $u^m(x,o) = u^m(x,b) = 0$ for $x \in \Omega$ when $m < 0$ or $|u^m(x,o)| < \infty$ and $u^m(x,b) = 0$ for $x \in \Omega$ when $m \geq 0$. Then $u^m \equiv 0$ if and only if $J_p(\sqrt{\lambda_n}b) \neq 0$ where the λ_n are the eigenvalues of the elliptic problem in the statement of Theorem 4.5 and $p = |m|$.

<u>Remark 4.10</u> Theorem 4.9 (rephrased for $u^m = 0$ on $\Gamma \times [o,b)$) is independent of Theorem 4.3 for example even when $- 1/2 < m < 0$. In Theorem 4.3 one assumes u_t^m is continuous in $\overline{\Omega} \times [o,b]$ whereas Theorem 4.9 assumes only continuity of u_t^m on $\overline{\Omega} \times (o,b)$. Consider for example $u^m(x,t) = t^{-m}J_p(t) \sin x$ for $- 1/2 < m < 0$ which satisfies $u_{tt}^m + \frac{2m+1}{t}u_t^m - u_{xx}^m = 0$ with $u^m(o,t) = u^m(\pi,t) = 0$ and $u^m(x,o) = 0$. But $u_t^m(x,t) = O(t^{-2m-1})$ as $t \to 0$ so the hypotheses of Theorem 4.3 do not hold whereas Theorem 4.9 (rephrased) will apply. On the other hand Theorem 4.3 does not require conditions at $t = b$ for $m > -\frac{1}{2}$. Similarly Travis [1] states a result for a Neumann problem ($u_t(x,o) = u_t(x,b) = 0$)

related to a theorem of Young [4].

4.5 Miscellaneous problems. Various Cauchy problems with boundary data for EPD equations have been studied and we will mention a few results. First consider (cf. Fusaro [4])

Problem 5.1 Solve $L_m u^m = u^m_{tt} + \frac{2m+1}{t} u^m_t - u^m_{xx} = 0$, where $m > -1/2$, $0 < x < \pi$, $t > 0$, $u^m(x,o) = f(x)$ ($f \in C^2$), $u^m_t(x,o) = 0$, and $u^m(o,t) = u^m(\pi,t) = 0$. One assumes also $f(o) = f(\pi) = 0$.

Then by separation of variables Fusaro [4] obtains a solution in the form

$$(5.1)\qquad u^m(x,t) = \Gamma(m+1) \sum_{n=1}^{\infty} b_n \left(\frac{nt}{2}\right)^{-m} J_m(nt) \sin nx$$

where $b_n = \left(\frac{2}{\pi}\right) \int_0^\pi f(x) \sin nx \, dx$. Using the formula (1.3.7) with a change of variables $\phi = \pi/2 - \theta$ and putting this in (5.1) one obtains

$$(5.2)\qquad u^m(x,t) = \beta\left(\frac{1}{2}, m + \frac{1}{2}\right)^{-1} \int_0^{\pi/2} [f(x+t \sin \phi) + f(x-t \sin \phi)] \cdot \cos^{2m} \phi \, d\phi$$

where β denotes the Beta function. There results (cf. Fusaro [4])

Theorem 5.2 A unique solution of Problem 5.1 is given by (5.2).

The proof of Theorem 5.2 is straightforward and will be omitted. If one sets $\xi = \sin \phi$ and $M(x,\sigma,f) = \frac{1}{2} [f(x+\sigma) + f(x-\sigma)]$ we can write (5.2) in the form

$$(5.3)\qquad u^m(x,t) = 2\beta\left(\frac{1}{2}, m + \frac{1}{2}\right)^{-1}\int_0^1 (1-\xi^2)^{m-1/2} M(x,\xi t,f)\,d\xi$$

which coincides with (1.6.3). We remark here that f is extended from (o,π) as an odd periodic function of period 2π.

Another type of such "mixed" problems is solved by Young [9]; here mixed problem means mixed initial-boundary value problem and does not mean that the problem changes type. Thus consider

<u>Problem 5.3</u> Solve $L_m u^m = u^m_{tt} + \frac{2m+1}{t} u^m_t - u^m_{xx} - u^m_{yy} = f(x,y,t)$ in the quarter space $x > 0$, $t > 0$, $-\infty < y < \infty$, where $m > -1/2$, $u^m(x,y,o) = u^m_t(x,y,o) = 0$, and $u^m(o,y,t) = g(y,t)$ with $f \in C^2$, $g \in C^2$, and $g(y,o) = g_t(y,o) = 0$.

At points (x,y,t) where $x \geq t > o$ the presence of g is not felt and Young takes the solution in the form given by Diaz-Ludford [3]

$$(5.4)\qquad u^m(x,y,t) = \frac{2^{2m}}{\pi}\int_D \frac{\tau^{2m+1} f(\xi,\eta,\tau)}{R^{1/2}\hat{R}^{m+1/2}} F\left(m+\frac{1}{2}, m, \frac{1}{2}, \frac{R}{\hat{R}}\right) d\xi\, d\eta\, d\tau$$

where D is the domain bounded by $\tau = 0$ and the retrograde characteristic cone $R = (t-\tau)^2 - (x-\xi)^2 - (y-\eta)^2 = 0$ $(t-\tau>0)$ with vertex at (x,y,t) while $\hat{R} = (t+\tau)^2 - (x-\xi)^2 - (y-\eta)^2$. In the region where $0 < x < t$ formula (5.4) no longer gives the solution since it does not involve g and Young uses a method of M. Riesz as adapted by Young [1; 2] and Davis [1; 2] to adjoin additional terms to the u^m of (5.4). We will not go into the derivation of the final formula but simply state the result; for suitable m the technique can be extended to more space variables. The solution is given by the formula

$$(5.5)\qquad u^m(x,y,t) = \frac{2^{2m}}{\pi}\int_D \frac{\tau^{2m+1}f(\xi,\eta,\tau)}{R^{1/2}\hat{R}^{m+1/2}} F(m+\tfrac{1}{2}, m, \tfrac{1}{2}, \tfrac{R}{\hat{R}})d\xi d\eta d\tau$$

$$- \frac{2^{2m}}{\pi}\int_{D^*} \frac{\tau^{2m+1}f(\xi,\eta,\tau)}{R_*^{1/2}\hat{R}_*^{m+1/2}} F(m+\tfrac{1}{2}, m, \tfrac{1}{2}, \tfrac{R_*}{\hat{R}_*})d\xi d\eta d\tau$$

$$- \frac{2^{2m+1}}{\pi}\frac{\partial}{\partial x}\int_T \frac{\tau^{2m+1}g(\eta,\tau)}{R_o^{1/2}\hat{R}_o^{m+1/2}} F(m+\tfrac{1}{2}, m, \tfrac{1}{2}, \tfrac{R_o}{\hat{R}_o})d\eta d\tau$$

where $R_* = (t-\tau)^2 - (x+\xi)^2 - (y-\eta)^2$, R_o (resp. $\hat{R}_o$) denotes the value of R (resp. $\hat{R}$) for $\xi = 0$, $\hat{R}_* = (t+\tau)^2 - (x+\xi)^2 - (y-\eta)^2$, D^* is the domain where $\xi > 0$, $\tau > 0$, bounded by the retrograde cone with vertex at $(-x,y,t)$ (thus $D^* \subset D$), and T is the region intercepted on the plane $\xi = 0$ by the cone $R = 0$ ($\tau > 0$). Note that for $x \geq t > 0$ D^* and T are empty while as $x \to 0$ the integrals over D and D^* cancel each other since $D^* \to D$ and $(R,\hat{R}) \to (R_*,\hat{R}_*)$.

Theorem 5.4 The solution of Problem 5.3 is given by (5.5).

It is interesting to note what happens when all of the data in an EPD type problem are analytic. Thus for example Fusaro [3] considers for $t > 0$, $1 \leq k \leq n$, and $x = (x_1, \ldots, x_n)$

$$(5.6)\qquad u^m_{tt} + \frac{2m+1}{t}u^m_t = F(x,t,u^m,D_k u^m,u^m_t,D_k u^m_t,D_k D_\ell u^m);$$

$$u^m(x,o) = f(x);\quad u^m_t(x,o) = 0$$

where $D_k = \partial/\partial x_k$ while F and f are analytic in all arguments.

<u>Theorem 5.5</u> The problem (5.6) has a unique analytic solution for $m \neq -1, -3/2, -2, \ldots$.

<u>Corollary 5.6</u> If $F(x,t,u,\ldots) = G(x,u,D_k u,D_k D_\ell u) + g(x,t) + h(t)u_t$ with g even in t and h odd in t, then the problem (5.6) has for all $m \neq -1, -3/2, -2, \ldots$ a unique analytic solution and it is even in t. If $m = -3/2, -5/2, \ldots$ there exists a unique solution in the class of analytic functions which are even in t.

Fusaro [3] uses a Cauchy-Kowalewski technique to prove these results and we refer to his paper for details. One should note also the following results of Lions [1], obtained by transmutation, which illustrate further the role played by even and C^∞ functions of t in the uniqueness question for EPD equations. Let E_* be the space of C^∞ functions f on $[o,\infty)$ such that $f^{(2n+1)}(o) = 0$ (thus $f \in E_*$ admits an even C^∞ extension to $(-\infty,\infty)$) and give E_* the topology induced by E on $(-\infty,\infty)$. Let H be a Hilbert space and Λ a self adjoint operator in H such that for all $h \in D(\Lambda)$ $(\Lambda h,h) + \alpha||h||^2 \geq 0$ for some real α. Setting $L_m = D^2 + \frac{2m+1}{t} D$ and putting the graph topology on $D(\Lambda)$ Lions proves

<u>Theorem 5.7</u> For any $m \in \mathbb{C}$, $m \neq -1, -2, -3, \ldots$, there exists a unique solution $u^m \in E_* \hat{\otimes}_\pi D(\Lambda)$ of $L_m u^m + \Lambda u^m = 0$ for

$t \geq 0$ with $u^m(o) = h \in D(\Lambda^\infty)$ and $u_t^m(o) = 0$.

Here π denotes the projective limit topology of Grothendieck (cf. Schwartz [5]) which coincides with the ε topology since E_* is nuclear. If $m \neq -1, -3/2, -2, \ldots$ and E_+ is the space of C^∞ functions on $[o,\infty)$ with the standard E-type topology then any solution $u^m \in E_+ \hat{\otimes}_\pi D(\Lambda)$ of $L_m u^m + \Lambda u^m = 0$ satisfying $u^m(o) = h \in D(\Lambda^\infty)$ and $u_t^m(o) = 0$ is automatically in $E_* \hat{\otimes}_\pi D(\Lambda)$ so the corresponding problem in $E_+ \hat{\otimes}_\pi D(\Lambda)$ has a unique solution. For the exceptional values $m = -1, -2, -3, \ldots$ Lions [1] proves

<u>Theorem 5.8</u> Let $m = -(p+1)$ with $p \geq 0$ an integer so that $2m + 1 = -(2p+1)$. There exists a unique solution $u^m \in E_+ \hat{\otimes}_\pi D(\Lambda)$ of $L_m u^m + \Lambda u^m = 0$, $u^m(o) = u_t^m(o) = \ldots = D_t^{2p+1} u^m(o) = 0$, $D_t^{2p+2} u^m(o) = h \in D(\Lambda^\infty)$, and $D_t^{2p+3} u^m(o) = 0$.

In connection with the uniqueness question we mention also the following recent result of Donaldson-Goldstein [2]. Let H be a Hilbert space and S a self adjoint operator in H with no nontrivial null space. Let $L_m = D^2 + \frac{2m+1}{t} D$ again and define u^m to be a type A solution of

$$(5.7) \qquad L_m u^m + S^2 u^m = 0; \quad u^m(o) = h \in D(S^2); \quad u_t^m(o) = 0$$

if $u^m \in C^2([o,\infty),H)$ and a type B solution if $u^m \in C^1([o,\infty),H) \cap C^2((o,\infty),H)$.

<u>Theorem 5.9</u> The problem (5.7) can have at most one type A solution for $m \geq -1$.

Further, by virtue of (1.4.10) and the formula $u^m = t^{-2m}u^{-m}$ (cf. Remark 1.4.7), for $m < -1/2$ not a negative integer there is no uniqueness for type B solutions. We note here that for $-1 < m < -1/2$ we have $1 < -2m < 2$ and $u^m = t^{-2m}u^{-m} = O(t^{-2m}) \to 0$ as $t \to 0$ with $u_t^m = O(t^{-2m-1}) \to 0$ but $u_{tt}^m = O(t^{-2m-2}) \to \infty$ so such a u^m is not of type A. For $m = -1$ we look at $u^{-1} = t^2u^1$ with $u^{-1}(o) = u_t^{-1}(o) = 0$ and $u_{tt}^{-1} \in C^{\circ}([o,\infty),H)$. However from Remark 1.4.8 we see that the solution u^{-1} arising from the resolvant R^{-1} (which could be spectralized as in Section 1.5) does not have a continuous second derivative in t unless the logarithmic term vanishes, which corresponds to $S^2h = 0$ in the present situation. Thus in general (5.7) does not seem to have a type A solution for $m = -1$ and if it does the solution arises when $S^2h = 0$. Indeed if $L_{-1}u^{-1} + S^2u^{-1} = 0$ with $u^{-1}(o) = h$ and $u_{tt}^{-1}(o)$ is well defined then $u_{tt}^{-1}(o) = \lim 1/t\, u_t^{-1}(t)$ as $t \to 0$ so $L_{-1}u^{-1} = u_{tt}^{-1} - \frac{1}{t}u_t^{-1} = -S^2u^{-1} \to 0$ which means that $S^2u^{-1}(o) = S^2h = 0$ (cf. Weinstein [3]), which is precluded unless $h = 0$. Note that if $h \neq 0$ with $S^2h = 0$ then $u^{-1}(t) = h$ and $u^{-1}(t) = h + t^2h$ will both be solutions of (5.7).

We go now to some work of Bragg [2; 3; 4; 5] on index shifting relations for singular problems. This is connected to the idea of "related" differential equations developed by Bragg and Dettman, loc. cit., and described briefly in Chapter 1 (cf. Section 1.7). Consider first for example ($a \geq 0$, $\mu \geq 1$)

$$(5.8) \qquad u_{tt}^{(a,\mu)} + \frac{a}{t}u_t^{(a,\mu)} = u_{rr}^{(a,\mu)} + \frac{\mu-1}{r}u_r^{(a,\mu)}$$

This is a radially symmetric form of the EPD equation when $\mu = n$ for example since the radial Laplacian in $\mathbf{R}^n$ has the form $\Delta_R = D_r^2 + \frac{n-1}{r} D_r$. As initial data we take

$$(5.9) \qquad u^{(a,\mu)}(r,0) = T(r); \qquad u_t^{(a,\mu)}(r,0) = 0$$

From the recursion relations of Chapter 1 (cf. (1.3.11), (1.3.12), (1.3.54), (1.3.55), (1.4.6), (1.4.10), and Remark 1.4.7) we see that there are certain "automatic" shifts in the a index. For example the Sonine formula (1.4.6) with $a_1 = 2p + 1$ and $a_1 + a_2 = 2m + 1$ yields for $a_1 \geq 0$, $a_2 > 0$

$$(5.10) \qquad u^{(a_1+a_2,\mu)}(r,t) = c_{1,2} t^{1-a_1-a_2} \int_0^t (t^2-\eta^2)^{\frac{a_2-2}{2}} \eta^{a_1} \cdot u^{(a_1,\mu)}(r,\eta)d\eta$$

where $c_{1,2} = 2/\beta(\frac{a_1+1}{2}, \frac{a_2}{2})$. Bragg [5] obtains this result using Laplace transforms and a related radial heat equation

$$(5.11) \qquad v_t^{\mu} = v_{rr}^{\mu} + \frac{\mu-1}{r} v_r^{\mu}; \qquad v^{\mu}(r,0) = T(r)$$

The connection formula for $a \geq 1$ is given under suitable hypotheses by (cf. Bragg [5], Bragg-Dettman [6])

$$(5.12) \qquad u^{(a,\mu)}(r,t) = t^{1-a}\Gamma(\tfrac{a+1}{2})L_s^{-1}\{s^{-\frac{(a+1)}{2}} v^{\mu}(r,\tfrac{1}{4s})\}_{s\to t^2}$$

(here L_s^{-1} is the inverse Laplace transform with kernel e^{st^2}) and (5.10) follows directly by a convolution argument. Similarly, setting $I_\nu(z) = \exp(-\frac{1}{2}\nu\pi i)\, J_\nu(iz)$ and $K_\mu(r,\xi,t) =$

$(2t)^{-1}r^{1-\mu/2}\xi^{\mu/2}\exp[-(r^2+\xi^2)/4t]I_{\mu/2-1}(r\xi/2t)$, one can show that $v^\mu(r,t) = \int_0^\infty K_\mu(r,\xi,t)T(\xi)d\xi$ satisfies (5.11), and using (5.12) there results from known properties of I_ν

Theorem 5.10 Let $a \geq 0$ with $u^{(a,\mu)}$ a continuous solution of (5.8) - (5.9). Then

$$(5.13)\qquad u^{(a,\mu+2)}(r,t) = u^{(a,\mu)}(r,t) + \frac{1}{r}\frac{\partial}{\partial r}\int_0^t u^{(a,\mu)}(r,\eta)\eta d\eta$$

satisfies (5.8) = (5.9) with index $\mu + 2$. On the other hand a direct substitution shows that if $u^{(a,4-\mu)}(r,t)$ satisfies (5.8) with index $4 - \mu$ and $u^{(a,4-\mu)}(r,o) = r^{\mu-2}T(r)$ then $u^{(a,\mu)}(r,t) = r^{2-\mu}u^{(a,4-\mu)}(r,t)$ satisfies (5.8) - (5.9) except possibly at $r = 0$.

Using the index shifting relations indicated on a and μ Bragg [5] proves a uniqueness theorem for (5.8) - (5.9) when $a \geq 0$ and $\mu \geq 1$ and constructs fundamental solutions. The idea of related differential problems such as (5.8) - (5.9) and (5.11) (or (5.8) - (5.9) with index changes), having connnection formulas like (5.12) (or (5.10)), has been exploited by Bragg [2; 3; 4] to produce a variety of index shifting relations involving at times the generalized hypergeometric functions

$$(5.14)\qquad {}_pF_q(\alpha,\beta,z) = 1 + \sum_{n=1}^\infty \frac{\Pi_1^p(\alpha_j)_n}{\Pi_1^q(\beta_j)_n}\frac{z^n}{n!}$$

where $(\sigma)_n = \sigma(\sigma+1)\ .\ .\ .\ (\sigma+n-1)$, $\Pi_1^p(\gamma_j)_n = 1$ if $p = 0$, and none of the β_j are nonpositive integers (cf. Appell-Kampé de

Fériet [1]). We will mention a few results in this direction without giving proofs and here we recall that if $\theta = zD_z$ then $w = {}_pF_q$ satisfies the differential equation $\{\theta\Pi_1^q(\theta+\beta_j-1)-z\Pi_1^p(\theta+\alpha_j)\}w = 0$. Thus let $P(x,D) = \sum a_\lambda(x)D_1^{\lambda_1} \ldots D_n^{\lambda_n}$ where $D_k = \partial/\partial x_k$ and $|\lambda| = \sum\lambda_i \leq m$. Let $\theta_1(q,\beta,t,D_t) = tD_t\ \Pi_1^q(tD_t+\beta_j-1)$ and $\theta_2(p,\alpha,t,D_t) = t\Pi_1^p(tD_t+\alpha_j)$. Then given $P(x,D)$ and $Q(x,D)$ of the type indicated with orders ℓ_1 and ℓ_2 respectively and $r = \max\ (p+\ell_1,q+\ell_2)$ with $p \leq q$ one can prove for example (cf. Bragg [4])

<u>Theorem 5.11</u> Let $u \in C^r$ in (x,t) for $t > 0$ with u and all its derivatives through order r bounded and $\lim u(x,t) = T(x)$ as $t \to 0$. Suppose $Q(x,D)\theta_1(q,\beta,t,D_t)u - P(x,D)\theta_2(p-1,\alpha,t,D_t)u = 0$. Then for $\alpha_p > 0$ the function $v(x,t) = \Gamma(\alpha_p)^{-1}\int_0^\infty e^{-\sigma}\sigma^{\alpha_p-1}u(x,t\sigma)d\sigma$ satisfies $Q\theta_1(q,\beta,t,D_t)v - P\theta_2(p,\alpha,t,D_t)v = 0$ and $v(x,o) = T(x)$.

Similarly, under suitable hypotheses, if $[Q\theta_1(q-1,\beta,t,D_t) - P\theta_2(p,\alpha,t,D_t)]u = 0$ and $w(x,t) = t^{1-\beta_q}\Gamma(\beta_q)L_s^{-1}[s^{-\beta_q}u(x,1/s)]_{s\to t}$ then one has $[Q\theta_1(q,\beta,t,D_t) - P\theta_2(p,\alpha,t,D_t)]w = 0$. Index shift formulas in (p,q), α, and β are also proved and when applied to EPD equations for example will yield some of the standard recursion relations. We note that under the change of variables $\xi = \frac{1}{4}t^2$ the EPD equation $u_{tt} + \frac{a}{t}u_t = Pu$ becomes $[\xi D_\xi(\xi D_\xi + \frac{a+1}{2} - 1) - \xi P]u = 0$. Using the R-L integral, continuous index shifts in α and β for similar abstract problems are treated in Bragg [3] where a more systematic theory is developed; nonhomogeneous problems are studied in Bragg [2].

We conclude with a few remarks about nonlinear EPD equations. In Keller [1] it is shown that for $m > 0$ and certain nonlinear f the Cauchy problem

$$(5.15)\qquad L_m u = u_{tt} + \frac{2m+1}{t} u_t - c^2 \Delta u = f(u);$$

$$u(x,o) = u_o(x); \qquad u_t(x,o) = 0$$

does not have global solutions in $\mathbb{R}^2 \times (o,\infty)$ for arbitrary initial data u_o. Levine [2] studies this problem for $-\frac{1}{2} < m \leq 0$ and $x \in \Omega \subset \mathbb{R}^n$ with $u = 0$ on $\Gamma \times [o,T)$ where Ω is bounded with sufficiently smooth boundary Γ. Let f be a real valued C^1 function and for each real $\phi \in C^2(\overline{\Omega})$ define $G(\phi) = \int_\Omega (\int_0^{\phi(x)} f(z)dz)dx$. Set $(\phi,\psi) = \int_\Omega \phi(x)\psi(x)dx$ and assume there is a constant $\alpha > 0$ such that for all $\phi \in C^2(\overline{\Omega})$

$$(5.16)\qquad 2(2\alpha+1)G(\phi) \leq (\phi,f(\phi))$$

It can be shown that if, for some $\alpha > 0$ and some monotone increasing function h, $f(z) = |z|^{4\alpha+1}h(z)$, then (5.16) is valid. Assuming the existence of a local solution for each $u_o \in C^2(\overline{\Omega})$ vanishing on Γ Levine proves

<u>Theorem 5.12</u> Let $-\frac{1}{2} < m \leq 0$ and assume (5.16) holds. If $u : \overline{\Omega} \times [o,T) \to \mathbb{R}^1$ is a classical solution of $L_m u = f(u)$ with $u(x,o) = u_o(x)$, $u_t(x,o) = 0$, and $u = 0$ on $\Gamma \times [o,T)$, while $G(u_o) > \frac{1}{2}\int_\Omega |\nabla u_o|^2 dx$, then necessarily $T < \infty$ and $\int_\Omega u^2(x,t)dx \to \infty$ as $t \to T^-$.

REFERENCES

Abdul-Latif, A. and Diaz, J.
1. Dirichlet, Neumann, and mixed boundary value problems for the wave equation $u_{xx}-u_{yy} = 0$ for a rectangle, Jour. Appl. Anal., 1 (1971), 1-12.

Adler, G.
1. Un type nouveau des problèmes aux limites de la conduction de la chaleur, Magyar Tud. Akad. Mat. Kutato Int. Kőzl., 4 (1959), 109-127.

Agmon, S., Nirenberg, L. and Protter, M.
1. A maximum principle for a class of hyperbolic equations and applications to equations of mixed elliptic - hyperbolic type, Comm. Pure Appl. Math., 6 (1953), 455-470.

Agmon, S.
2. Boundary value problems for equations of mixed type, Conv. Int. Eq. Lin. Der. Parz., Trieste, Cremonese, Rome, 1955, pp. 54-68.

Agmon, S. and Nirenberg, L.
3. Lower bounds and uniqueness theorems for solutions of differential equations in a Hilbert space, Comm. Pure Appl. Math., 20 (1967), 207-229.

Albertoni, S. and Cercigrani
1. Sur un problème mixte dans la dynamique des fluides, C. R. Acad. Sci. Paris, 261 (1965), 312-315.

Amos, D. E.
1. On half-space solutions of a modified heat equation, Quart. Appl. Math., 27 (1969), 359-369.

Appell, P. and Kampé de Fériet, J.
1. Fonctions hypergéométriques et hypersphériques. Polynomes d'Hermite, Gauthier - Villars, Paris, 1926.

Arens, R.
1. The analytic functional calculus in commutative topological algebras, Pacific Jour. Math., 11 (1961), 405-429.

Arens, R. and Calderón
2. Analytic functions of several Banach algebra elements, Annals Math., 62 (1955), 204-216.

Aronson, D.
1. Regularity properties of flows through porous media, SIAM Jour. Appl. Math., 17 (1969), 461-467.

REFERENCES

Asgeirsson, L.

1. Uber eine Mittelwerteigenschaft von Losungen homogener linearen partiellen Differentialgleichungen zweiter Ordnung mit konstanten Koeffizienten, Math. Annalen, 113 (1937), 321-346.

Babalola, V.

1. Semigroups of operators on locally convex spaces, to appear.

Babenko, K.

1. On the theory of equations of mixed type, Usp. Mat. Nauk, 8 (1953), 160.

Baiocchi, C.

1. Sulle equazioni differenziali lineari del primo e del secondo ordine negli spazi di Hilbert. Ann. Mat. Pura Appl. (4), 76 (1967), 233-304.

Baker, B. and Copson, E.

1. The mathematical theory of Huygens' principle, Oxford, 1939.

Baouendi, M. and Goulaouic, C.

1. Cauchy problems with characteristic initial hypersurface, Comm. Pure Appl. Math., 26 (1973), 455-475.

Baranovskij, F.

1. The Cauchy problem for an equation of Euler-Poisson-Darboux type and a degenerate hyperbolic equation, Izv. Vis. Uč. Zaved., Matematika, 6 (19), 1960, 11-23.

2. The Cauchy problem for a second order hyperbolic equation that is strongly degenerate (case A = 1), Dopovidi Akad. Nauk Ukrain. RSR, 91 (1971), 11-16.

3. On the Cauchy problem for a strongly degenerate hyperbolic equation, Sibirsk. Mat. Žur., 4 (1963), 1000-1011.

4. The mixed problem for a linear hyperbolic equation of the second order degenerating on the initial plane, Uč. Zap. Leningrad. Ped. Inst., 183 (1958), 23-58.

5. The mixed problem for a hyperbolic degenerate equation, Izv. Vis. Uč. Zaved., Matematika, 3 (16), 1960, 30-42.

6. On differential properties of the solution of the mixed problem for degenerate hyperbolic equations, Izv. Vis. Uč. Zaved., Matematika, 6 (137), 1963, 15-24.

7. The Cauchy problem for a linear hyperbolic equation degenerating on the initial plane, Uč. Zap. Leningrad Ped. Inst., 166 (1958), 227-253.

Barantzev, R.
1. Transzvukovoi gazodinamike, Izd. Leningrad. Univ., 1965.

Barbu, V.
1. Existence theorems for a class of two point boundary problems. Jour. Diff. Eqs., 17 (1975), 236-257.

Bardos, C. and Brézis H.
1. Sur une classe de problèmes d'évolution nonlinéaires, Jour. Diff. Eqs., 6 (1969), 345-394.

Bardos, C., Brézis, D. and Brézis, H.
2. Pertubations singulieres et prolongements maximaux d'opérateurs positifs, Arch. Rat'l. Mech. Anal. 53 (1973), 69-100.

Barenblatt, G., Zheltov, I. and Kochina, I.
1. Basic concepts in the theory of seepage of homogeneous liquids in fissured rocks, Jour. Appl. Math. Mech., 24 (1960), 1286-1303.

Barenblatt, G. and Chernyi, G.
2. On moment relations of surface discontinuity in dissipative media, Jour. Appl. Math. Mech., 27 (1963), 1205-1218.

Bargmann, V.
1. Irreducible unitary representations of the Lorentz group, Annals Math., 48 (1947), 568-640.

Bargmann, V. and Wigner, E.
2. Group theoretical discussion of relativistic wave equations, Proc. Nat'l. Acad. Sci., 34 (1948), 211-223.

Benjamin, T. B.
1. Lectures on nonlinear wave motion, in Nonlinear wave motion, Lect. Appl. Math, 15, Amer. Math. Soc., 1972.

Benjamin, T. B., Bona, J. L., Mahoney, J. J.
2. Model equations for long waves in nonlinear dispersive systems, Phil. Trans. Roy. Soc. London, Ser. A, 272 (1972), 47-78.

Berezanskij, Yu.
1. Existence of weak solutions of some boundary problems for equations of mixed type, Ukrain. Mat. Žur., 15 (1963), 347-364.

2. The expansion of selfadjoint operators in terms of eigenfunctions, Izd. Nauk. Dumka, Kiev, 1965.

Berezin, I.
1. On the Cauchy problem for linear equations of the second order with initial data on the parabolic line, Mat. Sbornik, 24 (1949), 301-320.

Bers, L.
1. Mathematical aspects of subsonic and transonic gas dynamics, Wiley, N.Y., 1958.

2. On the continuation of a potential gas flow across the sonic line, NACA Tech. Note 2058, 1950.

Bhanu Murti, T.
1. Plancherel's formula for the factor space SL $(n, \mathbb{R})/SO(n)$, Dokl. Akad. Nauk SSSR, 133 (1960), 503-506.

Bhatnager, S. C.
1. Higher order pseudoparabolic partial differential equations, Thesis, Indiana Univ., 1974.

Bitsadze, A.
1. Equations of mixed type, Izd. Akad. Nauk SSSR, Moscow, 1959.

2. On the problem of equations of mixed type, Trudy Steklov Mat. Inst., Akad. Nauk, SSSR, 41.

Blum, E.
1. The Euler-Poisson-Darboux equation in the exceptional cases, Proc. Amer. Math. Soc., 5 (1954), 511-520.

2. The solutions of the Euler-Poisson-Darboux equation for negative values of the parameter, Duke Math. Jour., 21 (1954), 257-269.

Blyumkina, I.
1. On the solution by the Fourier method of the boundary problem for the equation $u_{yy}-k(y)\,u_{xx}=0$ with data on a characteristic and on the degenerate line, Vestnik LGU, 1962, No. 1, pp. 111-115.

Bochner, S. and Von Neumann, J.
1. On compact solutions of operational-differential equations I, Annals Math., 36 (1936), 255-291.

Bona, J.
1. On the stability of solitary waves, Proc. Roy. Soc. London, A., 344 (1975), 363-374.

Bona, J., Bose, D. and Benjamin, T. B.
2. Solitary wave solutions for some model equations for waves in nonlinear dispersive media, to appear

Bona, J. L. and Bryant, P. J.
3. A mathematical model for long waves generated by wavemakers in nonlinear dispersive systems, Proc. Camb. Phil. Soc., 73 (1973), 391-405.

Bona, J. and Smith, R.
4. The initial-value problem for the KdV equation, Phil. Trans. Roy. Soc. London, A, 278 (1975), 555-604.

5. A model for the two-way propogation of water waves in a channel, to appear

Bourbaki, N.
1. Espaces vectoriels topologiques, Chaps. 1-2, Hermann, Paris, 1953.

2. Espaces vectoriels topologiques, Chaps. 3-5, Hermann, Paris, 1955.

3. Intégration, Chaps. 1-4, Hermann, Paris, 1952.

4. Intégration vectorielle, Hermann, Paris, 1959.

5. Groupes et algèbres de Lie, Chaps. 1-6, Hermann, Paris 1960, 1968, 1972.

Bourgin, D.
1. The Dirichlet problem for the damped wave equation, Duke Math. Jour., 7 (1940), 97-120.

Bourgin, D. and Duffin, R.
2. The Dirichlet problem for the vibrating string equation, Bull. Amer. Math. Soc., 45 (1939), 851-858.

Boussinesq, M.
1. Théorie génerale des mouvements qui sont propagés dans un canal rectangulaire horizontal, C. R. Acad. Sci. Paris, 73 (1871), 256.

REFERENCES

Bragg, L. and Dettman, J.

1. Multinomial representations of solutions of a class of singular initial value problems, Proc. Amer. Math. Soc., 21 (1969), 629-634.

Bragg, L.

2. Related nonhomogeneous partial differential equations, Jour. Appl. Anal., 4 (1974), 161-189.

3. The Riemann-Liouville integral and parameter shifting in a class of linear abstract problems, SIAM Jour. Math. Anal., to appear

4. Hypergeometic operator series and related partial differential equations, Trans. Amer. Math. Soc., 143 (1969), 319-336.

5. Fundamental solutions and properties of solutions of the initial value radial Euler-Poisson-Darboux problem, Jour. Math. Mech., 18 (1969), 607-616.

Bragg, L. and Dettman, J.

6. An operator calculus for related partial differential equations, Jour. Math. Anal. Appl., 22 (1968), 261-271.

7. Expansions of solutions of certain hyperbolic and elliptic problems in terms of Jacobi polynomials, Duke Math. Jour., 36 (1969), 129-144.

8. Related partial differential equations and their applications, SIAM Jour. Appl. Math., 16 (1968), 459-467.

Bragg, L.

9. Singular nonhomogeneous abstract Cauchy and Dirichlet problems related by a generalized Stieltjes transform, Indiana Univ. Math. Jour., 24 (1974), 183-195.

10. Linear evolution equations that involve products of commutative operators, SIAM Jour. Math. Anal., 5 (1974), 327-335.

Brauer, F.

1. Spectral theory for the differential equation $Lu = \lambda Mu$, Canad. Jour. Math., 10 (1958), 431-436.

Bresters, D.

1. On the equation of Euler-Poisson-Darboux, SIAM Jour. Math. Anal., 4 (1973), 31-41.

Brézis, H.
1. On some degenerate non-linear parabolic equations, Proc. Symp. Pure Math., 18, Amer. Math. Soc., 1970, pp. 28-38.

Brézis, H., Rosenkrantz, W. and Singer, B.
2. On a degenerate elliptic-parabolic equation occurring in the theory of probability, Comm. Pure Appl. Math., 24 (1971), 395-416.

Brézis, H.
3. Opérateurs maximaux monotones et semi-groupes de contractions dans les espaces de Hilbert, North Holland, Amsterdam, 1973.

4. Equations et inéquations non-linéaires dans les espaces vectoriels en dualité, Ann. Inst. Fourier, 18 (1968), 115-175.

Brézis, H. and Goldstein, J.
5. Liouville theorems for some improperly posed problems, to appear

Brill, H.
1. A semilinear Sobolev evolution equation in a Banach space, Jour. Diff. Eqs., to appear

Browder, F.
1. Non-linear maximal monotone operators in Banach spaces, Math. Annalen, 175 (1968), 89-113.

2. Nonlinear initial value problems, Annals Math. (2), 82 (1965), 51-87.

3. Existence and uniqueness theorems for solutions of non-linear boundary value problems, Proc. Symp. Appl. Math., 17, Amer. Math. Soc., 1965, pp. 24-49.

4. Parabolic systems of differential equations with time dependent coefficients, Proc. Nat'l. Acad. Sci., 42 (1956), 914-917.

5. On the unification of the calculus of variations and the theory of monotone nonlinear operators in Banach spaces, Proc. Nat'l. Acad. Sci., 56 (1966), 419-425.

6. Problèmes non-linéaires, Univ. Montréal, 1966.

7. Nonlinear operators and nonlinear equations of evolution in Banach spaces, Proc. Symp. Pure Math., 28, Part 2, Amer. Math. Soc., 1976.

Brown, P. M.
1. Constructive function-theoretic methods for fourth order pseudo-parabolic and metaparabolic equations, Thesis, Indiana Univ., 1973.

Buckmaster, J., Nachman, A., and Ting, L.
1. The buckling and stretching of a viscida, Jour. Fluid Mech,. 69 (1975), 1-20.

Bureau, F.
1. Divergent integrals and partial differential equations, Comm. Pure Appl. Math., 8 (1955), 143-202.

2. Problems and methods in partial differential equations, Ann. Mat. Pura Appl., 51 (1960), 225-299.

Calvert, B.
1. The equation $A(t, u(t))' + B(t, u(t)) = 0$, Math. Proc. Camb. Phil. Soc., to appear

Cannon, J. and Meyer, G.
1. On diffusion in a fractured medium, SIAM Jour. Appl. Math., 20 (1971), 434-448.

Cannon J. and Hill, C.
2. A finite-difference method for degenerate elliptic-parabolic equations, SIAM Jour. Numer. Anal., 5 (1968), 211-218.

Cannon, J., Ford, W. and Lair, A.
3. Quasilinear parabolic systems, Jour. Diff. Eqs., 20 (1976), 441-472.

Carroll, R.
1. L'équation d'Euler-Poisson-Darboux et les distributions sousharmoniques, C. R. Acad. Sci. Paris, 246 (1958), 2560-2562.

2. On some generalized Cauchy problems and the convexity of their solutions, AFOSR-TN-59-649, University of Maryland, 1959.

3. Quelques problèmes de Cauchy singuliers, C. R. Acad. Sci. Paris, 251 (1960), 498-500.

4. Some singular mixed problems, Proc. Nat'l. Acad. Sci., 46 (1960), 1594-1596.

5. Some singular Cauchy problmes, Ann. Mat. Pura Appl., 56 (1961), 1-31.

6. Sur le problème de Cauchy singulier, C. R. Acad. Sci. Paris, 252 (1961), 57-59.

7. Quelques problèmes de Cauchy dégénérés avec des coefficients opérateurs, C. R. Acad. Sci. Paris, 253 (1961), 1193-1195.

8. On the singular Cauchy problem, Jour. Math. Mech., 12 (1963), 69-102.

9. Some degenerate Cauchy problems with operator coefficients, Pacific Jour. Math., 13 (1963), 471-485.

10. Some quasi-linear singular Cauchy problems, Proc. Symp. Nonlinear Probs., Univ. Wisconsin, 1963, pp. 303-306.

11. On some singular quasilinear Cauchy problems, Math. Zeit. 81 (1963), 135-154.

Carroll, R. and Wang, C.

12. On the degenerate Cauchy problem, Canad. Jour. Math., 17 (1965), 245-256.

Carroll, R.

13. Some growth and convexity theorems for second order equations, Jour. Math. Anal. Appl., 17 (1967), 508-518.

14. Abstract methods in partial differential equations, Harper-Row, N.Y., 1969.

Carroll, R. and Silver, H.

15. Suites canoniques de problèmes de Cauchy singuliers, C. R. Acad. Sci. Paris, 273 (1971), 979-981.

16. Canonical sequences of singular Cauchy problems, Jour. Appl. Anal., 3 (1973), 247-266.

17. Growth properties of solutions of certain canonical hyperbolic equations with subharmonic initial data, Proc. Symp. Pure Math., 23, Amer. Math. Soc., 1973, pp. 97-104.

Carroll, R.

18. On a class of canonical singular Cauchy problems, Proc. Colloq. Anal., Univ. Fed. Rio de Janeiro, 1972, Anal. fonct. appl., Act. Sci. Ind., Hermann, Paris, 1975, pp. 71-90.

19. On some hyperbolic equations with operator coefficients, Proc. Japan Acad., 49 (1973), 233-238.

Carroll, R. and Donaldson, J.

20. Algunos resultados sobre ecuaciones diferenciales abstractas relacionadas, Portug. Math., to appear.

Carroll, R.

21. Eisenstein integrals and singular Cauchy problems, Proc. Japan Acad., 51 (1975), 691-695.

22. Singular Cauchy problems in symmetric spaces, Jour. Math. Anal. Appl., 55 (1976), to appear.

23. Lectures on Lie groups, University of Illinois, 1970.

24. Operational calculus and abstract differential equations, Notas de Matematica, North Holland Pub. Co., in preparation.

25. Growth theorems for some hyperbolic differential equations with operator coefficients, in preparation

26. A uniqueness theorem for EPD type equations in general spaces, to appear.

27. Local forms of invariant differential operators, Rend. Accad. Lincei, 48 (1970), 292-297.

28. Local forms of invariant operators, I, Ann. Mat. Pura Appl. 86 (1970), 189-216.

29. On the spectral determination of the Green's operator, Jour. Math. Mech., 15 (1966), 1003-1018.

30. Some differential problems related to spectral theory in several operators, Rend. Accad. Lincei, 39 (1965), 170-174.

31. Quelques problèmes différentials abstraits, Eqs. derivées part., Univ. Montréal, 1966, pp. 9-46.

Carroll, R. and Neuwirth, J.

32. Some uniqueness theorems for differential equations with operator coefficients, Trans. Amer. Math. Soc., 110 (1964), 459-472.

33. Quelques théorèmes d'unicité pour des équations différentielles opérationnelles, C. R. Acad. Soc. Paris, 255 (1962), 2885-2887.

Carroll, R.
34. Some remarks on degenerate Cauchy problems, to appear.

Carroll, R. and Cooper, J.
35. Remarks on some variable domain problems in abstract evolution equations, Math. Annalen, 188 (1970), 143-164.

Carroll, R and State, E.
36. Existence theorems for some weak abstract variable domain hyperbolic problems, Canad. Jour. Math., 23 (1971), 611-626.

Chen, P. and Gurtin, M.
1. On a theory of heat conduction involving two temperatures, Zeit. Angew. Math. Phys., 19 (1968), 614-627.

Chi Min-you
1. On the Cauchy problem for a class of hyperbolic equations with data given on the degenerate parabolic line, Acta Math. Sinica, 8 (1958), 521-529.

2. On the Cauchy problem for second order hyperbolic equations in two variables with initial data on the parabolic degenerating line, Acta Math. Sinica, 12 (1962), 68-76.

Cibrario, M.
1. Sui teoremi di esistenza e di unicitá per le equazioni lineari alle derivate parziali del secondo tipo misto iperbolico-paraboliche $x^{2m} Z_{xx} - Z_{yy} = 0$, Rend. Cir. Mat. Palermo, 58 (1934), 217-284.

2. Sulla reduzione a forma canonica dalle equazioni lineari alle derivate parziali di secondo ordine di tipo misto iperbolico-paraboliche, Rend. Cir. Palermo, 56, 1932.

3. Sulla reduzione a forma canonica dalle equazioni lineari alle derivate parziali di secondo ordine di tipo misto, Rend. Lombardo, 65 (1932), 889-906.

Cinquini-Cibrario, M.
1. Primi studii alle equazione lineari alle derivate parziali del secondo ordine di tipo misto iperbolico-paraboliche, Rend. Cir. Mat. Palermo, 56 (1932), 1-34.

2. Sui teoremi di esistenza e di unicitá per le equazioni lineari alle derivate parziali de secondo tipo iperbolico-paraboliche, Rend. Cir. Mat. Palermo, 58 (1934), 1-68.

REFERENCES

Coddington, E. and Levinson, N.

1. Theory of ordinary differential equations, McGraw-Hill, N.Y., 1955.

Coifman, R. and Weiss, G.

1. Analyse harmonique non-commutative sur certains espaces homogènes, Lect. Notes, 242, Springer, N.Y., 1971.

Coirier, Jr.

1. On an evolution problem in linear acoustics of viscous fluids, Joint IUTAM/IMU Symp. Appl. Fnl. Anal. Probs. Mech., Marseille, Sept., 1975.

Coleman, B. and Noll, W.

1. An approximation theorem for functionals, with applications to continuum mechanics, Arch. Rat'l. Mech. Anal., 6 (1960), 355-370.

Coleman, B. D., Duffin, R. J. and Mizel, V. J.

2. Instability, uniqueness and nonexistence theorems for the equation $U_t = U_{xx} - U_{xtx}$ on a strip, Arch. Rat'l. Mech. Anal., 19 (1965), 100-116.

Colton, D.

1. Integral operators and the first initial boundary value problem for pseudoparabolic equations with analytic coefficients, Jour. Diff. Eqs., 13 (1973), 506-522.

2. On the analytic theory of pseudoparabolic equations, Quart. Jour. Math., Oxford (2), 23 (1972), 179-192.

3. Pseudoparabolic equations in one space variable, Jour. Diff. Eqs., 12 (1972), 559-565.

Conti, R.

1. Sul problema di Cauchy per l'equazioni di tipo misto $x^2 Z_{xx} - y^2 Z_{yy} = 0$, Rend. Accad. Lincei, 6 (1949), 579-582.

2. Sul problema di Cauchy per l'equazioni di tipo misto $y^2 Z_{xx} - x^2 Z_{yy} = 0$, Ann. Scuola Norm. Sup., Pisa, 3 (1950), 105-130.

3. Sul problema di Cauchy per l'equazioni $y^{2\alpha} k^2(x,y) \cdot z_{xx} - z_{yy} = f(x,y,z,z_x,z_y)$ con i dati sulla linea parabolica, Ann. Math. Pura Appl., 31 (1950), 303-326.

Copson, E.
1. On a singular boundary value problem for an equation of hyperbolic type, Arch. Rat'l. Mech. Anal., 1 (1958), 349-356.

Copson, E. and Erdelyi, A.
2. On a partial differential equation with two singular lines, Arch. Rat'l. Mech. Anal., 2 (1959), 76-86.

Courant, R. and Hilbert, D.
1. Methods of mathematical physics, Vol. 2, Interscience-Wiley, N.Y., 1962.

2. Methods of mathematical physics, Vol. 1, Interscience-Wiley, N.Y., 1953.

Crandall, M. and Pazy, A.
1. Nonlinear evolution equations in Banach spaces, Israel Jour. Math., 11 (1972), 57-94.

Crandall, M. and Liggett, T.
2. Generation of semigroups of nonlinear transformations in general Banach spaces, Amer. Jour. Math., 93 (1971), 265-298.

Crandall, M.
3. Semigroups of nonlinear transformations in Banach spaces, Contrib. Nonlinear Fnl. Anal, Academic Press, N.Y., 1971, pp. 157-179.

Darboux, G.
1. Leçons sur la théorie génerale des surfaces, second edition, Gauthier-Villars, Paris, 1915.

Davis, P. L.
1. A quasilinear and a related third order problem, Jour. Math. Anal. Appl., 40 (1972), 327-335.

2. On the hyperbolicity of the equations of the linear theory of heat conduction for materials with memory, SIAM Jour. Appl. Math., 30 (1976), 75-80.

Davis, R.
1. The regular Cauchy problem for the Euler-Poisson-Darboux equation, Bull. Amer. Math. Soc., 60 (1954), 338.

2. On a regular Cauchy problem for the Euler-Poisson-Darboux equation, Ann. Mat. Pura Appl., 42 (1956), 205-226.

REFERENCES

Delache, S. and Leray, J.

1. Calcul de la solution élémentaire de l'opérateur d'Euler-Poisson-Darboux et de l'opérateur de Tricomi-Clairaut, hyperbolique, d'ordre 2, Bull. Soc. Math. France, 99 (1971), 313-336.

Delsarte, J.

1. Sur certains transformations fonctionnelles relatives aux équations linéaires aux derivées partielles du second ordre, C. R. Acad. Sci. Paris, 206 (1938), 1780-1782.

Delsarte, J. and Lions, J.

2. Transmutations d'opérateurs differéntiels dans le domaine complexe, Comm. Math. Helv., 32 (1957), 113-128.

3. Moyennes généralisées, Comm. Math. Helv., 33 (1959), 59-69.

Dembart, B.

1. On the theory of semigroups of operators on locally convex spaces, to appear.

Derguzov, V. I.

1. Solution of an anticanonical equation with periodic coefficients, Dokl. Akad. Nauk SSSR 193 (1970), 834-838.

Dettman, J.

1. Initial-boundary value problems related through the Stieltjes transform, Jour. Math. Anal. Appl., 15 (1969), 341-349.

Diaz, J. and Young, E.

1. A singular characteristic value problem for the Euler-Poisson-Darboux equation, Ann. Mat. Pura Appl., 95 (1973), 115-129.

Diaz, J. and Weinberger, H.

2. A solution of the singular initial value problem for the EPD equation, Proc. Amer. Math. Soc., 4 (1953), 703-718.

Diaz, J. and Ludford, G.

3. On the singular Cauchy problem for a generalization of the Euler-Poisson-Darboux equation in two space variables, Ann. Mat. Pura Appl., 38 (1955), 33-50.

4. On the Euler-Poisson-Darboux equation, integral operators, and the method of descent, Proc. Conf. Diff. Eqs., Univ. Maryland, 1956, pp. 73-89.

Diaz, J.
5. On singular and regular Cauchy problems, Comm. Pure Appl. Math., 9 (1956), 383-390.

6. Solution of the singular Cauchy problem for a singular system of partial differential in the mathematical theory of dynamical elasticity, etc., Theory of distributions, Proc. Summer Inst., Lisbon, 1964, pp. 75-131.

Diaz, J. and Kiwan, A.
7. A remark on the singular Cauchy problem for all values of time for the Euler-Poisson-Darboux equation, Jour. Math. Mech., 16 (1966), 197-202.

Diaz, J. and Young, E.
8. On the characteristic initial value problem for the wave equation in odd spatial dimensions with radial initial data, Ann. Mat. Pura Appl., 94 (1972), 161-176.

Diaz, J. and Martin, M.
9. Riemann's method and the problem of Cauchy, II. The wave equation in n dimensions, Proc. Amer. Math. Soc., 3 (1952), 476-484.

Diaz, J. and Young E.
10. Uniqueness of solutions of certain boundary value problems for ultrahyperbolic equations, Proc. Amer. Math. Soc., 29 (1971), 569-574.

Dieudonné, J.
1. Foundations of modern analysis, Academic Press, N.Y., 1960.

Dixmier, J.
1. Les algèbres d'opérateurs dans l'espace Hilbertien, Gauthier-Villars, Paris, 1957.

Donaldson, J.
1. A singular abstract Cauchy problem, Proc. Nat'l. Acad. Sci., 66 (1970), 269-274.

Donaldson, J. and Goldstein, J.
2. Some remarks on uniqueness for a class of singular abstract Cauchy problems, Proc. Amer. Math. Soc., 54 (1976), 149-153.

Donaldson, J.
3. An operational calculus for a class of abstract operator equations, Jour. Math. Anal. Appl., 37 (1972), 167-184.

4. The Cauchy problem for a first order system of abstract operator equations, Bull. Amer. Math. Soc., 81 (1975), 576-578.

5. New integral representations for solutions of Cauchy's problem for abstract parabolic equations, Proc. Nat'l. Acad. Sci., 68 (1971), 2025-2027.

Donaldson, J. and Hersh, R.
6. A perturbation series for Cauchy's problem for higher order abstract parabolic equations, Proc. Nat'l. Acad. Sci., 67 (1970), 41-44.

Donaldson, J.
7. The abstract Cauchy problem and Banach space valued distributions, to appear.

Douglis, A.
1. A criterion for the validity of Huygens' principle, Comm. Pure Appl. Math., 9 (1956), 391-402.

Dubinsky, J.
1. Weak convergence in nonlinear elliptic and parabolic equations, Mat. Sbornik, 67 (109) (1965), 609-642; Amer. Math. Soc. Transl. (2), 67 (1968), 226-258.

Duff. G.
1. Sur une classe de solutions élémentaires positives, Eqs. derivées part., Univ. Montréal, 1966, pp. 47-60.

2. Positive elementary solutions and completely monotonic functions, Jour. Math. Anal. Appl., 27 (1969), 469-494.

Dunninger, D. and Levine, H.
1. Uniqueness criteria for solutions of singular boundary value problems, to appear.

Dunninger, D. and Zachmanoglou, E.
2. The condition for uniqueness of the Dirichlet problem for hyperbolic equations in cylindrical domains, Jour. Math. Mech., 18 (1969), 763-766.

Dunninger, D. and Weinacht, R.
3. Improperly posed problems for singular equations of the fourth order, Jour. Appl. Anal., 4 (1975), 331-341.

Ehrenpreis, L. and Mautner, F.
1. Some properties of the Fourier transform on semisimple Lie groups, I-III, Annals Math., 61 (1955), 406-439; Trans. Amer. Math. Soc., 84 (1957), 1-55 and 90 (1959), 431-484.

Ehrenpreis, L.

2. Analytic functions and the Fourier transform of distributions, I, Annals Math., 63 (1956), 129-159.

Ekeland, I. and Teman, R.

1. Analyse convexe et problèmes variationnels, Dunod, Paris, 1974.

Eskin, G.

1. Boundary value problems for equations with constant coefficients on the plane, Mat. Sbornik (N.S.), 59 (101) (1962), suppl., 69-104.

2. A boundary value problem for the equation $\partial/\partial t\ P(\partial/\partial t, \partial/\partial x)u = 0$, where P is an elliptic operator, Sibirsk, Mat. Žur., 3 (1962), 882-911.

Euler, L.

1. Institiones calculi integralis, III, Petropoli, 1770.

Ewing, R.

1. The approximation of certain parabolic equations backward in time by Sobolev equations, SIAM Jour. Math. Anal., 6 (1975), 283-294.

2. Numerical solution of Sobolev partial differential equations, SIAM Jour. Numer. Anal., 12 (1975), 345-363.

3. Determination of a physical parameter in Sobolev equations, Int. Jour. Eng. Sci., to appear.

Fattorini, H.

1. Ordinary differential equations in linear topological spaces, I and II, Jour. Diff. Eqs., 5 (1968), 72-105 and 6 (1969), 50-70.

Feller, W.

1. Two singular diffusion problems, Annals Math., 54 (1951), 173-182.

2. The parabolic differential equations and the associated semigroups of transformations, Annals Math., 55 (1952), 468-519.

Ferrari, C. and Tricomi, F.

1. Aerodinamica transonica, Cremonese, Rome, 1962--English translation, Academic Press, N.Y., 1968.

Fichera, G.
1. Sulle equazioni differenziali lineari ellitico-paraboliche del secondo ordine, Atti Accad. Lincei, 5 (1956), 1-30.

2. On a unified theory of boundary value problems for elliptic-parabolic equations of second order, Boundary problems in differential equations, Univ. Wisconsin, 1960, pp. 97-120.

Filipov, B.
1. Solution of the Cauchy problem for hyperbolic equations with initial data on the parabolic line, Nauč. Dokl. Vys. Šk., 4 (1958), 69-73.

Flensted-Jensen, M.
1. On the Fourier transform on a symmetric space of rank one, Inst. Mittag-Leffler, Djursholm, 1970.

Foias, C., Gussi, G., and Poenanu, V.
1. Une méthode directe dans l'étude des équations aux derivées partielles hyperboliques, quasilinéaires en deux variables, Math. Nachr., 15 (1956), 89-116.

2. Sur les solutions généralisées de certaines équations linéaires et quasi-linéaires dans l'espace de Banach, Rev. Math. Pures Appl., Acad. Rep. Pop. Roumaine, 3 (1958), 283-304.

Foais, C.
3. La mesure harmonique-spectrale et la théorie spectral des opérateurs généraux d'un espace de Hilbert, Bull. Soc. Math. France, 85 (1957), 263-282.

Ford, W.
1. The first initial boundary value problem for a nonuniform parabolic equation, Jour. Math. Anal. Appl., 40 (1972), 131-137.

Ford, W. and Waid, M.
2. On maximum principles for some degenerate parabolic operators, Jour. Math. Anal. Appl., 40 (1972), 271-277.

Ford, W. H.
1. Numerical solution of pseudoparabolic partial differential equations, Thesis, University of Illinois, 1972.

Ford, W. H. and Ting, T. W.
2. Uniform error estimates for difference approximations to nonlinear pseudoparabolic partial differential equations, SIAM Jour. Numm. Anal., 11 (1974), 155-169.

Fox, D.
1. The solution and Huygens' principle for a singular Cauchy problem, Jour. Math. Mech., 8 (1959), 197-219.

Fox, D. and Pucci, C.
2. The Dirichlet problem for the wave equation, Ann. Mat. Pura Appl., 45 (1958), 155-182.

Fox, D.
3. Transient solutions for stratified fluid flows, to appear.

Frankl, F.
1. On Cauchy's problem for equations of mixed elliptic-hyperbolic type with initial data on the transition line, Izv. Akad. Nauk SSSR, 8 (1944), 195-224.

Friedlander, F. and Heins, A.
1. On a singular boundary value problem for the Euler-Darboux equation, Jour. Diff. Eqs., 4 (1968), 460-491.

Friedman, A. and Schuss, Z.
1. Degenerate evolution equations in Hilbert space, Trans. Amer. Math. Soc., 161 (1971), 401-427.

Friedman, A.
2. Generalized functions and partial differential equations, Prentice-Hall, Englewood Cliffs, N.J., 1963.

Friedman, A. and Shinbrot, M.
3. The initial value problem for the linearized equations of water waves, Jour. Math. Mech., 19 (1967), 107-180.

Friedman, B.
1. An abstract formulation of the method of separation of variables, Proc. Conf. Diff. Eqs., Univ. Maryland, 1956, pp. 209-226.

Fuglede, B.
1. A commutativity theorem for normal operators, Proc. Nat'l. Acad. Sci., 36 (1950), 35-40.

Fuks, B.
1. Introduction to the theory of analytic functions of several complex variables, Gos. Izd. Fiz.-Mat. Lit., Moscow, 1962.

Furstenberg, H.
1. A Poisson formula for semisimple Lie groups, Annals Math., 77 (1963), 335-386.

REFERENCES

Fusaro, B.

1. Spherical means in harmonic spaces, Jour. Math. Mech., 18 (1969), 603-606.

2. A correspondence principle for the Euler-Poisson-Darboux equation in harmonic space, Glasnik Mat., 21 (1966), 99-102.

3. The Cauchy-Kowalewski theorem and a singular initial value problem, SIAM Review, 10 (1968), 417-421.

4. A solution of a singular mixed problem for the equation of Euler-Poisson-Darboux, Amer. Math. Monthly, 73 (1966), 610-613.

5. A nonuniqueness result for an Euler-Poisson-Darboux problem, Proc. Amer. Math. Soc., 18 (1967), 381-382.

6. A simple existence and uniqueness proof for a singular Cauchy problem, Proc. Amer. Math. Soc., 19 (1968), 1504-1505.

Gagneux, G.

1. Thèse 3me cycle, Paris, to appear.

Gajewski, H. and Zacharius, K.

1. Zur regularization einer Klasse nichtkorrekter Probleme bei Evolutionsgleichungen, Jour. Math. Anal. Appl. 38 (1972), 784-789.

2. Uber eine Klasse nichtlinearen Differentialgleichungen im Hilbertraum, Jour. Math. Anal. Appl., 44 (1973) . 71-87.

3. Zur starken Konvergenz des Galerkinverfahrens bei einer Klasse pseudoparabolischer partieller Differentialgleichungen, Math. Nachr., 47 (1970), 365-376.

Galpern, S.

1. Cauchy's problem for an equation of S. L. Sobolev's type, Dokl. Akad. Nauk SSSR, 104 (1955), 815-818.

2. The Cauchy problem for general systems of linear partial differential equations, Usp. Mat. Nauk, 18 (1963), 239-249.

3. The Cauchy problem for Sobolev equations, Sibirsk. Mat. Žur., 4 (1963), 758-774.

4. The Cauchy problem for general systems of linear partial differential equations, Trudy Mosk. Mat. Obšč., 9 (1960), 401-423.

5. Cauchy problem for systems of linear differential equations, Dokl. Akad. Nauk SSSR, 119 (1958), 640-643.

Gangolli, R.
1. Spherical functions on semisimple Lie groups, to appear.

Garabedian, P.
1. Partial differential equations, Wiley, N.Y., 1964.

Garding, L.
1. Application of the theory of direct integrals of Hilbert spaces to some integral and differential operators, Lectures, Univ. Maryland, 1954.

Garsoux, J.
1. Espaces vectoriels topologiques et distributions, Dunod, Paris, 1963.

Gelbart, S.
1. Fourier analysis on matrix space, Mem. Amer. Math. Soc., 108, 1971.

Gelfand, I. and Naimark, M.
1. Unitary representations of the classical groups, Moscow, 1950.

Gelfand, I., Graev, M. and Vilenkin, N.
2. Generalized functions, Vol. 5, Integral geometry and related questions of representation theory, Moscow, 1962.

Gelfand, I., Graev, M. and Pyatetskij-Shapiro, I.
3. Generalized functions, Vol. 6, Representation theory and automorphic functions, Moscow, 1966.

Gelfand, I., Minlos, R. and Shapiro, Z.
4. Representations of the rotation and Lorentz groups and their applications, Moscow, 1958.

Gelfand, I. and Šilov, G.
5. Generalized functions, Vol. 1, Generalized functions and operations on them, Moscow, 1958.

6. Generalized functions, Vol. 2, Spaces of basic and generalized functions, Moscow, 1958.

7. Generalized functions, Vol. 3, Some questions in the theory of differential equations, Moscow, 1958.

Gelfand, I., Raikov, I. and Šilov, G.
8. Commutative normed rings, Moscow, 1960.

Germain, P.
1. Remarks on the theory of partial differential equations of mixed type and applications to the study of transonic flow, Comm. Pure Appl. Math., 7 (1954), 143-177.

Germain, P. and Bader, R.
2. Sur quelques problèmes aux limites, singuliers pour une equation hypérbolique, C. R. Acad. Sci. Paris, 231 (1950), 268-270.

3. Sur le problème de Tricomi, Rend. Cir. Mat. Palermo, 2 (1953), 53-70.

Ghermanesco, M.
1. Sur les moyennes successives des fonctions, Bull. Soc. Math. France, 62 (1934), 245-246.

2. Sur les valeurs moyennes des fonctions, Math. Annalen, 119 (1944), 288-320.

Gilbert, R.
1. Function theoretic methods in partial differential equations, Academic Press, N.Y., 1969.

Gindikin, S. and Karpelevič, F.
1. Plancherel measure of Riemannian symmetric spaces of nonpositive curvature, Dokl. Akad. Nauk, SSSR, 145 (1962), 252-255.

Gluško, V. and Krein, S.
1. Degenerating linear differential equations in Banach space, Dokl. Akad. Nauk SSSR, 181 (1968), 784-787.

Godement, R.
1. A theory of spherical functions, Trans. Amer. Math. Soc., 73 (1952), 496-556.

Gordeev, A.
1. Certain boundary value problems for a generalized Euler-Poisson-Darboux equation, Volz. Mat. Sbor. Vyp., 6 (1968), 56-61.

Grabmüller, H.
1. Relativ akkretive operatoren und approximation von Evolutionsgleichungen 2, Math. Nachr., 66 (1975), 67-87 and 89-100.

Grange, O. and Mignot, F.
1. Sur la résolution d'une équation et d'une inéquation paraboliques non-linéaires, Jour. Fnl. Anal. 11 (1972), 77-92.

Gunning, R. and Rossi, H.
1. Analytic functions of severalcomplex variables, Prentice-Hall, Englewood Cliffs, N.J., 1965.

Günther, P.
1. Uber einige spezielle Probleme aus der Theorie der linearen partiellen Differentialgleichungen zweiter Ordnung, Berichte Verh. Sachs. Akad. Wiss., Leipzig, 102 (1957), 1-50.

2. Uber die Darbouxsche Differentialgleichung mit variablen Koeffizienten, Math. Nachr., 22 (1960), 285-321.

3. Das iterierte Anfangswertproblem bei der Darbouxschen Differentialgleichung, Math. Nachr., 25 (1963), 293-310.

4. Huygenssche Differentialgleichungen, die zur Wellengleichung infinitesimal benachbart sind, Arch. Math., 16 (1965), 465-475.

5. Einige Sätze uber Huygenssche Differentialgleichungen, Wiss. Zeit. Univ. Leipzig, 17 (1965), 497-507.

Haack, W. and Hellwig, G.
1. Lineare partielle Differentialgleichungen zweiter Ordnung von gemischtem Typus, Math. Zeit., 61 (1954), 26-46.

Hadamand, J.
1. Lectures on Cauchy's problem in linear partial differential equations, Dover, N.Y., 1952.

Hairullina, S.
1. The solution of a completely hyperbolic equation with initial data on a line of degeneracy, Diff. Urav., 3 (1967), 994-1001.

2. Solution of a hyperbolic equation with data on the line of degenerate type, Uč. Zap. Kabardino-Balkarsk. Univ., 19 (1963), 277-283.

3. Solution of the Cauchy problem for a hyperbolic equation with data on the line of degenerate type, Dokl. Akad. Nauk SSSR, 8 (1964), 361-364.

Hairullin, H. and Nikolenko, V.

1. A mixed problem for an equation of hyperbolic type that degenerates on the boundary of the region, Volz. Mat. Sbor. Vyp., 7 (1969), 190-193.

Harish-Chandra

1. Spherical functions on a semisimple Lie group, I, Amer. Jour. Math., 80 (1958), 241-310.

Hausner, M. and Schwartz, J.

1. Lie groups; Lie algebras, Gordon-Breach, N.Y., 1968.

Helgason, S.

1. Differential geometry and symmetric spaces, Academic Press, N.Y., 1962.

2. A duality for symmetric spaces with applications to group representations, Advances Math., 5 (1970), 1-154.

3. Lie groups and symmetric spaces, Battelle Rencontres, 1967, Benjamin, N.Y., 1968, pp. 1-71.

4. Differential operators on homogeneous spaces, Acta Math., 102 (1960), 239-299.

5. Eigenspaces of the Laplacian; integral representations and irreducibility, Jour. Fnl. Anal., 17 (1974), 328-353.

6. Duality and Radon transform for symmetric spaces, Amer. Jour. Math., 85 (1963), 667-692.

7. Analysis on Lie groups and homogeneous spaces, Amer. Math. Soc. Reg. Conf. Ser., 14, Providence, 1971.

8. Fundamental solutions of invariant differential operators on symmetric spaces, Amer. Jour. Math., 86 (1964), 565-601.

9. Paley-Wiener theorems and surjectivity of invariant differential operators on symmetric spaces and Lie groups, Bull. Amer. Math. Soc., 79 (1973), 129-132.

10. Solvability of invariant differential operators on homogeneous manifolds, CIME Seminar, 1975.

11. The eigenfunctions of the Laplacian on a two point homogeneous space; integral representations and irreducibility, Proc. Amer. Math. Soc. Sum. Inst., Stanford, 1973.

Hellwig, G.

1. Anfangs-und Randwertprobleme bei partiellen Differentialgleichungen von wechselndem Typus auf dem Rändern, Math. Zeit., 58 (1953), 337-357.

2. Anfangswertprobleme bei partiellen Differentialgleichungen mit Singularitaten, Jour. Rat'l. Mech. Anal., 5 (1956), 395-418.

3. Partielle Differentialgleichungen, Teubner, Stuttgart, 1960.

Hermann, R.

1. Lie groups for physicists, Benjamin, New York, 1966.

Hersh, R.

1. Explicit solution of a class of higher order abstract Cauchy problems, Jour. Diff. Eqs., 8 (1970), 570-579.

2. Direct solution of general one dimensional linear parabolic equation via an abstract Plancherel formula, Proc. Nat'l. Acad. Sci., 63 (1969), 648-654.

3. The method of transmutations, Part. diff. eqs. and related topics, Lect. Notes, 446, Springer, N.Y., 1975, pp. 264-282.

Hille, E.

1. The abstract Cauchy problem, Jour. Anal. Math., 3 (1953-54), 81-196.

Hille, E. and Phillips, R.

2. Functional analysis and semigroups, Amer. Math. Soc. Colloq. Pub., Vol. 31, 1957.

Hochschild, S.

1. The structure of Lie groups, Holden-Day, San Francisco, 1965.

Horgan, C. O. and Wheeler, L. T.

1. A spatial decay estimate for pseudoparabolic equations, Lett. Appl. Eng. Sci., 3 (1975), 237-243.

Hörmander, L.
1. Linear partial differential operators, Springer, Berlin, 1963.

2. Complex analysis in several variables, Von Nostrand, Princeton, 1966.

Horváth, J.
1. Topological vector spaces and distributions, Addison-Wesley, Reading, Mass., 1967.

Huilgol, R.
1. A second order fluid of the differential type, Int. Jour. Nonlinear Mech., 3 (1968), 471-482.

Ilin, A.
1. Sur le comportement de la solution d'un problème aux limites pour $t \to \infty$, Mat. Sbornik, 81 (1972), 530-553.

Infeld, L. and Hull, T.
1. The factorization method, Rev. Mod. Physics, 23 (1951), 21-68.

Jacobson, N.
1. Lie algebras, Interscience, N.Y., 1962.

Jacquet, H. and Langlands, R.
1. Automorphic forms on GL(2), Springer Lect. Notes 114, N.Y., 1970.

Jehle, H. and Parke, W.
1. Covariant spinor formulation of relativistic wave equations under the homogeneous Lorentz group, Lect. Theoret. Physics, Vol. 7a, Univ. Colorado, 1965, pp. 297-386.

John, F.
1. Plane waves and spherical means, Interscience, N.Y., 1955.

2. The Dirichlet problem for a hyperbolic equation, Amer. Jour. Math., 63 (1941), 141-154.

Kamber, F. and Tondeur, Ph.
1. Invariant differential operators and the cohomology of Lie algebra sheaves, Mem. Amer. Math. Soc., 113, 1971.

Kamenomostskaya, S. L.
1. On the Stefan problem, Mat. Sbornik, 53 (1961), 489-514.

Kapilevič, M.
1. Equations of mixed elliptic and hyperbolic type, Linear eqs. math. physics, Ed. S. Mikhlin, Izd. Nauka, Moscow, 1964, chap. 8.

2. On an equation of mixed elliptic-hyperbolic type, Mat. Sbornik, 30 (1952), 11-38.

3. On the analytic continuation of the complementary solutions of an equation of hyperbolic type with singular coefficients, Dokl. Akad. Nauk SSSR, 116 (1957), 167-170.

4. On the theory of linear differential equations with two nonperpendicular lines of parabolicity, Dokl. Akad. Nauk SSSR, 125 (1959), 251-254.

5. The solution of singular Cauchy problems in operator series and basis series, Comment. Math. Univ. Carolinae, 9 (1968), 27-40.

Karapetyan, K.
1. On the Cauchy problem for equations of hyperbolic type, degenerating on the initial plane, Dokl. Akad. Nauk, SSSR, 106 (1956), 963-966.

Karimov, D. and Baikuziev, K.
1. A mixed problem for a hyperbolic equation degenerating on the boundary of the region, Nauč. Tr. Taškent. Inst., 208 (1962), 90-97.

2. A second mixed problem for a hyperbolic equation degenerating on the boundary of the region, Izv. Akad. Nauk Uzb. SSR 6 (1964), 27-30.

Karmanov, V.
1. On the existence of solutions of some boundary problems for equations of mixed type, Izv. Akad. Nauk, SSSR, 22 (1958), 117-134.

Karol, I.
1. On the theory of equations of mixed type, Dokl. Akad. Nauk SSSR, 88 (1953), 297-300.

2. On the theory of boundary problems for equations of mixed elliptic-hyperbolic type, Mat. Sbornik, 38 (1956), 261-282.

REFERENCES

Karpelevič, F.

1. Geometry of geodesics and eigenfunctions of the Laplace-Beltrami operator on symmetric spaces, Trudy Mosk. Mat. Obšč., 14 (1965), 48-185.

Kato, T.

1. Perturbation theory for linear operators, Springer, Berlin, 1966.

2. Nonlinear semigroups and evolution equations, Jour. Math. Soc. Japan, 19 (1967), 508-520.

Keller, J.

1. On solutions of nonlinear wave equations, Comm. Pure Appl. Math., 10 (1957), 523-530.

Kimura, M. and Ohta, T.

1. Probability of fixation of a mutant gene in a finite population when selective advantage decreases in time, Genetics, 65 (1970), 525-534.

Kluge, R. and Bruckner, G.

1. Uber einige Klassen nichtlinearen Differentialgleichungen und -ungleichungen im Hilbert-Raum, Math. Nachr., 64 (1974), 5-32.

Konishi, Y.

1. On the nonlinear semigroups associated with $u_t = \Delta\beta(u)$ and $\phi(u_t) = \Delta u$, Jour. Math. Soc. Japan, 25 (1973), 622-628.

Knapp, A. and Stein, E.

1. Singular integrals and the principal series, Proc. Nat'l. Acad. Sci., 63 (1969), 281-284.

Kobayashi, S. and Nomizu, K.

1. Foundations of differential geometry, I, Interscience-Wiley, N.Y., 1963.

2. Foundations of differential geometry, II, Interscience-Wiley, N.Y., 1969.

Kohn, J. and Nirenberg, L.

1. Degenerate elliptic-parabolic equations of second order, Comm. Pure, Appl. Math., 20 (1967), 797-872.

Komatsu, H.

1. Semigroups of operators in locally convex spaces, Jour. Math. Soc. Japan, 16 (1964), 230-262.

Komura, T.
1. Semigroups of operators in locally convex spaces, Jour. Fnl. Anal., 2 (1968), 258-296.

Kononenko, V.
1. Fundamental solutions of singular partial differential equations with variable coefficients, Sov. Mat. Dokl., 8 (1967), 71-74.

Kostant, B.
1. On the existence and irreducibility of certain series of representations, Bull. Amer. Math. Soc., 75 (1969), 627-642.

Kostyučenko, A. and Eskin, G.
1. The Cauchy problem for an equation of Sobolev-Galpern type, Trudy Mosk. Mat. Obšč., 10 (1961), 273-284.

Köthe, G.
1. Topologische lineare Raume, Springer, Berlin, 1960.

Krasnov, M.
1. Mixed boundary value problems for degenerate linear hyperbolic differential equations of second order, Mat. Sbornik, 91 (1959), 29-84.

2. The mixed boundary problem and the Cauchy problem for degenerate hyperbolic equations, Dokl. Akad. Nauk, SSSR, 107 (1956), 789-792.

3. Mixed boundary problems for degenerate hyperbolic equations, Trudy Mosk. Energ. Inst., 28 (1956), 25-45.

Krein, M. G.
1. Introduction to geometry of indefinite J-spaces, Amer. Math. Soc. Transl. (2), Vol. 93 (1970), 103-199.

Krivenkov, Yu.
1. On a representation of the solutions of the Euler-Poisson-Darboux, Dokl. Akad. Nauk SSSR, 116 (1957), 351-354.

2. Some problems for the Euler-Poisson-Darboux equation, Dokl. Akad. Nauk SSSR, 123 (1958), 397-400.

Kunze, R. and Stein, E.
1. Uniformly bounded representations and harmonic analysis of the 2 x 2 real unimodular group, Amer. Jour. Math., 82 (1960), 1-62.

2. Uniformly bounded representations, 3, Amer. Jour. Math., 89 (1967), 385-442.

Lacomblez, C.
1. Une équation d'évolution du second ordre en t à coefficients dégénérés ou singuliers, Pub. Math. Bordeaux, Fasc. 4 (1974), 33-64.

Ladyženskaya, O.
1. Mixed problems for hyperbolic equations, Gos. Tek. Izd., Moscow, 1953.

Ladyženskaya, O., Solonnikov, V. and Uralceva, N.
2. Linear and quasilinear equations of parabolic type, Izd. Nauka, Moscow, 1967.

Lagnese, J.
1. General boundary value problems for differential equations of Sobolev type, SIAM Jour. Math. Anal., 3 (1972), 105-119.

2. Exponential stability of solutions of differential equations of Sobolev type, SIAM Jour. Math. Anal., 3 (1972), 625-636.

3. Rate of convergence in a class of singular perturbations, Jour. Fnl. Anal., 13 (1973), 302-316.

4. Approximation of solutions of differential equations in Hilbert-space, Jour. Math. Soc. Japan, 25 (1973), 132-143.

5. Perturbations in a class of nonlinear abstract equations, SIAM Jour. Math. Anal., 6 (1975), 616-627.

6. Rate of convergence in singular perturbations of hyperbolic problems, Indiana Univ. Math. Jour., 24 (1974), 417-432.

7. Existence, uniqueness and limiting behavior of solutions of a class of differential equations in Banach space, Pacific Jour. Math., to appear.

8. The final value problem for Sobolev equations, Proc. Amer. Math. Soc., to appear.

9. Singular differential equations in Hilbert space, SIAM Jour. Math. Anal., 4 (1973), 623-637.

10. Perturbations in variational inequalities, Jour. Math. Anal. Appl., to appear.

11. The structure of a class of Huygens' operators, Jour. Math. Mech., 18 (1969), 1195-1201.

12. A solution of Hadamards' problem for a restricted class of operators, Proc. Amer. Math. Soc., 19 (1968), 981-988.

13. The fundamental solution and Huygens' principle for decomposable differential operators, Arch. Rat'l. Mech. Anal., 19 (1965), 299-307.

Lagnese, J. and Stellmacher, K.

14. A method of generating classes of Huygens' operators, Jour. Math. Mech., 17 (1967), 461-472.

Lavrentiev, M. and Bitsadze, A.

1. On the problem of equations of mixed type, Dokl. Akad. Nauk SSSR, 70 (1950), 373-376.

Lavrentiev, M.

2. Some improperly posed problems of mathematical physics, Springer, N.Y., 1967.

Lax, P. and Phillips, R.

1. Scattering theory, Academic Press, N.Y., 1967.

Lebedev, V. I.

1. The use of nets for the Sobolev type of equation, Dokl. Akad. Nauk SSSR, 114 (1957), 1166-1169.

Leray, J. and Lions, J. L.

1. Quelques résultats de Višik sur les problèmes elliptiques non linéaires par les méthodes de Minty-Browder, Bull. Soc. Math. France, 93 (1965), 97-107.

Levikson, B. and Schuss, Z.

1. Nonhomogeneous diffusion approximation to a genetic model, Jour. Math. Pures Appl., to appear.

Levine, H.

1. Some uniqueness and growth theorems in the Cauchy problem for $Pu'' + Mu' + Nu = 0$ in Hilbert space, Math. Zeit., 126 (1972), 345-360.

2. On the nonexistence of global solutions to a nonlinear EPD equation, Jour. Math. Anal. Appl., 48 (1974), 646-651.

3. Instability and nonexistence of global solutions to nonlinear wave equations of the form $Pu_{tt} = -Au + F(u)$, Trans. Amer. Math. Soc., 192 (1974), 1-21.

4. Nonexistence theorems for the heat equation with nonlinear boundary conditions and for the porous medium equation backward in time, Jour. Diff. Eqs., 16 (1974), 319-334.

5. Logarithmic convexity and the Cauchy problem for some abstract second order differential inequalities, Jour. Diff. Eqs., 8 (1970), 34-55.

6. Logarithmic convexity, first order differential inequalities and some applications, Trans. Amer. Math. Soc., 152 (1970), 299-320.

7. Some nonexistence and instability theorems for solutions to formally parabolic equations of the form $Pu_t = - Au + F(u)$, Arch. Rat'l. Mech. Anal., 51 (1973), 371-386.

8. Nonexistence of global weak solutions to some properly and improperly posed problems of mathematical physics: the method of unbounded Fourier coefficients, Math. Annalen. 214 (1975), 205-220.

9. Some additional remarks on the nonexistence of global solutions to nonlinear wave equations, SIAM Jour. Math. Anal., 5 (1974), 138-146.

10. Uniqueness and growth of weak solutions to certain linear differential equations in Hilbert space, Jour. Diff. Eqs., 17 (1975), 73-81.

11. Uniqueness and growth of weak solutions to certain linear differential equations in Hilbert space, Jour. Diff. Eqs., 17 (1975), 73-81.

Levine, H. and Murray, A.

12. Asymptotic behavior and lower bounds for semilinear wave equations in Hilbert space with applications, SIAM Jour. Math. Anal., 6 (1975), 846-859.

Lezhnev, V. G.

1. The decreasing property of a solution for Sobolev's equation, Diff. Eqs., 9 (1973), 389.

Lieberstein, H.
1. A mixed problem for the Euler-Poisson-Darboux equation, Jour. Anal. Math., 6 (1958), 357-379.

2. On the generalized radiation problem of A. Weinstein, Pacific Jour. Math., 7 (1959), 1623-1640.

Lighthill, M. J.
1. Dynamics of rotating fluids: a survey, Jour. Fluid Mech. 26 (1966), 411-431.

Lions, J.
1. Opérateurs de Delsarte et problèmes mixtes, Bull. Soc. Math. France, 84 (1956), 9-95.

2. Equations d'Euler-Poisson-Darboux généralisées, C. R. Acad. Sci. Paris, 246 (1958), 208-210.

3. Opérateurs de transmutation singuliers et équations d'Euler-Poisson-Darboux généralisées, Rend. Sem. Mat. Fis. Milano, 28 (1959), 3-16.

4. On the generalized radiation problem of Weinstein, Jour. Math. Mech., 8 (1959), 873-888.

5. Equations différentielles-opérationnelles, Springer, Berlin, 1961.

6. Sur la regularité et l'unicité des solutions turbulentes des équations de Navier-Stokes, Rend. Sem. Mat. Padova, 30 (1960), 16-23.

7. Sur les semigroupes distributions, Portug. Math., 19 (1960), 141-164.

Lions, J. and Strauss, W.
8. Some nonlinear evolution equations, Bull. Soc. Math. France, 93 (1965), 43-96.

Lions, J.
9. Boundary value problems Tech. Report, Univ. Kansas, 1957.

10. Quelques méthodes de résolution des problèmes aux limites non-linéaires, Dunod, Paris, 1969.

Lojasiewicz, S.
1. Sur la fixation des variables dans une distribution, Studia Math., 17 (1958), 1-64.

REFERENCES

Loos, O.

1. Symmetric spaces, Vols. 1 and 2, Benjamin, N.Y., 1969.

Love, A. E. H.

1. A treatise on the mathematical theory of elasticity, Dover, N.Y., 1944.

Luke, Y.

1. The special functions and their approximations, Vols. 1 and 2, Academic Press, N.Y., 1969.

Lyubarskij, G.

1. The theory of groups and its applications in physics, Moscow, 1958.

Magnus, W., Oberhettinger, F. and Soni, R.

1. Formulas and theorems for the special functions of mathematical physics, Springer, N.Y., 1966.

Makarov, I.

1. Uniqueness theorems for the solution of problem D, problem E, and a problem of type N, Volz. Mat. Sbor. Vyp., 6 (1968), 142-144.

2. Solution of boundary value problems for equations of hyperbolic type, Volz. Mat. Sbor. Vyp., 6 (1968), 131-141.

Makusina, R.

1. A local extremum principle for an equation with a first order singularity on the boundary of the domain, Volz. Mat. Sbor. Vyp., 6 (1968).

Martin, M.

1. Riemann's method and the problem of Cauchy, Bull. Amer. Math. Soc., 57 (1951), 238-249.

Maslennikova, V.

1. An explicit representation and asymptotic behavior for $t \to \infty$ of the solution of the Cauchy problem for a linearized system of a rotating compressible fluid, Soviet Math. Dokl. 10 (1969), 978-981.

2. Solution in explicit form of the Cauchy problem for a system of partial differential equations, Izv. Akad. Nauk, SSSR., 22 (1958), 135-160.

3. The rate of decrease, for large time, of the solution of a Sobolev system with allowance made for viscosity, Mat. Sbornik 92 (134), 1973, 584-610.

4. L_p-estimates and the asymptotic behavior as $t \to \infty$ of a solution of the Cauchy problem for a Sobolev system, Proc. Stekov Inst. Math., 103 (1969), 123-150.

5. Construction of a solution of Cauchy's problem for a system of partial differential equation, Dokl. Akad. Nauk. SSSR 102 (1955), 885-888.

6. Mixed problems for a system of partial differential equations of first order, Izv. Akad. Nauk, SSSR, 22 (1958), 271-278.

Maurin, I.

1. General eigenfunction expansions and unitary representations of topological groups, Monogr. mat., 48, Panst. Wydaw. Nauk, Warsaw, 1968.

2. Uber gemischte Rand und Anfangswertprobleme im Grossen für gewisse systeme von Differentialgleichungen auf differenzierbaren Mannigfaltigkeiten, Studia Math., 15 (1956), 314-327.

3. Metody przestrzeni Hilberta, Monogr. mat., 36, Warszawa, 1959; Methods of Hilbert spaces, Warsaw, 1967.

Mazumdar, T.

1. Existence theorems for noncoercive variable domain evolution problems, Thesis, University of Illinois, 1971.

2. Generalized projection theorem with application to linear equations and some nonlinear situations, Jour. Math. Anal. Appl., 43 (1973), 72-100.

3. Generalized projection theorem and weak noncoercive evolution problems in Hilbert space, Jour. Math. Anal. Appl., 46 (1974), 143-168.

4. Regularity of solutions of linear coercive evolution equations with variable domain, I, Jour. Math. Anal. Appl., 52 (1975), 615-647.

McKerrell, A.

1. Covariant wave equations for massless particles, Annals Physics, 40 (1966), 237-267.

McShane, E.

1. Integration, Princeton Univ. Press, Princeton, N.J., 1947.

Medeiros, L. and Menzala, G.
1. On a modified KdV equation, to appear.

Mikhlin, S. et al.
1. Linear equations of mathematical physics, Izd. Nauka, Moscow, 1964.

Mikhlin, S.
2. Numerical performance of variational methods, Wolters-Noordhoff, Groningen, The Netherlands, 1971.

Miles, E. and Williams, E.
1. A basic set of polynomial solutions for the Euler-Poisson-Darboux and Beltrami equations, Amer. Math. Monthly, 63 (1956), 401-404.

Miles, E. and Young, E.
2. Basic sets of polynomials for generalized Beltrami and Euler-Poisson-Darboux equations and their iterates, Proc. Amer. Math. Soc., 18 (1967), 981-986.

3. On a Cauchy problem for a generalized Euler-Poisson-Darboux equation with polyharmonic data, Jour. Diff. Eqs., 2 (1966), 482-487.

Miller, W.
1. Lie theory and special functions, Academic Press, N.Y., 1968.

2. On Lie algebras and some special functions of mathematical physics, Mem. Amer. Math. Soc., 50, 1964.

Milne, E. A.
1. The diffusion of imprisoned radiation through a gas, Jour. London Math. Soc., 1 (1926), 40-51.

Miranda, M. M.
1. Weak solutions of a modified KdV equation, Bol. Soc. Brasileira Mat., to appear.

Miyadera, I.
1. Semigroups of operators in Frechet space and applications to partial differential equations, Tohoku Math. Jour., 11 (1959), 162-183.

Morawetz, C.
1. A weak solution for a system of equations of elliptic-hyperbolic type, Comm. Pure Appl. Math., 11 (1958), 315-331.

2. Note on a maximum principle and a uniqueness theorem for an elliptic-hyperbolic equation, Proc. Roy. Soc., 236A (1956), 141-144.

Nachbin, L.
1. Lectures on the theory of distributions, Textos do Matematica, 15, Inst. Fis. Mat., Univ. Recife, 1964.

2. The Haar integral, Van Nostrand, Princeton, 1965.

Naimark, M.
1. Linear representations of the Lorentz group, Moscow, 1958.

Nersesyan, A.
1. The Cauchy problem for degenerating hyperbolic equations of second order, Dokl. Akad. Nauk SSSR, 166 (1966), 1288.

Neves, B. P.
1. Sur un problème non linéaire d'évolution, C. R. Acad. Sci. Paris, to appear.

Nevostruev, L.
1. The local extremum for an equation of second kind, Volz. Mat. Sbor. Vyp., 6 (1968), 190-197.

Nosov, V.
1. The Cauchy problem and the second Cauchy-Goursat problem for a certain equation of hyperbolic type, Volz. Mat. Sobr. Vyp., 6 (1968), 198-203.

Ogawa, H.
1. Lower bounds for solutions of hyperbolic inequalities, Proc. Amer. Math. Soc., 16 (1965), 853-857.

2. Lower bounds for solutions of hyperbolic inequalities on expanding domains, Jour. Diff. Eqs., 13 (1973), 385-389.

Oleinik, O.
1. On the smoothness of solutions of degenerate elliptic-parabolic equations, Dokl. Akad. Nauk SSSR, 163 (1965), 577-580.

2. On linear equations of the second order with a nonnegative characteristic form, Mat. Sbornik, 69 (1966), 111-140.

Olevskij, M.
1. Quelques théorèmes de la moyenne dans les espaces a courbure constante, Dokl. Akad. Nauk SSSR, 45 (1944), 95-98.

2. On the solution of Cauchy's problem for a generalized Euler-Poisson-Darboux equation, Dokl. Akad. Nauk, SSSR, 93 (1953), 975-978.

Ossicini, A.
1. Problema singolare di Cauchy, relativo ad una generalizazione dell equatione di Eulero-Poisson-Darboux, Rend. Accad. Lincei, 35 (1963), 454-459.

2. Problema singolare di Cauchy, relativo ad una generalizzazione dell equatione di Eulero-Poisson-Darboux, II, caso k < p-1, Rend. Mat. Appl., 23 (1964), 40-65.

Oucii, S.
1. Semigroups of operators in locally convex spaces, Jour. Math. Soc. Japan, 25 (1973), 265-276.

Ovsyanikov, L.
1. On the Tricomi problem in a class of generalized solutions of the Euler-Darboux equation, Dokl. Akad. Nauk SSSR, 91 (1953), 457-460.

Payne, L.
1. Representation formulas for solutions of a class of partial differential equations, Jour. Math. Phys., 38 (1959), 145-149.

Payne, L. and Sather, D.
2. On a singular hyperbolic operator, Duke Math. Jour., 34 (1967), 147-162.

Payne, L.
3. Improperly posed problems in partial differential equations, SIAM Reg. Conf. series in Appl. Math., 22, 1975.

Peregrine, D.
1. Calculations of the development of an undular bore, Jour. Fluid Mech., 25 (1966), 321-330.

2. Long waves on a beach, Jour. Fluid Mech., 27 (1967), 815-827.

Phillips, R. S.
1. Dissipative operators and hyperbolic systems of partial differential equations, Trans. Amer. Math. Soc., 90 (1959), 193-254.

Poincaré, H.
1. Sur l'equilibre d'une masse fluide animée d'un mouvement de rotation, Acta Math., 7 (1885), 259-380.

Poisson, S.
1. Memoirs sur l'intégration des équations linéaires aux derivées partielles, Jour. Ecole Polytech., 12 (1823), No. 19.

Poritsky, H.
1. Generalizations of the Gauss law of spherical means, Trans. Amer. Math. Soc., 43 (1938), 147.

Prokopenko, L.
1. The Cauchy problem for Sobolev's type of equation, Dokl. Akad. Nauk SSSR, 122 (1968), 990-993.

Protter, M. and Weinberger, H.
1. Maximum principles in differential equations, Prentice-Hall, New Jersey, 1967.

Protter, M.
2. The Cauchy problem for a hyperbolic second order equation with data on the parabolic line, Canad. Jour. Math., 6 (1954), 542-553.

3. A boundary value problem for an equation of mixed type, Trans. Amer. Math. Soc., 71 (1951), 416-429.

4. The two noncharacteristics problem with data partly on the parabolic line, Pacific Jour. Math., 4 (1954), 99-108.

5. Uniqueness theorems for the Tricomi problem, I-II, Jour. Rat'l. Mech. Anal., 2 (1953), 107-114; 4 (1955), 721-732.

6. On partial differential equations of mixed type, Proc. Conf. Diff. Eqs., Univ. Maryland, 1956, pp. 91-106.

7. A maximum principle for hyperbolic equations in a neighborhood of an initial line, Trans. Amer. Math. Soc., 87 (1958), 119-129.

Pucci, C. and Weinstein, A.
1. Sull'equazione del calore con dati subarmonici e sue generalizzazioni, Rend. Accad. Lincei, 24 (1958), 493-496.

Pucci, C.
2. Discussione del problema di Cauchy per le equazioni di tipo ellitico, Ann. Mat. Pura Appl., 46 (1958), 131-153.

REFERENCES

Pukansky, L.

1. On the Plancherel theorem of the real unimodular group, Bull. Amer. Math. Soc., 69 (1963), 504-512.

Gopala Rao, V. and Ting, T.

1. Solutions of pseudo-heat equations in the whole space, Ann. Mat. Pura Appl., 49 (1972), 57-78.

2. Initial value problems for pseudo-parabolic partial differential equations, Indiana Univ. Math. Jour., 23 (1973), 131-153.

Gopala Rao, V.

3. Sobolev-Galpern equations of order n + 2 in $R^m \times R$, $m \geq 2$, Trans. Amer. Math. Soc., 210 (1975), 267-278.

4. A Cauchy problem for pseudo-parabolic partial differential equations in the whole space, Thesis, Univ. Illinois, 1972.

Gopala Rao, V. and Ting, T.

5. Pointwise solutions of pseudo-parabolic equations in the whole space, Jour. Diff. Eqs., to appear.

Raviart, P.

1. Sur la résolution et l'approximation de certaines équations paraboliques nonlinéaires dégénérées, Arch. Rat'l. Mech. Anal., 25 (1967), 64-80.

2. Sur la résolution de certaines équations paraboliques non-linéaires, Jour. Fnl. Anal., 5 (1970), 299-328.

de Rham, G.

1. Variétés différentiables, Hermann, Paris, 1955.

Rickart, C.

1. Banach algebras, Van Nostrand, Princeton, 1960.

Riesz, M.

1. L'intégrale de Riemann-Liouville et le problème de Cauchy, Acta Math., 81 (1949), 1-223.

Robin, L.

1. Fonctions sphériques de Legendre et fonctions sphéroidales, Vols. 1-3, Gauthier-Villars, Paris, 1957-1959.

Rosenbloom, P.

1. Numerical analysis and partial differential equations, coauthor, Wiley, N.Y., 1958.

2. Singular partial differential equations, Proc. Symp. Fluid Dyn. Appl. Math., Gordon-Breach, N.Y., 1962, pp. 67-77.

3. An operational calculus, Bull. Amer. Math. Soc., 61 (1955), 73.

Rotman, J.
1. The theory of groups; an introduction, Allyn-Bacon, Boston, 1965.

Rühl, W.
1. The Lorentz group and harmonic analysis, Benjamin, N.Y., 1970.

Rundell, W.
1. Initial boundary value problems for pseudo-parabolic equations, Thesis, Univ. of Glasgow, 1974.

Rundell, W. and Stecher, M.
2. A method of ascent for parabolic and pseudoparabolic partial differential equations, SIAM Jour. Math. Anal., to appear.

3. A Runge approximation and unique continuation theorem for pseudoparabolic equations, to appear.

Ruse, H., Walker, A., and Willmore, T.
1. Harmonic spaces, Cremonese, Rome, 1961.

Sally, P.
1. Intertwining operators and the representations of SL(2,R), Jour. Fnl. Anal., 6 (1970), 441-453.

2. Analytic continuation of the irreducible unitary representations of the universal covering group of SL(2,R), Mem. Amer. Math. Soc., 69, 1967.

Sansone, G. and Conti, R.
1. Equazioni differenziali non lineari, Cremonese, Rome, 1956.

Sather, D.
1. Maximum properties of Cauchy's problem in three dimensional space time, Arch. Rat'l. Mech. Anal., 18 (1965), 14-26.

2. A maximum property of Cauchy's problem in n-dimensional space time, Arch. Rat'l. Mech. Anal., 18 (1965), 27-38.

3. A maximum property of Cauchy's problem for the wave operator, Arch. Rat'l. Mech. Anal., 21 (1966), 303-309.

Sather, D. and Sather, J.
4. The Cauchy problem for an elliptic parabolic operator, Ann. Mat. Pura Appl., 80 (1968), 197-214.

Schaeffer, H.
1. Topological vector spaces, Macmillan, New York, 1966.

Schuss, Z.
1. Regularity theorems for solutions of a degenerate evolution equation, Arch. Rat'l. Mech. Anal., 46 (1972), 200-211.

2. Backward and degenerate parabolic equations, to appear.

Schwartz, L.
1. Théorie des distributions, Edition "Papillon", Hermann, Paris, 1966.

2. Espaces de fonctions différentiables à valeurs vectorielles, Jour. Anal. Math., 4 (1954-55), 88-148.

3. Les équations d'évolution liées au produit de composition, Ann. Inst. Fourier, 2 (1950), 19-49.

4. Un lemme sur la dérivation des fonctions vectorielles d'une variable réele, Ann. Inst. Fourier, 2 (1950), 17-18.

5. Théorie des distributions à valeurs vectorielles, Ann. Inst. Fourier, 7 and 8, 1957-58, 1-141 and 1-209.

6. Lectures on mixed problems in partial differential equations and representations of semigroups, Tata Institute, Bombay, 1957.

Selezneva, F.
1. An initial value problem for general systems of partial differential equations with constant coefficients, Soviet Math. Dokl., 9 (1968), 595-598.

Serre, J.
1. Algèbres de Lie semisimples complexes, Benjamin, New York, 1966.

2. Lie algebras and Lie groups, Harvard lectures, Benjamin, N.Y., 1964.

Showalter, R.

1. Pseudo-parabolic differential equations, Thesis, Univ. Illinois, 1968.

2. Degenerate evolution equations, Indiana Univ. Math. Jour., 23 (1974), 655-677.

3. Existence and representation theorems for a semilinear Sobolev equation in Banach space, SIAM Jour. Math. Anal., 3 (1972), 527-543.

4. A nonlinear parabolic-Sobolev equation, Jour. Math. Anal. Appl., 50 (1975), 183-190.

5. Nonlinear degenerate evolution equations and partial differential equations of mixed type, SIAM Jour. Math. Anal., 6 (1975), 25-42.

6. The Sobolev equation, I, Jour. Appl. Anal., 5 (1975), 15-22; II, 81-99.

7. Partial differential equations of Sobolev-Galpern type, Pacific Jour. Math., 31 (1969), 787-793.

Showalter, R. and Ting, T.

8. Asymptotic behavior of solutions of pseudo-parabolic differential equations, Ann. Mat. Pura Appl., 90 (1971), 241-248.

9. Pseudo-parabolic partial differential equations, SIAM Jour. Math. Anal., 1 (1970), 1-26.

Showalter, R.

10. Hilbert space methods for partial differential equations, Pittman Press, London, 1977 (to appear).

11. Local regularity of solutions of partial differential equations of Sobolev-Galpern type, Pacific Jour. Math., 34 (1970), 781-787.

12. Well-posed problems for a differential equation of order 2m + 1, SIAM Jour. Math. Anal., 1 (1970), 214-231.

13. Weak solutions of nonlinear evolution equations of Sobolev-Galpern type, Jour. Diff. Eqs., 11 (1972), 252-265.

14. Equations with operators forming a right angle, Pacific Jour. Math., 45 (1973), 357-362.

15. Global perturbation of generators of semi-groups, Proc. Second Annual U.S.L. Math. Conf., Univ. S.W. Louisiana, Lafayette, La., 1972.

16. The final value problem for evolution equations, Jour. Math. Anal. Appl., 47 (1974), 563-572.

17. Regularization and approximation of second order evolution equations, SIAM Jour. Math. Anal., 7 (1976), to appear.

18. A Sobolev equation for long waves in nonlinear dispersive systems, to appear.

19. Energy estimates for perturbations of evolution equations, Proc. 12th Annual Mtg. Soc. Eng. Sci. (1975), 275-278.

20. Quasi-reversibility of first and second order evolution equations, in Improperly posed boundary value problems, Ed. A. Carasso and A. Stone, Pitman Publ., Carlton, 1975.

21. Well-posed problems for some nonlinear dispersive waves, Jour. Math. Pures Appl., to appear.

Sigillito, V. G.

1. A priori inequalities and the dirichlet problem for a pseudo-parabolic equation, SIAM Jour. Math. Anal., to appear.

2. Exponential decay of functionals of solutions of a pseudoparabolic equation, SIAM Jour. Math. Anal., 5 (1974), 581-585 .

Sigillito, V. G. and Pickle, J. C., Jr.

3. A priori inequalities and norm error bounds for solutions of a third order diffusion-like equation, SIAM Jour. Appl. Math., 25 (1973), 69-71.

Sigillito, V. G.

4. On the uniqueness of solutions of certain improperly posed problems, Proc. Amer. Math. Soc., 24 (1970), 828-831.

Silver, H.

1. Canonical sequences of singular Cauchy problems, Thesis, University of Illinois, 1973.

Simms, D.

1. Lie groups and quantum mechanics, Springer, Berlin, 1968.

Smirnov, M.
1. Degenerate elliptic and hyperbolic problems, Izd. Nauka, Glav. Red. Fiz-Mat. Lit., Moscow, 1966.

2. The Cauchy problem for degenerate hyperbolic equations of the second order, Vest. Leningrad. Inst., 3 (1950), 50-58.

3. Equations of mixed type, Izd. Nauka, Glav. Red. Fiz.-Mat. Lit., Moscow, 1970.

4. Degenerate hyperbolic and elliptic equations, Linear equations of mathematical physics, Ed. S. Mikhlin, Izd. Nauka, Moscow, 1964, chap. 7.

Smoke, W.
1. Invariant differential operators, Trans. Amer. Math. Soc., 127 (1967), 460-494.

Snow, C.
1. Hypergeometric and Legendre functions with applications to integral equations of potential theory, Nat'l. Bur. Stand., Appl. Math. Ser., 19, 1952.

Sobolev, S.
1. Some new problems in mathematical physics, Izv. Akad. Nauk, SSSR, 18 (1954), 3-50.

Solomon, J.
1. Huygens' principle for a class of singular Cauchy problems, Jour. Diff. Eqs., 10 (1971), 219-239.

2. The solution of some singular Cauchy problems, US Naval Ord. Lab., NOLTR 68-105, 1968.

Sommerfeld, A.
1. Partial differential equations, Academic Press, N.Y., 1964.

Stecher, M. and Rundell, W.
1. Maximum principles for pseudoparabolic partial differential equations, Jour. Math. Anal. Appl., to appear.

Stein, E.
1. Analysis in matrix space and some new representations of SL(N,C), Annals Math., 86 (1967), 461-490.

Stellmacher, K.
1. Eine Klasse Huyghenscher Differentialgleichungen und ihre Integration, Math. Annalen, 130 (1955), 219-233.

Strauss, W. A.

1. Evolution equations nonlinear in the time derivative, Jour. Math. Mech., 15 (1966), 49-82.

2. The initial-value problem for certain non-linear evolution equations, Amer. Jour. Math., 89 (1967), 249-259.

3. The energy method in nonlinear partial differential equations, Notas de Matematica, 47, IMPA, Rio de Janeiro, Brasil, 1969.

Sun, Hu-Hsien

1. On the uniqueness of the solution of degenerate equations and the rigidity of surfaces, Dokl. Akad. Nauk SSSR, 122 (1958), 770-773.

Suschowk, D.

1. On a class of singular solutions of the Euler-Poisson-Darboux equation, Arch. Rat'l. Mech. Anal., 11 (1962), 50-61.

Takahashi, R.

1. Sur les représentations unitaires des groupes de Lorentz généralisés, Bull. Soc. Math. France, 91 (1963), 289-433.

Talman, J.

1. Special functions; a group theoretic approach, Benjamin, N.Y., 1968.

Taylor, D.

1. Research on consolidation of clays, Mass. Inst. Tech. Press, Cambridge, 1952.

Tersenov, S.

1. On the theory of hyperbolic equations with data on the degenerate line, Sibirsk. Mat. Žur., 2 (1961), 913-935.

2. On an equation of hyperbolic type degenerating on the boundary, Dokl. Akad. Nauk SSSR, 129 (1959), 276-279.

3. A singular Cauchy problem, Dokl. Akad. Nauk SSSR, 196 (1971), 1032-1035.

4. On a problem with data on the degenerate line for systems of equations of hyperbolic type, Dokl. Akad. Nauk SSSR, 155 (1964), 285-288.

Thyssen, M.

1. Opérateurs de Delsarte particuliers, Bull. Soc. Roy. Sci. Liège, 26 (1957), 87-96.

2. Sur certains opérateurs de transmutation particuliers, Mem. Soc. Roy. Sci. Liège, 6 (1961), 7-32.

Ting, T.
1. Parabolic and pseudo-parabolic partial differential equations, Jour. Math. Soc. Japan, 21 (1969), 440-453.

2. Certain non-steady flows of second-order fluids, Arch. Rat'l. Mech. Anal., 14 (1963), 1-26.

3. A cooling process according to two-temperature theory of heat conduction, Jour. Math. Anal. Appl., 45 (1974), 23-31.

Tinkham, M.
1. Group theory and quantum mechanics, McGraw-Hill, N.Y., 1964.

Titchmarsh, E.
1. The theory of functions, Oxford Univ. Press, London, 1932.

2. Introduction to the theory of Fourier integrals, Oxford Univ. Press, London, 1937.

3. Eigenfunction expansions, Part I, Oxford Univ. Press, London, 1962.

Ton, B. A.
1. Nonlinear evolution equations of Sobolev-Galpern type, to appear.

Tondeur, Ph.
1. Introduction to Lie groups and transformation groups, Lect. Notes, 7, Springer, N.Y., 1965.

Tong, Kwang-Chang
1. On singular Cauchy problems for hyperbolic partial differential equations, Sc. Rev., 1, No. 5, 1957.

Travis, C.
1. On the uniqueness of solutions to hyperbolic boundary value problems, Trans. Amer. Math. Soc., 216 (1976), 327-336.

Treves, F.
1. Topological vector spaces, distributions, and kernels, Academic Press, N.Y., 1967.

Tricomi, F.
1. Sulle equazioni lineari alle derivate parziali di secondo ordine di tipo misto, Rend. Accad. Lincei, 14 (1923), 134-247.

Vilenkin, N.
1. Special functions and the theory of group representations, Izd. Nauka, Moscow, 1965.

Višik, M.
1. The Cauchy problem for equations with operator coefficients. . ., Mat. Sbornik, 39 (1956), 51-148.

Volkodavov, V.
1. The principle of local extremum for a general Euler-Poisson-Darboux equation and its applications, Volz. Mat. Sbor. Vyp., 7 (1969), 6-10.

Volpert, A. I. and Hudjaev, S. I.
1. On the Cauchy problem for composite systems of nonlinear differential equations, Mat. Sbornik, 87 (129) (1972); transl. Math. USSR Sbornik 16, (1972), 517-544.

Waelbroeck, L.
1. Les semigroupes différentiables, Deux. Colloq. Anal. Fonct., CBRM (1964), pp. 97-103.

2. Le calcul symbolique dans les algèbres commutatives, Jour. Math. Pures Appl., 33 (1954), 147-186.

Wahlbin, L.
1. Error estimates for a Galerkin method for a class of model equations for long waves, Numer. Math., 23 (1975), 289-303.

Walker, W.
1. The stability of solutions to the Cauchy problem for an equation of mixed type, to appear.

2. A stability theorem for a real analytic singular Cauchy problem, Proc. Amer. Math. Soc., 42 (1974), 495-500.

3. Bounds for solutions to ordinary differential equations applied to a singular Cauchy problem, Proc. Amer. Math. Soc., 54 (1976), 73-79.

Wallach, N.
1. Harmonic analysis on homogeneous spaces, Dekker, N.Y., 1973.

Walter, W.
1. Uber die Euler-Poisson-Darboux Gleichung, Math. Zeit., 67 (1957), 361-376.

2. Mittelwertsätze und ihre Verwendung zur Lösung von Randwertaufgaben, Jahresber. Deut. Math. Ver., 59 (1957), 93-131.

Wang, C.
1. On the degenerate Cauchy problem for linear hyperbolic equations of the second order, Thesis, Rutgers, 1964.

2. A uniqueness theorem on the degenerate Cauchy problem, Canad. Math. Bull., 18 (1975), 417-421.

Warner, G.
1. Harmonic analysis on semisimple Lie groups, I, Springer, N.Y., 1972.

2. Harmonic analysis on semisimple Lie groups, II, Springer, N.Y., 1972.

Watson, G.
1. A treatise on the theory of Bessel functions, Cambridge Univ. Press, Second ed., Cambridge, 1944.

Weil, A.
1. L'intégration dans les groupes topologiques et ses applications, Act. Sci. Ind., Hermann, Paris, 1953.

Weinberger, H.
1. A maximum property of Cauchy's problem, Annals Math., 64 (1956), 505-513.

Weinstein, A.
1. Sur le problème de Cauchy pour l'équation de Poisson et l'équation des ondes, C. R. Acad. Sci. Paris, 234 (1952), 2584-2585.

2. On the Cauchy problem for the Euler-Poisson-Darboux equation, Bull. Amer. Math. Soc., 59 (1953), p. 454.

3. On the wave equation and the equation of Euler-Poisson, Proc. Fifth Symp. Appl. Math., Amer. Math. Soc., McGraw-Hill, New York, 1954, pp. 137-147.

4. The singular solutions and the Cauchy problem for generalized Tricomi equations, Comm. Pure Appl. Math., 7 (1954), 105-116.

5. On a class of partial differential equations of even order, Ann. Mat. Pura Appl., 39 (1955), 245-254.

6. The generalized radiation problem and the Euler-Poisson-Darboux equation, Summa Brasil. Math., 3 (1955), 125-147.

7. Sur un problème de Cauchy avec des données sousharmoniques, C. R. Acad. Sci. Paris, 243 (1956), 1193.

8. Generalized axially symmetric potential theory, Bull. Amer. Math. Soc., 59 (1953), 20-38.

9. Singular partial differential equations and their applications, Proc. Symp. Fluid Dyn. Appl. Math., Gordon-Breach, New York, 1962, pp. 29-49.

10. On a Cauchy problem with subharmonic initial values, Ann. Mat. Pura Appl., 43 (1957), 325-340.

11. On a singular differential operator, Ann. Mat. Pura Appl., 49 (1960), 349-364.

12. Spherical means in spaces of constant curvature, Ann. Mat. Pura Appl., 60 (1962), 87-91.

13. The method of axial symmetry in partial differential equations, Conv. Int. Eq. Lin. Der. Parz., Trieste, Cremonese, Rome, 1955, pp. 1-11.

14. On Tricomi's equation and generalized axially symmetric potential theory, Acad. Roy. Belg., Bull. Cl. Sci, 37 (1951), 348-358.

15. Hyperbolic and parabolic equations with subharmonic data, Proc. Rome Symp., 1959, 74-86.

16. Elliptic and hyperbolic axially symmetric problems, Proc. Int. Cong. Math. (1954), 3, 1956, 264-269.

17. Applications of the theory of functions in continuum mechanics, Int. Symp., Tbilisi, 1963, 440-453.

18. Sur une classe d'équations aux derivées partelles singulières, Proc. Int. Colloq. Part. Diff. Eqs., Nancy, 1956, pp. 179-186.

19. Subharmonic functions and generalized Tricomi equations, Tech. Note BN-88, AFOSR-TN-56-574, AD-110396, Univ. Maryland, 1956.

20. Discontinuous integrals and generalized potential theory, Trans. Amer. Math. Soc., 63 (1948), 342-354.

Whitham, G. B.
1. Linear and nonlinear waves, Wiley, N.Y., 1974.

Wigner, E.
1. On unitary representations of the inhomogeneous Lorentz group, Annals Math., 40 (1939), 149-204.

Williams, W.
1. Cauchy problem for the generalized radially symmetric wave equation, Mathematika, 8 (1961), 66-68.

Willmore, T.
1. Mean value theorems in harmonic Riemannian spaces, Jour. London Math. Soc., 25 (1950), 54-57.

Yosida, K.
1. Functional analysis, Springer, Berlin, 1965.

Young, E.
1. On a method of determining the Riesz kernel for the Euler-Poisson-Darboux operator, Jour. Math. Anal. Appl., 16 (1966), 355-362.

2. A solution of the singular Cauchy problem for the non-homogeneous Euler-Poisson-Darboux equation, Jour. Diff. Eqs., 3 (1967), 522-545.

3. A Cauchy problem for a semi-axially symmetric wave equation, Compositio Math., 23, 1971, pp. 297-306.

4. Uniqueness of solutions of the Dirichlet and Neumann problems for hyperbolic equations, Trans. Amer. Math. Soc., 160 (1971), 403-409.

5. Uniqueness theorems for certain improperly posed problems, Bull. Amer. Math. Soc., 77 (1971), 253-256.

6. A characteristic initial value problem for the Euler-Poisson-Darboux equation, Ann. Mat. Pura Appl. 85 (1970), 357-367.

7. Uniqueness of solutions of the Dirichlet problem for singular ultrahyperbolic equations, Proc. Amer. Math. Soc., 36 (1972), 130-136.

8. Uniqueness of solutions of improperly posed problems for singular ultrahyperbolic equations, Jour. Austral. Math. Soc., 18 (1974), 97-103.

9. A mixed problem for the Euler-Poisson-Darboux equation in two space variables, to appear.

10. On a generalized EPD Equation, Jour. Math. Mech., 18 (1969), 1167-1175.

11. Uniqueness of solutions of the Neumann problem for singular ultrahyperbolic equations, Portug. Math., to appear.

12. The characteristic initial value problem for the wave equation in n dimensions, Jour. Math. Mech., 17 (1968), 885-890.

Zalcman, L.

1. Mean values and differential equations, to appear.

2. Analyticity and the Pompeiu problem, Arch. Rat'l. Mech. Anal., 47 (1972), 237-254.

Zalenyak, T.

1. The behavior as $t \to \infty$ of solutions of a problem of S. L. Sobolev, Soviet Math., 2, No. 4 (1961), 956-958.

2. A problem of Sobolev, Soviet Math., 3, No. 6 (1962), 1756-1759.

2. Mixed problem for an equation not solvable for the highest time derivative, Dokl. Akad. Nauk SSSR, 158 (1964), 1268-1270.

4. On the asymptotic behavior of the solutions of a mixed problem, Diff. Urav., 2 (1966), 47-64.

Želobenko, D.

1. Compact Lie groups and their representations, Izd. Nauka, Moscow, 1970.

Žitomirski, Ya.

1. Cauchy's problem for systems of linear partial differential with differential operators of Bellel type, Mat. Sbornik, 36 (1955), 299-310.

Index

A 6
B 7
C 8
D 9
E 0
F 1
G 2
H 3
I 4
J 5